S0-BPI-648

ALLE ZEIT WACH
1842

Åke E. Andersson · David F. Batten
Charlie Karlsson (Eds.)

Knowledge and Industrial Organization

With Contributions by

Å. E. Andersson, C. Anderstig, R. Artle, R. U. Ayres, S. Bass Warner Jr.
D. F. Batten, L. Bodström, D. L. Bosworth, E. J. Davelaar
M. M. Fischer, H. Folmer, H. W. Herzog Jr., B. Hotz-Hart, J. Howells
T. Hutten, Y. M. Ioannides, C. Karlsson, K. Kobayashi
T. R. Lakshmanan, G. Larsson, E. Mansfield, A. Markusen, P. Nijkamp
A. M. Schlottmann, P. Varaiya, K. Velupillai, R. A. Wilson
M. Wiseman, W. Zegveld

Springer-Verlag Berlin Heidelberg New York
London Paris Tokyo Hong Kong

Professor Dr. Åke E. Andersson, Director, Institute for Futures Studies, Hagagatan 23 A, 3tr, 113 47 Stockholm, Sweden

Professor Dr. David F. Batten, Department of Economics, University of Umeå, 901 87 Umeå, Sweden

Dr. Charlie Karlsson, Department of Economics, University of Karlstad, 650 09 Karlstad, Sweden

With 33 Figures

ISBN 3-540-51529-1 Springer-Verlag Berlin Heidelberg New York Tokyo
ISBN 0-387-51529-1 Springer-Verlag New York Berlin Heidelberg Tokyo

Library of Congress Cataloging-in-Publication Data
Transportation for the future / David F. Batten, Roland Thord (eds).; with contributions by Å. E. Andersson ... [et al.].
p. cm.
"This volume contains a colletion of papers presented at the First International Conference on Transportation for the Future, held at the Hotel Skogshöjd in Södertälje, Sweden from 24-25th May 1988"-Pref.
Sponsored by the Swedish Board for Technical Development, the Swedish National Road Administration, and the Community of Södertälje.
ISBN (invalid) 0-387-22809-0 (U.S.)
1. Transportation-Forecasting-Congresses. I. Batten, David F. II. Thord, Roland, 1949.
III. International Conference on Transportation for the Future (1st: 1988: Södertälje, Sweden).
IV. Styrelsen för teknisk utveckling. V. Sweden. Vägverket.
HE 11, T73 1989
388'.01'12-dc20

Printing: Weihert-Druck GmbH, Darmstadt
Bookbinding: J. Schäffer GmbH u. Co. KG., Grünstadt
2142/7130-543210

FOREWORD

This volume contains a selection of papers presented at an international symposium on research and development, industrial change and economic policy organized and hosted by the University of Karlstad, Värmland, Sweden.

Situated about halfway between Stockholm and Oslo, Karlstad stands on the River Klara, which reaches north into the mountains of Norway. Founded by King Charles IX of Sweden, whose statue stands in the city centre beside the river, the city celebrated its 400th anniversary in 1984. For many decades the wealth of Karlstad has been based on traditional industries such as iron, timber and paper, and throughout the province of Värmland there are a considerable number of industrial communities which grew up around mines, ironworks, sawmills and papermills. Even the cultural structure of these communities is heavily marked by the industrial environment in which they developed.

However, for over a decade now a major structural reorientation has been taking place and the old industrial structures have been disappearing. For various reasons the importance of large scale, manufacturing companies has declined and as a result of intense development work with new ideas, new entrepreneurs and new technology, we have seen the rapid rise of small companies.

In this context, recent research has shown that universities playing an increasingly central role in regional development. Thus the continued development of the University of Karlstad is of significant regional and national interest. Although fairly small in an international perspective (approximately 4000 students) the university is expanding rapidly.

It is natural that regional research should be an important field in an environment such as the one sketched above, so a special regional science research unit has been established at the university, organized along the lines of what is termed the 'Karlstad approach'. This means that in its research work Karlstad partly draws on the academic competence of the larger universities in the country. For instance, the regional science research unit, includes among its scientific advisers Professor Åke E. Andersson, now head of the Institute for Futures Studies in Stockholm, and Professor Börje Johansson of Umeå University, both of whom played a key role in promoting the symposium.

The above background helps to explain why Karlstad was a suitable venue for the meeting. However, the decisive factor for hosting a symposium on this theme was that research and development, industrial change and economic policy have become major research fields at the international level. I am very pleased, therefore, that the organizers succeeded in creating a meeting place for prominent researchers from many parts of the world. A stimulating international spirit was the keynote of the symposium. We were honoured by the presence of participants and observers from universities, major companies, trade unions, government departments, various national boards, research institutes and regional organizations. It gave us particular pleasure that Mr. Lennart Bodström, the Swedish Minister of Education, was able to accept the invitation to open the proceedings.

That the symposium was successful was a tribute to the efficiency of the organizing committee and secretariat. Further, we would like to express our appreciation to those research councils, authorities, organizations and companies in Sweden who provided the necessary funds. The main funding agencies were the National Swedish Board for Technical Development and the Bank of Sweden Tercentenary Foundation.

Last but not least, I would like to thank the editors for their work in preparing this volume. My sincere thanks also go to Michael Cooper and Ann-Britt Höglund who made invaluable contributions to the editorial process and to Ingrid Lindqvist for her extraordinarily conscientious secretarial work. I am firmly convinced that the volume they have produced is a major contribution to research in the field.

Lennart Andersson
Vice Chancellor
University of Karlstad

CONTENTS

CHAPTER 1

From the Industrial Age to the Knowledge Economy

Åke E. Andersson, David F. Batten and Charlie Karlsson

1.1 INDUSTRIAL ORGANIZATION AND THE KNOWLEDGE ECONOMY

1.1.1 Economic Life in the Industrial Age

Some aspects of economic life were taken for granted in the industrial age. Surprisingly enough, one of the scarcest resources of all - knowledge - was treated as given and inflexible in its supply, just like minerals and other natural resources.

During the 19th and 20th centuries, two opposing assumptions have emerged out of this basic idea. The first assumption is associated with David Ricardo. In his world different regions were assumed to be endowed with differing sets of technological and commercial knowledge. Thus the availability of natural resources, agricultural traditions, and technological knowledge were assumed to have given Portugal an absolute advantage over England in the production of oats, wine and cloth. His riddle was then: would there be any mutual advantage in engaging in trade between these two potential trading partners, who were so unequal in productivity? Ricardo gave an astonishing answer. The absolute productivity differential, however great, would never matter. As long as there were *comparative* productivity advantages in any field of production, there would be gains from trade for the sum of countries involved in international trade.

Almost a century later, Eli Heckscher and Bertil Ohlin made the second, much more heroic assumption of identical technological knowledge ruling everywhere. In this world where knowledge differences were completely ruled out, it was possible to show that comparative advantages in terms of basic resource endowments were sufficient to generate comparative advantages from trade. Then it would even be possible to compensate for shortages in some immobile factors, such as sun or rain, by exchanging goods intensive in water and sunrays in the process of cultivation and other production. Heckscher and Ohlin disregarded knowledge as an important factor determining trade and thus productive specialization of different countries.

This lack of interest in knowledge as a productive factor was not so surprising. The division of labour between individuals, regions, and nations was the most obvious aspect of 19th century development among the North Atlantic economies. The economic potentials of linking resource-rich regions in North America with the labour-rich countries of Western Europe were significant enough to generate an enormous increase of trade, specialized production in new manufacturing cities, and massive increases in income per capita.

The idea of specialization by division of labour, first formulated by Smith and Ricardo, became a macro and micro theory of economic organization. In the practical form advocated by Taylor, it provided guidelines for exploiting the comparative advantage principle from the individual factory floor to the industrial corporation as a home. The more limited the number of tasks performed by each worker the more comparative advantages would be used and the higher would be productivity and profitability of production. Marx had foreseen some of the political consequences of unrestricted division of labour. In the long run nobody would be able to set up competitive industries due to the complete fragmentation of the production process. Two classes would emerge - the fragmented and specialized labour class and the capitalist class with a complete overview of the whole production process. Limitless application of the principal of division of labour and the use of indivisible capital equipment would create a complete hierarchy of monopolistic production within each country. Only the size of the market would prevent such a system from becoming completely monopolistic. This amounts to the structuring of society by static principles alone.

Schumpeter was the first economist to radically question this view of economic development. His is a much more dynamic approach to the problem. He willingly admitted to all of the principles of division of labour at any conceivable level of economic life. He also willingly admitted that a general equilibrium with a given structure can exist, but only as a very limiting case where all the natural dynamics of real economies are precluded. To him, knowledge was a dynamic factor of production potentially available in abundance. The synergetic combination of entrepreneurs and such potential technological knowledge was enough to make the general equilibrium unstable. When the economic system is approaching a state of general equilibrium, there is by definition also a decline of profitability that would sooner or later force the entrepreneur capitalist into a frantic search for new ideas - "innovations". Thus, a new method of production would decrease the long run marginal cost of production, or a new product would be introduced, shifting demand in favour of the new product. The closer we come to the general equilibrium the stronger would be the disequilibrating forces. In modern terminology, the general equilibrium would be dynamically repulsive. Two principles were sufficient to generate this conclusion - the existence of an entrepreneurial class, and the availability of non-innovated knowledge.

In the static theories of Adam Smith, David Ricardo, Eli Heckscher and Bertil Ohlin, knowledge was not progress but corresponded to a given set of capacities to be exploited through the division of labour. In Schumpeter's theory, knowledge was not the outcome of an economic process but a good to be exploited by entrepreneurs in their innovative economic processes.

1.1.2 Knowledge as a Productive Factor

Since Arrow's learning-by-doing process, knowledge has been introduced as a productive factor endogenous to the production process. According to this theory, any predetermined division of labour will influence the pattern of accumulation of knowledge. Learning-by-doing implies that comparative advantages within an industrial organization, as well as between industrial organizations, will shift over time as a reflection of the initial choice of division of labour.

But knowledge can also be introduced into economic theory in a much deeper dynamic fashion. The last three decades of development within "cognitive science" have credited the process of interaction with the fundamentally important role of knowledge accumulation in industrial and other organizations. It has also become more important to clarify the precise meaning of knowledge in contrast to information. Some of the conceptual issues delimiting the capabilities of computers from those of human intelligence are discussed in the contribution to this volume by *Vellupilai*.

From a purely empirical point of view, economists must seriously engage themselves in an analysis of the knowledge creation process. More than ten percent of the GNP of most OECD-countries is currently allocated to formal education and research and development expenditures. Countries are now engaged in a qualitative form of competition, where knowledge investments are the key long-term policy instruments. The pattern of employment is rapidly changing. The most rapid expansion is occurring in the occupations involved in knowledge handling. Here we have teachers, scientists, research and development personnel, journalists and other mass media occupations, consultants and more.

Using identical statistical databases, it has been possible to estimate the structural changes proceeding within labour markets in Sweden and Japan. This study shows that the restructuring of labour markets in the Stockholm and Tokyo regions are qualitatively similar. The cognitive, knowledge-oriented jobs are expanding rapidly (together with the interactive, service jobs), while information-oriented jobs and physically demanding occupations are stagnating or decreasing in absolute and relative importance. This recent tendency may be contrasted with a declining commitment to education and R&D occupations during the seventies in the USA (see Table 1.1).

In the analysis of the division of labour and other aspects of industrial organisation, it was assumed that the productivity of one individual is independent of others involved in the production process. Labour was assumed to interact with machinery and raw materials, but the interaction patterns between individuals within the labour force were not assumed to be significant for their respective productivities. This is an unreasonable assumption in a truly dynamic economy with endogenously shifting production technology and a changing spectrum of products. In his contribution to this volume, *Varaiya* shows that the whole doctrine of the division of labour by comparative advantage is misleading if applied to high technology and other knowledge-intensive industries. When dynamic complementarities or *synergistic* conditions prevail - as they do in research and development activities - it is fruitless to apply the static principles of the division of labour. In the creation of knowledge, dynamic self-organization during the creative period seems to be a much more viable procedure than a pre-organisation into tasks according to measured comparative advantages.Varaiya's analysis points in the general direction of a new theory of industrial organization.

As we noted earlier, the classical theory of interregional and international trade has disregarded the dynamic role of knowledge and communication possibilities for the organization of production in space. In their chapter, *Batten, Kobayashi and Andersson* have laid the foundations for a dynamic analysis of the location process in which knowledge accumulation and communication are explicitly considered. Their analysis shows that the synergistic potentials of knowledge can be analysed in a spatial equilibrium context, manifesting themselves in the form of a dramatic structural change when slowly changing variables are transgressing certain critical values. This implies that smooth changes in policy variables should not necessarily be expected to always lead to correspondingly smooth changes in economic structure. As soon as knowledge and communication systems are introduced into economics, one should allow for the possibility of dramatic industrial reorganization as an embodied dynamic characteristic. A more general conclusion would be the following: that in the dynamic evolution of a knowledge-intensive corporation or region, there will be a need for complete reorganization at certain stages in order to break out of an outmoded equilibrium structure and establish a new equilibrium at a higher level of total system efficiency.

1.2 TECHNOLOGICAL AND ECONOMIC INTERACTIONS - EMPIRICAL STUDIES

1.2.1 Long Cycles of Technological Progress

In addition to the realization that knowledge ought to be regarded as an endogenous factor in the production process, a growing number of economists have become fascinated with the idea that the world economy has been experiencing a structural crisis that in some ways parallels earlier major crises in economic activity (Mensch, 1975; Forrester, 1977; Rostow, 1978; Mandel 1980; Freeman, 1981; Freeman, Clark and Soete, 1982; van Duijn, 1983). However, this idea has also been the subject of major criticism, see e.g. Rosenberg and Frischtak, 1983). The more or less regular occurrence of structural crises is generally associated with the work of the Russian economist Kondratieff, who presented his theory of long waves or cycles in the 1920s (Kondratieff, 1926). In reality, the idea had been formulated earlier in some detail by the Dutch economist van Gelderen (van Gelderen, 1913) and even by Marx who in the third volume of *Capital* wrote about "fluctuations extending over long periods" (Kuczynski, 1987).

Schumpeter also advanced the notion of radical technological innovations as a major factor in recurrent structural crises (Schumpeter, 1939). He stressed the uniqueness of each long cycle, pointing to important exogeneous factors such as wars and harvest failures. He also wrote explicitly about technological innovations as being the driving force of economic growth (see also Section 1.1.1). The quasi-monopolistic profits gained by the initiators led to a swarming effect where imitators jumped on the entrepreneurial bandwagon. This generated a wave of new investment, which in turn generated the boom conditions associated with the rising segment of the long wave.

Schumpeter compared technological innovations with a series of explosions and claimed that these explosions were sufficient to engender long cycles of the Kondratieff type. He based his conclusion on the following argument: (i) innovations are not distributed over the whole economic system at random, but tend to concentrate in certain sectors and their surroundings, (ii) the diffusion process is inherently a very uneven one because innovations do not remain isolated events, and are not evenly distributed in time, on the contrary, they tend to come in clusters or bunches, i.e. the initiators are usually followed by a large group of imitators in the wake of successful innovation, and (iii) the characteristics (i) and (ii) imply that the disturbances the diffusion process engenders are enough to disrupt the existing system and enforce a distinct process of adaptation.

In a Schumpeterian framework, causation runs from science via technology to the economy. The cumulative exploition of scientific and technological ideas is called the 'natural trajectory' (Freeman, 1982). Different industries have different capacities to exploit these natural trajectories (Nelson and Winter, 1977). It is disequilibrium and imperfect competition among entrepreneurs engaging in industrial innovation that provide the basis for economic development and the focus is on the supply side with induced investments. Mensch (1975) noted the importance of basic innovations in providing the basis of new industries. He claimed that his data revealed that innovation peaks precede the bottoms of Kondratieff cycles by two decades and so the beginning of a new boom is embedded in a previous downswing. This hypothesis has been severely criticized (see e.g. Freeman, Clark and Soete, 1982).

Leaving aside the difficult question of whether long cycles exist at all - 200 or so years of industrial capitalism are too short a time to establish the existence of a cycle from highly disturbed statistical series (Goodwin, 1987) - we would in any case face a problem with a number of competing explanations. A number of attempts have been made to verify the different theories empirically, but as far as we can see there are still many uncertainties.

In this volume, the chapter by *Ayres* provides one example of the type of information needed to improve our understanding of these cyclic possibilities. His perspective is essentially Schumpeterian, but places special emphasis on the role of perceived technological barriers for technological innovation and the opportunities created by a breakthrough (when a major barrier is overcome). An important implication of his model is that major inventions do not occur at random. His model also predicts a clustering of innovations following any major breakthrough. Thus, rapid progress in particular fields at particular times is not accidental. The identities of the inventors are unpredictable, along with the technical details of their inventions. On the other hand the technological functions (i.e. the problems solved) are often well understood in advance. Ayres' paper reviews the history of the electrical industry from the perspective of this barrier-breakthrough model.

The chapter by *Bass Warner* on "The evolution of high technology in the Boston Region 1920-1980" contrasts and complements the paper by *Ayres*. It is a contrast in the sense that its major theme is unpredictability. Bass Warner maintains that although the Boston metropolis seems to be prospering at the present moment on the high technology branches of the electronics industry, nobody in 1920 and probably nobody in 1950 could have predicted this prosperity. It is a complement in the sense that it presents an excellent case study of the transition from the electrical industry to the electronics industry at one specific location.

Mansfield also deals with high technology in his chapter. His purpose is to further our understanding of the evolution and dynamic behaviour of both the Japanese and American robot industries. To do this, he compares the speed and efficiency of Japanese and American robot producers as innovators, and the amounts spent by Japanese and American robot producers on various kinds of R&D and other innovative activities. He also carefully studies the process of firm growth in the robot industry in both countries and how a firm´s growth rate is related to its R&D and other innovative activities.

1.2.2 Spatial Innovation Diffusion

The diffusion of innovations is at the centre of the Schumpeterian framework, although the spatial dimension of innovation diffusion is often neglected. The papers by Ayres, Bass Warner and Mansfield demonstrate that innovative activities have definite spatial aspects. Thompson (1965) advanced the hypothesis that successful inventions tend to become innovations in the large metropolitan areas of 'high-technology' regions in the most advanced countries. Once inventions have become innovations, this knowledge diffuses to other countries and regions. The spatial patterns of diffusion are of great interest, since they determine the capacity of cities and regions to compete successfully in the world market. Although not all authors agree (Pred, 1974), several reasons for a hierarchical diffusion process can be offered (Karlsson, 1988):

A first reason can be found in the character of the information flows. In much of the pioneer work on the spatial diffusion of innovations (Hägerstrand, 1952 and 1967), the role of information availability in neighbourhood diffusion was afforded paramount importance (i.e. the diffusion of innovations was viewed as a learning process). But this approach is also valid for the problem of hierarchical diffusion, since that there are networks of social communications which connect certain nodes (central places in this case) to the exclusion of others. Innovations are transmitted through a hierarchy of information networks (international, interregional, intraregional, etc). Enterprises with spatially abundant information networks may learn of the relevance of a particular technique much earlier than enterprises which obtain technical information from a narrower base. "The diffusion of information therefore becomes a key variable in the diffusion of a new technique"(Thwaites, 1978, p.452). However, the strength of information flows may vary over space and information networks may be stronger in certain regions. Given this background, we may think that a hierarchical diffusion pattern is generated by a simple probability mechanism in which the probability of adoption depends upon the chance that an enterprise residing in a given centre will learn of the new innovation - a probability which declines with the size of the centre (Hudson, 1969). Information about the characteristics of innovations is transferred interpersonally and thus flows hierarchically, generating a tendency for innovations to diffuse down the urban hierarchy.

A second reason is concerned with the need to minimize risk. The introduction of an innovation in a new market is an inherently risky business for its propagator: "uncertainty has a spatial dimension" (Pred, 1966, p.99) . Problems of a technical nature, of factor supply or of marketing may arise, and each is likely to be more easily overcome the larger the city. The argument is not simply that technical expertise, skilled labour and a ready market are most abundant in cities, but also that there is more information. Thus, the argument is again one of abundance of information. Hierarchical diffusion may be the result of a "market-searching" process in which the enterprises propagating the innovations exploit market opportunities in a larger-to-smaller sequence. The idea is that the most profitable locations (=markets) are exploited first, and successively less profitable or more risky locations are exploited later.

A third reason as to why we may hypothesize a hierarchical diffusion pattern has to do with the availability of factors of production. To be able to adopt an innovation in a profitable way, an enterprise must have access to complementary inputs or factors of production. Two kinds of complementary inputs can be distinguished: inputs which are supplied in the marketplace and inputs from social overhead capital. An important example of the first type is the availability of a sufficient quantity of qualified labour to be capable of adopting the innovation. The argument here is that highly educated labour, capital and entrepreneurial skills are in short supply and most readily available in the largest urban areas. This means, for example, that the absolute number of people, ready, willing and able to perform a certain job is likely to vary with the size of urban centre. Also the supply of services necessary for adopting an innovation must be mentioned here. In a situation of excess demand for the innovation, one may expect that the development of the spatial pattern of supply of services is led by the development of the spatial pattern of adoption. This implies a certain advantage for potential adopters in regions where there are already many adopters compared with other regions (Nijkamp and Rietveld, 1987). Social overhead capital may also play a role as a complementary input to the adoption of innovations. Examples are research-oriented universities, transport infrastructure (e.g. airports) and high-speed/high-volume tele-communication facilities. Enterprises in regions with plenty of such facilities are candidates for an early adoption of innovations.

When a product embodying an innovation becomes standardized, it pays multi-locational enterprises to shift the production centre (or, at least, the manufacturing and assembly of standardized parts) to branch-establishments in the more peripheral areas of low-cost less skilled labour. Thus, the innovation diffusion process takes the form of a relocation process. This process is accelerated by price competition from other producers. Typically, manufacturing in an multi-locational enterprise is relocated to a branch-establishment in a smaller community, where labour and other production cost savings are possible. Oakey (1979) suggests that branch-establishments will receive mature products. Thus, the manufacturing of products containing the product innovations of yesterday is beting relocated to peripheral non-metropolitan areas. This means that, when products reach the later life cycle stages, the producers in metropolitan areas seek the industrial backwaters of non-metropolitan areas, where cheaper unskilled labour is available for the simplified manufacturing processes. This relocation process is known as the "trickle-down" process (Thompson, 1968).

As a fourth reason for hierarchical diffusion, we may hypothesize an "imitation" process in which independent entrepreneurs in smaller centres mimic the actions of enterprises in larger centres (Pedersen, 1970).

The chapter by *Fischer* is a survey attempting to summarize some of the major findings, to point to some gaps in our understanding of the conditions under which technical advance takes place, and to highlight spatial aspects of the innovation-diffusion process. He considers R&D in some detail, describes industry- and size-specific variations in R&D, and attempts in particular to explain the location of R&D. The diffusion of technological innovations is also dealt with and four major conceptual developments in diffusion research is briefly characterized. A third major purpose of the paper is to identify the key elements which affect the rate and degree of adoption and to develop an understanding of the temporal and spatial nature of the diffusion process. Some major problems arising in diffusion research are briefly discussed and the paper concludes with a discussion of some policy implications. The survey made by Fischer is complementary to that by *Lakshmanan* which surveys technological and institutional innovations in the service sector. His key point is that the analysis of the service sector requires models of change that are consistent with the special characteristics of the sector. Further, in the evolution of services, technical change and institutional change are important and have to be viewed together in a dynamic interactive manner.

The chapter by *Howells* also has the character of a survey but is backed by data from a case study of the pharmaceutical industry. His chapter is an attempt to provide some insights into industrial R&D organization and location, using the case study as a basis. It also examines some of the problems associated with developing a research and technological development strategy for regional economies. His results indicate that it is not only the location and level of R&D and technological capacity that is important in the transmission of research capacity into more specific benefits for a particular locality, but also its function and corporate context and its relationships with the wider technological/-institutional environment. Thus the attraction of R&D capacity into a region is not a sufficient measure for triggering a process of economic growth in the region.

Diffusion of technological change and economic growth is treated in the chapter by *Ioannides*. In the context of a standard neoclassical model of economic growth, he investigates the consequences of assuming that technological change does not affect the entire economy simultaneously. He assumes that the proportion of enterprises in the economy which have adopted embodied capital-augmenting technical progress follows a given diffusion sequence. He then derives a second-order differential equation for aggregate capital per capita, from which he concludes that the long-run growth rate of all per capita magnitudes is independent of the economy´s propensity to save. Thus, a key result of neoclassical growth theory still holds when technological change is not implemented uniformly throughout the economy.

The chapter by *Anderstig* and *Karlsson* presents an empirical investigation, by means of logit analysis, of the occurrence of hierarchical spatial diffusion patterns in Sweden. The study deals with the diffusion of applications of information technology in the production process, in administration and management, and in products in the engineering industry. Regarding the production process, the results indicate that applications associated with standardized production do not show a hierarchical diffusion pattern while applications associated with differentiated production do. As regards information technology components in products, a hierarchical diffusion pattern applies to 'advanced' IT-applications in terms of the relative value of the IT-content of the products. Generally, the use of information technology components in products is more frequent at lower levels in the urban hierarchy.

In the chapter by *Davelaar* and *Nijkamp*, spatial and temporal dimensions of technical innovation are dealt with in both a theoretical and an empirical sense. They provide a systematic typology of different innovations. As far as the regional context is concerned, they design a theoretical framework has been designed from which it is possible to deduce which regions will most likely 'specialize' in certain kinds of innovations. Their results indicate the important role of internal R&D for product innovations, while for process innovations external R&D appears to be of great importance. Furthermore, they indicate that small firms with their own R&D division appear to achieve excellent results in terms of product innovations, but poor results in terms of process innovations. On the other hand, large firms with their own R&D division, prove to perform very well in terms of both product and process innovations.

The chapter by *Hotz-Hart* concentrates on the following questions: What were the typical problems connected with innovative behaviour in Swiss industry during the last decade and what kind of conclusions for economic policy can be drawn from them? The author discusses findings from different studies concerning the innovative behaviour of Swiss industry point, in particular to (i) problems of exchange, linkage and transition in the innovation process, (ii) risk aversion in terms of a focus upon rationalization and cost reduction, (iii) low innovative behaviour in the small and medium sized firms, (iv) a sub-optimal framework of rules for company foundation, and (v) an increasing gap between qualifications available and needed in order to innovate. He suggests some starting points for economic policy promoting innovative development.

1.2.3 R&D and Property Rights

One central aspect of the Schumpeterian approach is the importance of imperfect competition among entrepreneurs as one foundation of economic development. When a major innovation occurs it will disequilibrate existing market structures and reward the successful innovator with exceptional growth and temporary monopoly profits. However, this monopoly will eventually give way to the cyclical phenomena already described. The onset of innovation is seen as the result of R&D investments creating an intellectual property. The duration of the monopoly profits depends on how well this intellectual property is protected.

The protection of intellectual property rights is discussed in the chapter by *Bosworth and Wilson*. The authors examine the nature of the current system for the protection of intellectual property in the UK. They also look at the implications of monopoly power, examining the inter-related goals of the Patent Office and the Patent Agents. Their results indicate that the current system in the UK appears least satisfactory for individual applicants and small firms. However, the evidence outlined is also indicative of the possible existence of monopoly power which may affect all users and, thereby, the operation of the patent system as a whole. The authors maintain that while the negative static welfare effects may be highly significant, the dynamic effects may be more important; not only affecting the current use of the system amongst existing inventors, but also reducing the incentive to invent and thereby affecting the dynamic performance of the British economy.

1.3 THE POLITICAL ECONOMY OF STRUCTURAL CHANGE

It is clear from the above that knowledge transfer and technology diffusion have significant temporal, spatial and organizational dimensions. The transfer of products, processes and knowhow in the form of patents, licences and other scientific ideas between different regions is in itself a catalyst of structural change. Thus we must study the consequences of R&D and technological change at the global level for effective policy making at the regional and industrial levels.

1.3.1 Interdependencies between Industrial and Regional Structural Change in the USA

According to Fritz Machlup (1962), economic activities associated with the knowledge industry accounted for some 29 percent of Gross National Product (GNP) in the United States in 1958. Slightly less than 32 percent of the U.S. workforce was engaged in knowledge-producing activities in 1959. Although the knowledge industry has expanded modestly up into the eighties, the rate of growth in knowledge production has been much slower than in the heady decade before 1958 (see Table 1.1). More significantly, there has been no clear pattern evident in the share of R&D expenditure as a percent of GNP, which seems to have stabilized within the range of 2.1-2.6%.

Table 1.1: Expenditure on knowledge-producing activities as a percent of adjusted GNP, United States 1958-1980

	1958	1963	1967	1972	1977	1980
Education	11.8	13.3	14.7	14.8	13.7	12.5
Research & development	2.2	2.6	2.6	2.2	2.1	2.2
Media communications	7.7	7.5	7.7	7.9	8.1	8.0
Information machines	2.0	2.4	2.6	2.3	2.7	3.2
Information services	4.9	5.2	5.7	6.7	7.6	8.4
TOTAL	28.6	31.0	33.3	33.9	34.2	34.3

Source: Rubin and Huber (1986)

By way of contrast, the level of R&D investments in a number of other industrialized countries has risen steadily throughout the seventies and the early eighties (see Figure 1.1). For example, Japan and Sweden currently spend a higher share of their GNP on R&D than does the USA.

What are the structural adjustments associated with this tapering off in levels of R&D growth in the USA? *Markusen* highlights the growing share of R&D financed by the federal government, the bulk of it going into defence and space-related activities. In her chapter, she argues that innovation has become heavily institutionalized in a group of national laboratories, defence-dependent corporations, and particular universities. The diversion of R&D resources away from the more commercially-oriented sectors has coincided with a suite of rather fundamental problems: a slowdown in productivity growth, an accelerated displacement of workers in the manufacturing sector, an expensive interregional resettlement program for the population, and a deepening trade and budget deficit.

In their chapter, *Herzog and Schlottman* concentrate on job creation in the U.S. manufacturing sector which is attributable to the recent growth in high-technology industries. Their study addresses the urban dimensions of the migration and residential choice decisions made by high-technology workers. Earlier work demonstrated that high-technology workers are more mobile than other workers, placing a greater emphasis on transportation access when considering relocation. Such workers tend to prefer large urban areas because of social diversity and increased employment opportunity. Herzog and Schlottman's present study reveals that a disproportionately high share of employees in high-technology industries choose suburban residences and workplaces rather than central city locations.

Wiseman examines parallel issues within a particularly dynamic regional economy, the Sacramento metropolitan area. His employer database permits scrutiny of the component processes of birth, death, expansion and contraction - each contributing to the observed evolution of employment structure in printing and publishing, high technology industries and business services. A negative relationship between firm size and rate of employment

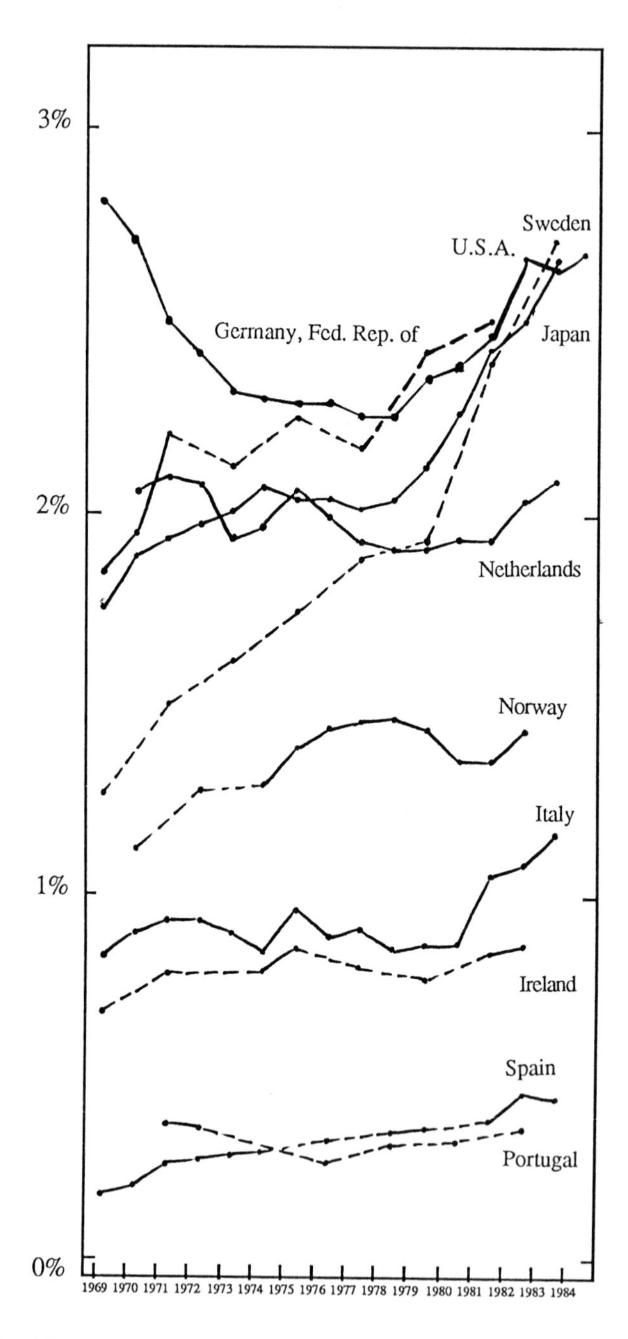

Figure 1.1: Expenditure on R&D as a percentage of GNP in 10 selected developed countries: Trends since 1969
Source: UNESCO Statistical Yearbook, 1986

is confirmed, albeit weakly. The unexpected high rate of job turnover calls for flexible training policies to complement greater mobility between jobs.

1.3.2 The Micro-Macro Policy Problem

One of the greatest challenges facing those responsible for the formulation of science and technology policy may be termed the "micro-macro policy problem". Whereas a macro perspective is generally needed to understand how the various organizations and agents combine to form science and technology policy, it is only at the microlevel of the individual agents that various processes of change can be properly comprehended. Table 1.2 illustrates the problem. The individual agents of knowledge production are not restricted to firms engaged in private sector manufacturing activity, but also include other firms engaged in knowledge-based service activity, some informal organizations and other entities or individuals providing knowledge. Technological changes generally result from processes of knowledge exchange between these various agents who are located in specific nodes (e.g. cities, towns, multinational firms) within large networks (containing many similar nodes). These nodes are linked together by transportation and communication systems, which facilitate interaction between agents and thus promote knowledge

Table 1.2: The micro-macro policy perspective

	MICROSYSTEMS		
MACROSYSTEMS	INDIVIDUAL AGENTS IN		
	Production & handling of goods	Mixed activities	Production of knowledge & knowledge-based services
NETWORKS OF INTRAREGIONAL, INTERNATIONAL AND MULTIREGIONAL NODES	PROCESSES: Creative processes Innovation activities related to industrials public policy Temporal and spatial organization of technology diffusion		
NETWORKS OF AGENTS & ORGANIZATIONS	Technology transfer related to imports & exports of technology intensive commodities & systems Consequences of technological change for policy making at the regional and industrial levels		

exchange (see also Chapter 4). As the instrumental agents, nodes and networks change, so does a nation's overall R&D propensity alter. But because the microlevel agents enter and exit the system at a relatively fast pace, it is very difficult to monitor the state of science and technology policy at the macrolevel.

In the final section of the book, various aspects of this micro-macro problem emerge as we discuss policy. *Zegveld* asks whether our current knowledge infrastructure and the traditional means by which politics, science and industry deal with knowledge are appropriate for fashioning today's science and technology policy. He alludes to the interactive, nonlinear and discontinuous character of the innovation process (see also Batten,

Casti and Johansson, 1987), citing works by Freeman (1982), Nelson and Winter (1982) and Dosi (1982) as pertinent to the formulation of a useful new theory of innovation. Such R&D developments are qualitatively different to those existing in the previous three decades, and may require new institutional arrangements to foster their fulfillment.

Reflecting on innovation-stimulating policy, *Folmer and Hutten* cite the preoccupation with "key" innovations instead of a broader set of policy targets. They assert that imitations lie at the very bottom of the innovations hierarchy, but the Japanese success suggests that they ought to play a more prominent role in science and technology policy. *Artle's* chapter looks at the nature of defence-led innovation policy in the USA (as did *Markusen* in a earlier chapter), and highlights the interdependencies between international flows of capital into a large nation and the international transfer of technologies and ideas. Interesting questions are raised about the effectiveness of targeted industrial policy and government subsidies for R&D activity.

Science policy in Sweden is examined in the last two chapters of the book. Sweden's research policy covers basic and applied research grants to the institutes of higher education, who also engage in mission-oriented research since there are few research institutes or laboratories outside the universities. The 1987 Bill on Research assured a substantial increase in State grants to research (4% in real terms) for the current three-year period, with much of this being targeted towards younger researchers and the creation of a stimulating research environment rather than the establishment of a greater number of research positions. *Minister Bodström's* key point is that it will always be difficult to develop research environments which are internationally competitive in a country of Sweden's size, but they should in any case be as creative as possible. *Gudmund Larsson's* ten-year review of Swedish science policy focuses on five years of crisis followed by five years of reconstruction. He argues that a general consensus exists in Swedish society about the importance of research. However stormy the debates over economic crises or nuclear power or whatever have been, the debate on research policy has glided smoothly along with few tremors. Researchers seem to have the nation's support. Perhaps the crucial role of creativity in social and industrial organization is widely appreciated in Sweden. It is hoped that this book may help to carry that same message to the remotest corners of the world.

REFERENCES

Dosi, G., 1982, "Technical Paradigms and Technical Trajectories", *Research Policy* 11:147

Fischer, M. M. and P. Nijkamp, eds.,1987, *Regional Labour Markets*. North-Holland, Amsterdam.

Forrester, J. W., 1977, Growth Cycles, *De Economist* 125:4, 525-43.

Freeman, C., ed., 1981, "Technical Innovation and Long Waves in World Economic Development", *Futures* 13:4, Special Issue.

Freeman, C., 1982, *The Economics of Industrial Innovation*. 2nd ed, Frances Pinter, London.

Freeman, C., J. Clark and L. Soete, 1982, *Unemployment and Technical Innovation*. Frances Pinter.

Goodwin, R.M., "The Economy as an Evolutionary Pulsator", in Vasko, ed., 1987, 27-34.

Hägerstrand, T., 1952, *The Propagation of Innovation Waves,* Lund Studies in Geography, Ser.B, Human Geography, No.4, Gleerup, Lund.

Hägerstrand, T., 1967, *Innovation Diffusion as a Spatial process*. (translated and postscript by A R Pred), University of Chicago Press, Chicago.

Karlsson, C., 1988, *Innovation Adoption and The Product Life Cycle,* Umeå Economic Studies No. 185, Umeå.
Kondratieff, N.O., 1926, "Die langen Wellen der Konjunktur", *Archiv für Sozialwissenschaft und Sozialpolitik* 56:573-609.
Kuczynski, T., 1987, "Marx and Engels on Long Waves", in Vasko, ed., 1987, 35-45.
Lonsdale, R.E. and Seyler H.L., 1979, *Nonmetropolitan Industrialization,* H.W.H. Einston and Sons, Washington, DC.
Machlup, F., 1962, *The Production and Distribution of Knowledge in the United States*, Princeton University Press, Princeton, N.J.
Mandel, E., 1975, *Long Waves of Capitalist Development,* Cambridge University Press, London.
Mensch, G., 1975, *Das Technologishe Patt. Innovationen überwinden die Depression.* Umschau Verlag, Frankfurt/M.
Nelson, R.R. and S.G. Winter, 1977, "In Search of a Useful Theory of Innovation", *Research Policy.* 6:36-76
Nelson, R.R. and S.G. Winter, 1982, *An Evolutionary Theory of Economic Change*, Harvard/Belknap, Cambridge, Mass.
Nijkamp, P. and P. Rietveld, 1987, "Technological Development and Regional Labour Markets", in Fisher and Nijkamp, eds., 1987,117-38.
Oakey, R. P., 1979, "Distribution of Significant British Innovations", *Discussion Paper No. 25.* University of Newcastle, CURDS, Department of Geography.
Perloff, H.S. and L. Vingo, eds., 1968, *Issues in Urban Economics*, The Johns Hopkins University Press, Baltimore.
Pred, A.R., 1966, *The Spatial Dynamics of U.S. Urban-Industrial Growth 1800-1914: Interpretive and Theoretical Essays.* The M.I.T. Press, Cambridge Mass.
Pred, A.R., 1974, *Major Job-providing Organizations and Systems of Cities.* Association of American Geographers, Washington, D.C.
Rosenberg, N. and C.R. Frischtak, 1983, "Long Waves and Economic Growth: A Critical Appraisal", *American Economic Review, Papers and Proceedings* 146-51.
Rostow, W.W. 1978,*The World Economy. History and Prospect.* Macmillan, New York.
Rubin, M.R. and M. T. Huber, 1986, *The Knowledge Industry in the United States: 1960-1980*, Princeton University Press, Princeton, N.J.
Schumpeter, J., 1939, *Business Cycles,* McGraw Hill, New York and London.
Thompson, W.R., 1965, *A Preface to Urban Economics,* The Johns Hopkins University Press, Baltimore.
Thompson, W.R., 1965, "Internal and External Factors in the Development of Urban Economics", in Perloff and Vingo, eds.,1968, 43-62.
Thwaites, A.T., 1978, "Technological Change, Mobile Plants, and Regional Development", *Regional Studies* 12:445-61.
van Duijn, J.J., 1983, *The Long Wave in Economic Life*, Van Gorcum, Assen.
van Gelderen, J., 1913, "Springvloed: Beschouwingen over industriele ontwickkeling en prijsbeweging", *De Nieuwe Tijd* 18.
Vasko, T., ed.,1987, *The Long Wave Debate.* Springer-Verlag, Berlin.

PART I

INDUSTRIAL ORGANIZATION IN THE KNOWLEDGE ECONOMY

CHAPTER 2

Productivity in Manufacturing and the Division of Mental Labor

Pravin Varaiya

2.1 INTRODUCTION

When we think of manufacturing we imagine a factory with assembly lines, sophisticated machines and workers whose pace of work is controlled by the machines. In our imagination, the factory of the future is different only in that it has many more robots and very few workers. The modern factory is the result of a two centuries long evolution of the interaction between manufacturing technology, the growth of markets and the labor process. The interaction of these three factors can be summarized as follows.

The development of manufacturing technology can be summed up as the progressive realization of the basic principles of Ford and Taylor.[1] Ford's principle is that a complex part should be built by assembling simpler, interchangeable parts. Taylor's principle is that the fabrication process should be broken down into a sequence of precise, very simple operations that can be efficiently executed by workers and equipment.

The implementation of these two principles requires a market of sufficient size so that the high fixed costs of specialized assembly lines and dedicated equipment can be spread over a large output. This high fixed cost but low average cost production technology was a major factor in the tremendous increase in the size of firms. In turn, this led to high barriers to entry of new firms and production was monopolized in many branches. Monopolies in turn often led to the stabilization of production volumes and prices, and to the internalization of technical innovation within the firm. This is sometimes called the virtuous circle: increase in the scale of production leads to higher monopoly profits, which leads to greater innovation and further increase in scale.

The principles of Ford and Taylor shaped the labor process. The subdivision of tasks, the introduction of expensive, special-purpose machinery to execute the tasks more efficiently and precisely, and the assembly line that controlled the flow of material and hence the pace of work, led to a highly ramified subdivision of labor. It is sometimes suggested that this subdivision of labor was associated with a specialization and deepening of skills. In fact, by and large the opposite is the case. The subdivision of tasks and the nature of the specialized machinery and automated assembly lines required a work force that was predictable, malleable, and easily substitutable, In other words, it required workers with a few generalized skills, namely a modest level of literacy and manual skills, and a high adaptibility to the work environment. The result has been a progressive reduction of skills among production workers in most branches of industry. This has

---

[1]The names of Ford and Taylor are used only to evoke the assembly line and time and motion studies.

become much more apparent in the last twenty years of so, as managers of clothing, automobile and electronics firms are finding that they can effectively move their production facilities to regions and countries where wages are low, low-skilled workers are plentiful, and unions are weak or absent.

This development of the interaction between production technology, the growth of markets and the process of work is summarized in the famous law of economics. "The division of labor is governed by the extent of the market." For present purposes, however, it is better to characterize the modern factory as the outcome of the fragmentation of the production process. This is a double fragmentation: the fragmentation of a product into individual parts that must be assembled and the fragmentation of a complex fabrication process into a sequence of material-transformation operations carried out by sophisticated machines and low-skilled workers.

2.2 DIVISION OF LABOR

This image of the modern factory as the site of manufacturing is highly misleading. The factory is only the most visible and no longer the most important part of manufacturing. It is only the location of the physical fabrication process, that is, of the material-transformation operations. But before this fabrication process is set in motion, it has to be conceived mentally, not by a single brain, but by the collective brain of a large number of mental workers. Before a single integrated circuit chip is produced in the factory or an automobile rolls off the assembly line, this mental work force must conceive a product, propose a preliminary design, assess the market response, design the product, design a fabrication process, build, test and refine a prototype, certify the final prototype, order the equipment used for fabrication, re-design the fabrication process and the product to suit the equipment, and then start production. Once production has started, this mental labor must develop production plans, control the fabrication process to meet those plans, intervene in that process when there are equipment failures and bottlenecks develop, schedule the work on each shift, keep track of inventories of supplies and work-in-progress, and provide field testing and customer service once the product is sold.

If the fabrication process is run by the collective hand of production workers, the innumerable decisions that constitute the planning, design and control of this fabrication process are carried out by the collective brain of these mental workers. If the fabrication process is subdivided into a sequence of material-transformation operations, this mental process is subdivided into a sequence of symbol-manipulation operations. If the daily progress of the fabrication process can be measured in the consumption of raw materials and supplies and the accumulation of inventories and finished products, the only tangible product of the daily progress of the mental process is "the paperwork that it creates- the records, drawings, information files, production orders, inspection plans, purchase orders, production schedules, and all of the communications associated with planning, controlling, and measuring" the fabrication process. In a way the paper trail provides a symbolic reconstruction of the whole manufacturing process and how it operates.

The productivity of manual labor can be measured in terms of the output per unit of labor, but there is no simple way of measuring the efficiency with which the data and information used to design the product, and plan, support and control the fabrication process, is created, maintained, modified, adapted, and transmitted. The reason for this is not that one cannot quantify the amount and quality of the data and information that is produced by mental labor. It is because the real work of this mental labor is the creation and communication of knowledge of various kinds - some of it scientific and of lasting value, most of it based on experience and short-lived - and the relation between this knowledge and its material representation on paper is by and large misunderstood. The productivity of manual labor has been growing steadily, but the productivity increases of manual and mental labor together in manufacturing has slowed down in the past twenty years. The reason for this is that the division of mental labor has been based on the same principles that were so successful in the division of manual labor. But those principles now appear to be quite unsuited for the efficient organization of mental labor.

2.3 DIVISION OF MENTAL LABOR

To understand how and why this happened, we must recapitulate briefly the development of the division of mental labor and its separation from manual labor. As we know, manufacturing began with artisans long before the Industrial Revolution. The artisan assessed the market, conceived and designed the product, procured and transformed raw material, sold the final product and repeated the cycle. In the days of the artisan, mental and manual labor was integrated and manufacturing was an indivisible whole.

The Industrial Revolution destroyed this unity, separated manual from mental labor, fragmented the fabrication process and relegated the functions of design, planning and control of the fabrication process to mental workers. The division and subdivision of manual labor and the coordination of the material-transformation operations were conducted on an objective basis provided by the development of machines specialized to execute each operation and the flow of materials between machines. Of course, this objectivity does not at all mean that the division of manual labor is "natural", since there are many alternative ways of fragmenting the fabrication process. Rather, objective basis means that once the fabrication process is decomposed into a sequence of operations and appropriate machines are chosen to execute each operation, then the task of each worker is immediately specified within fairly narrow limits, and the coordination of tasks and the performance of each worker are objectively and automatically determined by the flow of materials.

Mental workers were given the responsibility of decomposing the fabrication process into sequences of material-transformation operations, and of designing and selecting the associated specialized machines. As the range of decisions of manual labor narrowed and became specified in simple instructions, the range of decisions of mental labor increased correspondingly. Now, as long as production volumes were large and the market for the product was stable and predictable, the "overhead" represented by mental workers was spread over a large number of units, and the organization of mental labor was not a critical factor in total cost. However, the economic environment has changed considerably in the last twenty years. First, there is greater price competition. The initial reaction of management to this price competition was to reduce further the cost of manual workers, mainly through location in low cost manual labor markets and through greater automation. A much more important change in the economic environment, however, is the increase in the pace of the introduction of new products and new fabrication technology. As a result, product cycles have decreased significantly, and production volumes have declined. For example, the lifetime of new memory chips is about five years or less, and of this, up to three years may be taken up by product development. Thus there is a growing pressure to reduce product development costs, in other words, pressure to increase the productivity of mental labor.

Management responds to this pressure by applying the same principle of the fragmentation of the fabrication process to the mental process. Mental work is decomposed into a sequence of knowledge-transformation operations starting from product conception, through design, choice of fabrication technology, production planning and scheduling, to post-production tasks. Each specialized operation is assigned to separate departments which hire mental workers with skills needed to execute those operations. Finally, to control the process of mental labor, management applies quantitative methods that were developed for controlling manual labor, namely, control over the rate of production of paper or documents - which is the only tangible thing that mental workers produce. Thus, for example, we find that management techniques such as the critical path method originally created for the fabrication process, are now being used for the monitoring and coordination of mental workers. From this management perspective, an efficient organization is one in which each department prepares the appropriate documents at the specified time.

2.4 KNOWLEDGE AND COMMUNICATION PROBLEMS

This initial approach to the division of mental labor has proved not to lead to increased productivity. Two factors are responsible for this failure. First, the coordination of the knowledge-transformation tasks carried out by individual and groups of mental workers creates very difficult problems of communication. This is because most bodies of knowledge consist of a vast super-structure of empirical facts and rules-of-thumb built on top of a systematic, scientific base and expressed in dialects that are peculiar to the group. it is very difficult to record this knowledge and transmit it to people outside this group in a reliable and succinct manner. In other words, this knowledge is similar to folklore. Different groups may use similar words and phrases but they have different meanings because they share different experiences. Thus, for example, Macintosh buffs speak a different dialect from people who use IMB PCS. And, as everyone knows, it is very difficult to read and understand a reasonably complex computer program that has been written by someone else.

The second factor responsible for management's failure to organize the division of mental labor so as to increase its productivity is that there is no objective basis for this division. The decomposition of the mental process into a sequence of knowledge-transformation operations is, in fact, more or less arbitrary. Once you think about it, you realize that the neat division of responsibility into marketing and sales, research and development, fabrication process design, and customer services, seems natural only because that is the traditional way of management, and because these administrative divisions have made their way into our language and schools of management and engineering to such an extent that they seem self-evident principles of organization. In actual fact, the marketing people could have suggested a different product, the product design people could have changed the design in many ways, the fabrication process could have been different. That is, numerous choices are made at each stage, and all these choices interact in ways that cannot be predicted by any individual department, but this interaction can have a profound effect on how efficiently the product can be fabricated, its quality, safety, reliability and usefulness.

2.5 DECOMPOSITION OF MENTAL LABOR

Management has attempted this failure in two contrasting ways. The first attempt may be called the intensification of the fragmentation and objectification of mental labor. The second is an attempt to reintegrate manufacturing as a whole.

The first attempt strongly emphasizes the decomposition of the mental labor process into increasingly narrow tasks, documentation, and data collection. This attempt is greatly facilitated by the proliferation of computer and communication technology, which makes it very easy to hire low-skilled workers to enter data in prespecified formats, to store and manipulate massive amounts of data, and to summarize and re-present it in a variety of ways that can be easily incorporated in attractive documents. This accumulation and manipulation of massive amounts of data, while it is very useful for routine operations, has two shortcomings. First, it is very difficult to collect the more subtle and idiosyncratic information that people glean from their daily work experience. Both the routinized channels established for data collection, as well as the simple data formats that computers can allow, militate against the acquisition of informal, non-routine information.[2] The second shortcoming associated with task decomposition and computerized data is that each group assigned to carry out a particular task needs to represent the data in terms that are meaningful to that group's own private dialect. In other words, each group has its own view of the fabrication process. For example, the integrated circuit designer wants to know if the circuit will meet the design specifications; the process designer wants to know if the sequence of deposition, lithography, etching, ion implantation and photoresist strip operations will end up producing a chip with the electrical properties specified by the circuit designer, the production engineer is concerned with the fabrication process only to the extent of knowing the availability of equipment and workers; and so on. Thus the decomposition of task leads to a proliferation of views and specialized languages.

Computer technologists claim that new innovations will be able to accommodate idiosyncratic, informal information, and to integrate the use of data by groups that think in different languages. The new catchword is "computer-integrated manufacturing", or CIM. Central to CIM are integrated databases and expert systems. The claim is that integrated databases will allow different user groups to construct their own data objects and invent their own access methods and ways of manipulating these data objects. Thus different groups can view the same data from different perspectives and within different frames of knowledge or dialects. The database provides "bridges" between these dialect groups, maintains consistency and integrity. Expert systems are supposed to provide the methodology and the software for organizing the empirical facts and the the heuristic rules that mental workers have intuitively invented on the basis of their experience. The expert system operating alongside a new worker can assist this worker by providing, in an objective, easily learned from, a distillation of the experience of senior workers. Thus expert systems offer the hope of replacing more skilled workers with less-skilled workers.

While successful databases and expert systems applications do exist, at the moment these claims are largely promises and optimistic predictions. The best possible outcome is that they can relieve people from more routine tasks, assist communication across groups, support more creative work. The worst possible outcome is that these systems will stifle intuition and creative thought by forcing people to think and communicate in extremely restrictive, highly structured languages that computer systems can manipulate. We can gain a sense of this worst possible outcome by looking at office automation in which letters and memos are automatically synthesized by computer using a few key words. While the number of distinct letters that can be "composed" in this way seems very large because the number of combinations and permutations is large, in fact this number is minuscule in comparison with the potential of natural language. But it is not difficult to imagine that the potential will become more difficult to realize as workers are forced to adapt to the use of this restricted language. And as language becomes restrained, so will human thought.

As I mentioned, management has also responded to the productivity crisis of mental workers by seeking to reintegrate manufacturing as a whole. In this approach, a small team of workers is assembled and charged with the total manufacturing of a product, including its conception and design, fabrication and sales. The team consists of people with experience in these areas. The characteristic organizing principle is that there is no formal organization, there is intense informal interaction so that each specialized worker understands the language of others, and there is a virtual absence of formal documenta-

---

[2]A simple example will serve to illustrate this. A xerox machine manufacturer maintains an elaborate database of its fabrication process. There is an established procedure for inspecting machines before they leave the factory. Tracing back customer complaints, analysts found that machines inspected during one particular shift had a much lower rate of complaints than the average. When questioned, the worker who inspected the machines on that shift said that, in addition to conducting the prescribed test, he "listened" to each machine and if it did not "sound right" he did not release the machine. This kind of information cannot be captured by automated data acquisition systems.

tion, until the final stage. This is the reverse of the first approach. Where one approach emphasizes routinization, formal lines of authority, monitoring of work through documentation deadlines, the other abolishes formal authority and documentation, and deliberately destroys claims of expertise and bureaucratic turf. This approach has had some notable successes. For example, it was used by IBM to develop the Proprinter. "Candidates for the project were interviewed with an eye to how open they were to new ways of thinking, what their basic skill backgrounds were, and how entrepreneurial they were... The company assembled about 100 people - a product manager, engineers and programmers, a market and business planning team, procurement engineers, purchasing people, tooling engineers, an assembly engineering team, and a skeleton financial team... They virtually lived in the same complex". The Proprinter was developed within two years - half the normal time, and has been very successful in the market.

Even if this integrative approach proves successful in many other instances, there are obvious and great obstacles to implementing it on a large scale. Two of these seem particularly far-reaching. First, there is the loss of management control. During 200 years of manufacturing that control was consolidated by organizing the division of labor in such a way that workers can be readily replaced by others with clearly specified skills. Instead of reducing skills through routinization and specialization, the integrative approach seeks to generalize skills and reduce claims of expertise. Workers gain more knowledge in the process, and they become less replaceable. Second, the integrative approach can destabilize the market for mental workers. That market is now segmented in ways reminiscent of the old guild system. Engineering, for example, is divided and subdivided into numerous disciplines, and within each discipline there are different career tracks and pecking orders. For instance, research engineers have more prestige than development engineers who, in turn, are considered superior to engineers engaged in fabrication. Research engineers lose esteem in the eyes of their guild peers if they work in the factory. This guild system supports a differentiated wage structure. The integrative approach to manufacturing threatens to undermine this organization of the labor market, and hence one can expect resistance from workers themselves.

2.6 SUMMARY

Let me summarize my remarks. The historically unprecedented increase in manufacturing productivity was achieved by fragmenting the fabrication process and by dividing manual labor. Attention has now turned to increasing the productivity of mental labor. Attempts at relying on subdivision of mental labor in traditional ways have not led to expected increase in productivity, because mental work requires intense communication among groups with specialized knowledge represented in different dialects, and because manufacturing is an indivisible whole in which decisions taken in different stages interact in unforeseen ways. In responding to this failure, management has followed two approaches. One approach seeks to further subdivide mental work and to provide for the needs of communication and coordination by computer technology. The second approach seeks to reverse the trend towards the subdivision of mental work and to reintegrate the process of manufacturing. The first approach seems much more in vogue both because it reinforces management's tendency to increase control and because management is afraid to be left out of the computer revolution. The second approach threatens the stability of both management and the organization of mental labor markets.

The productivity crisis is now a subject of public debate in the United States. Every group has its own solution to the crises. Some recommend enforcing "fair" trade, others recommend permitting firms to pool their research and development effort; some advocate public subsidies for "high tech" industries, others would support "mature" industries, yet others see the solution in the growth of the "service" sector; some claim that this is a transitory problem due to the rapid growth of inexperienced, young and female workers, others see the problem in the reduction in the "work ethic". Implicit in each recommendation is a model of the manufacturing process. We should acknowledge that these models are based only on tradition and casual observation. They have little scientific value and so we have no sound basis for comparing alternative solutions to the productivity crisis. We should be engaged in understanding the nature of contemporary manufacturing. Perhaps as a first step we must give up identifying manufacturing with what goes on in the factory.

REFERENCES

Braverman, H., 1974, *Labor and Monopoly Capital*, Monthly Review Press, New York.
IEEE Spectrum, 1987, Special Report: On Good Design, May.
Lardner, J.F., 1986, "Computer Integrated Manufacturing and the Complexity Index", *The Bridge*, (Journal of the National Academy of Engineering), 10-16, Spring.

CHAPTER 3

What Have We Learned in the Path from Gödel and Turing to Artificial Intelligence?

Kumaraswamy Velupillai

3.1 PROLOGUE[1]

The distant connection with that romantic and mythical figure of nights of splendour - Harun Al Rashid - that the term ALGORITHM evokes should not be dismissed lightly. From Arithmetic and Geometry to Analysis; then Al Khowarizmi's Algebra; then again to Cartesian synthesis of Algebra and Geometry to the reign of SPACE, INTUITION *and* REASON; and finally the noble steps to the rehabilitation of logic in the analysis of reason and the founding of mathematics in the formalism of logic. It is as if we have traced a full circle from the analysis of number(s) to the number theoretic analysis of the reasoning processes of the mind - from the Pythagorean awe at the magic of numbers to the Gödelian numberings of our attempts to reason. Magic and Reason, the stuff that epics - be they Greek or Indian - are made of, have cause to be mediated by Decision. Economics, some say, is, partly at least, the science of optimal decision making under various constraints. And the making of decisions whether optimal or not, is almost life. The analysis of the making of decisions, the ubiquitous meta-level study, is where the confrontation between Magic and Reason must be resolved. That memorable resolution has been brought to the frontiers of pure and applied science by, first, Gödel introducing the notion of GENERAL RECURSIVENESS on the basic of a suggestion made by Herbrand; Turing making it fundamental with his codification by Turing Machines of Effective Computability and finally Putnam showing the equivalence of Rational Economic Man and the Universal Turing Machine. If this 'rational reconstruction' is even remotely plausible then the 'Turing Test', a turning point in Man's endless quest to understand the embodiment of mind and his sometimes tragic but often heroic attempts to build 'Artificially' an 'Intelligence', deserves the public discussion that moral and ethical issues have always had in the body politic. It will not do, as Popper will have us do, to ignore the 'Turing Test':

> "Turing said something like this: Specify the way in which you believe that a man is superior to a computer and I shall build a computer which refutes your belief. *Turing's challenge should not be taken up; for any sufficiently precise specification*

[1]These notes were put together after many memorable conversations with Daniel Heyman, Christina Bramuglia, Axel Leijonhufvud, John McCall and Stefano Zambelli. However, in the final stages, Ann Markusen's incisive comments helped clear some of the fog in my mind and its representation. If these worthies were responsible for this version the blemishes that remain may not have survived.

could be used in principle to programme a computer. Also the challenge was about behaviour - admittedly including verbal behaviour - rather than about subjective experience."

(Popper and Eccles, 1977. p. 208; Italics added).

The Rational Economic Man (REM) in standard economic theory has often been defined in terms of preference relations and initial endowments; the preference relations, in turn, are analyzed in terms of behaviour. Is Popper asking the economist to abdicate and should he do so simply because REM can be shown to be equivalent to some UTM? Is that the lesson to be learned from the long journey that brought us from Pythagoras to Turing? I think not; and it is part of my aim to try to show that Turing's challenge should not only be taken up, but mercifully, will always be taken up. For Popper forgets one important element in his advocacy of abdication: any definition of "precise" presupposes a meta-level study and, if we have learned anything at all from Gödel then it is that a metatheoretic ascent is itself endless. The day the human mind accepts Popper's injunction, that day will herald the end of speculative philosophy. Such an end is not only to be deplored but is also inconceivable.

3.2 THE 13 STEPS TO COMPLEXITY

From Boole to Turing is the path from the analysis of reason to the construction of reasoning automata. However, the path is really from Cantor to Turing. The 13-steps can be stylized as follows:

(1) The analysis of Infinite Series by Fourier methods to the elements of set theory.

(ii) The search for the 'rigorous' foundations of classical analysis and its LOGIC.

(iii) The Logic of Set Theory and the rediscovery of Antinomies.

(iv) The implications of the paradoxes of set theory; formalism vs. intuitionism. Brouwer's sustained programme of research to drive Hilbert, and everyone else, from Cantor's Paradise.

(v) The Hilbert Programme to remain in Cantor's Paradise: International Mathematical Congress, Bologna, 1927. Consistency, Completeness and Decidability.

(vi) The infeasibility of paradise or 'Paradise Lost': Gödel's results.

(vii) Forever Undecided: Turing Machines, Decidability and Church's Thesis.

(viii) Effective Computability, Algorithms and Recursiveness: Herbrand $\rightarrow$ Gödel $\rightarrow$ Turing-Church.

(ix) Von Neumann enters the scene. (M)ENIACS, Machines and the Brain: the first step by McCulloch and Pitts.

(x) Wiener and Shannon: Logic and Information. Not only (M)ENIACS and Machines but Brains and (Maxwellian) Demons.

(xi) The Turing Test and the birth of Artificial Intelligence.

(xii) The Recursive Universe.

(xiii) Computability, Complexity and Dynamics.

As an economist it is the last and 13th step that intrigues me most. That it is the 13th step - in the above 'rational reconstruction' - in a long and arduous programme of speculative philosophy is what must be remembered; that it will give rise to a 14th and 1729th[2] step is what makes it all worthwhile.

3.3 OF 'EMPTY ECONOMICS BOXES'

Why should an economist be interested in the esoteric speculations of Logicians, Mathematicians and Computer Scientists? It is easy to make a case. We have, in the 'one-hundred year diversion' from Walras to Debreu, accepted with little dissent - 'except in the under worlds of Silvio Gesell...' - the characterization of the economic agent as REM and define him (it, she!) in terms of a preference relation and initial endowments:

$$REM \equiv (\geq, e) \qquad (1)$$

where $\geq$: (usually binary) preference relation

e : initial endowments.

From 'Value and Capital' in 1939 to 'The Theory of Value' in 1959, with increasing sophistication mathematical economists imported Bourbaki - that mythical creature of much formal mischief. Thus it was that REM was studied in terms of the Logic of Relations, or more precisely the Logic of Preference Relations. Between 'Value and Capital' and the 'Theory of Value' there was Samuelson's 'Foundations': and so it was that between Preference and the Logic of Relations there was Behaviour. Bourbaki's world of Formal Elegance was the uneasy handmaiden of the messy and unpredictable Behavioural Agent. What is the connection, then between the Rational Agent and the Behavioural Agent? That is the crucial qeustion Herbert Simon has been asking ever since he and his colleagues faced Turing's challenge squarely in the face. It is satisfying, therefore, to know that even Lucas has come round to defining REM not simply in terms of preferences and initial endowments but also including decision rules. thus the 'new' REM is characterized as:

$$REM \equiv (Dr, \geq, e) \qquad (2)$$

where

Dr: decision rules

Or, in Lucas's own words:

> "In general terms, we view or model an individual as a collection of decision rules (rules that dictate the action to be taken in given situations) and a set of preferences used to evaluate the outcomes arising from particular situation - action combinations."

and then,

> "We use economic theory to calculate how certain variations in the situation are predicted to affect behavior, but these calculations obviously do not reflect or usefully model the adaptive process by which subjects have themselves arrived at

[2]It is, perhaps, fair to call 1729 Ramanujam's number.

the decision rules they use. Technically I think of economics as studying decision rules that are steady states of some adaptive process, decision rules that are found to work over a range of situations and hence are no longer revised appreciably as more experience accumulates."

(Lucas, (1986), p. s401 and s402)

Having, in the first instance, finally admitted Herbert Simon's important distinction between substantive (>) and procedural (DR) rationality, Lucas deftly parametrizes the latter to then proceed with 'business-as-usual'. But that won't do; simply because not only is there circularity in the reasoning but also because the wonderful implications of the 13th step provide no easy guarantee about the existence of 'steady states of some adaptive process'- unless they are very specially restricted processes. Interesting adaptive processes are, invariably, nonlinear processes: this is only the obverse of Morris Hirsch's succinct observation that:

"...only nonlinear differential equations have interesting dynamics."

(Hirsch (1984), p. 23)

Lucas's adaptive processes are obviously discrete. The combination of discreteness and nonlinearity unleashes a ferocious monster that calls forth the full power of recent results in Computability, Complexity and Dynamics -and even then without much hope of taming it. This Pandora's box, possibly inadvertently opened by Lucas, cannot but lead to even deeper questions about decidability, completeness and consistency - the Hilbertian Trinity - which, I conjecture, will restore Simon's distinction with a vengeance if there is to be any analysis at all.

In some senses the 'dual' to the problem of DR's and procedural rationality is the one faced by the poor Auctioneer attempting to equate supply and demand. Years ago Arrow perceptively noted that:

"Equality of supply and demand is a central tenet of ordinary economic theory, but only as the end result of a process, not as a state holding at every instant of time. On the contrary, inequalities between supply and demand are usually regarded as an integral part of the process by which the price on a market reaches its equilibrium position."

(Arrow and Hurwicz (1977), p. 352)

(cf. also Arrow (1986), esp. p. s387). If the Auctioneer is to interpret the disequilibrium behavior of rational agents to construct an iterative process of price adjustment which will lead to an equilibrium in some conventional sense (s)he has to face squarely - all over again - the ubiquitous XIIIth step (cf. Velupillai (1987)).

So much for some justification from the microeconomic point of view.

What about those of us who have never felt comfortable with Bourbakian formalisms, Utilitarianism and other 'unpleasant Monetarist' isms? Those of us who have embraced the 'Magnificent Dynamics' of the classical Economists and then, after Frisch, Kalecki, Tinbergen, Harrod and Domar made Macroeconomics bristle with Dynamics, accepted post-Keynesian (whatever that means) Macroeconomics, have been accused of adhockery in analysis. This latter charge is primarily directed at the lack of MICRO (or, for that matter ANY) Foundation for the Macroeconomics we were doing. Let that pass - or rather let us face our inquisitors squarely in their utilitarian faces and pass on to matters more interesting (but perhaps a little less useful).

Dynamics, in Macroeconomics, has meant basically two things: the Theory of Growth on the one hand and Trade Cycle Theory on the other. The formalism of either implies the study of Dynamical Systems (in the strict technical sense of the term). Where are the frontiers of Dynamical Systems? The XIIIth step appears once again even in this case.

Economists cannot ignore the implications of the steps that have led from Cantor to Turing (and now to Kolmogorov, Arnold, Smale and Mandelbrot).

Perhaps the interesting question is 'what should have been learned in the path from...' rather than 'what have we learned...' - mainly because economists seem not to have learned anything except blind obedience to the 'theoretical technology' of pure mathematics.

The morale of the 100-year history of the almost mad drive for rigour in proof and formalism in analysis - at least for economics and economists - seems to be many faceted. Let me try to disentangle what I believe to be the lessons, on the basis of the 13-steps from Cantor's Paradise to Kolmogorov's Complexities and Smale's Dynamics.

a) Ever since Hilbert's "proof" of Gordon's theorem (almost exactly 100 years ago) Existence proofs have been dominated by Formalistic mathematics (although Gordon himself called Hilbert's method 'theological'). This trend was taken over 'lock, stock and barrel' by the early developments in Mathematical Economics and Equilibrium Existence Proofs in the circle around Karl Menger. The trend has only been strengthened since Debreu's 'Theory of Value' - i.e., since about the mid-50's. No attention whatsoever was paid to Constructive Analysis, Intuitionism and Proof Theory.

b) Economic Theory, in its standard neo-classical microeconomic variety, pushed the assumption of REM in terms of the logic of relations to every feasible frontier.

c) Equally, on the basis of Arrow's celebrated result, Social Choice Theory paid homage to the mathematical underpinnings of the logic of relations.

d) Developments in Logic itself - towards Modal and Deontic Logic - were totally ignored - except, as always, by Herbert Simon and those inspired by his fertile mind. Whether such interesting developments in Logic can be circumvented by reducing imperatives to indicatives is, in my opinion, an open question. Distinguished Logicians have discussed this issue since Jørgensen's pioneering works in the heydays of Logical Positivism.[3]

e) Macrodynamics in both variants, i.e. growth and cycle theories was based upon the Theory of Ordinary Differential Equations and more recently on formal developments in Dynamical Systems Theory.

Given all this, none of these areas paid any attention to the deeper questions being asked in Mathematics, Foundations of Mathematics and Computer Science. Existence of Equilibrium under more and more general conditions, proof of dynamical economic systems with limit cycles, strange attractors, etc., REM without completeness and so on. Where was the subject heading? Increasing sophistication in the use of the 'theoretical technology' provided by the 'hard sciences' and made to work on concepts that were unable to bear the heavy load of the technical machinery thrust upon them.

At this point some seem to have culled out Popperian Lessons: do not accept Tuning's challenge. If the Popperian challenge has to be admitted now, why should it not have been admitted when Cantor went from Fourier analysis to set theory and Pioncaré from Differential Equations to Topology and Global Dynamics?

The answer is simple. The inherent curiosity of the Natural Mind cannot be encapsulated by formal methods. Every attempt must lead to metatheoretically higher levels of abstraction and hence deeper and wider concepts in our comprehension of

---

[3]However see also Carnap (1935).

ourselves and the setting in which we theorize, speculate, dream and weave the magic wands.

3.4 EPILOGUE

The XIII-th step, Computability, Complexity and Dynamics ought to widen the horizon of economists to take research in Artificial Intelligence seriously. The ever widening spiral network of feedback relations between the last three steps - XI, XII and XIII - cannot be ignored for too long if economic theorizing is to progress meaningfully. On the one hand it will help us to escape from the straitjacket of utilitarian REM's. It will force ethical issues that have been abstracted away from the formal setting in which REM has been considered in the long neoclassical diversion. On the other hand, Macrodynamic formalism, particularly Trade Cycle Theory must come to terms with the ramification of complexity theory. it is no longer clear whether deterministic or stochastic processes are meaningful in the classic measure theoretic framework of Kolmogorov. If an algorithmic definition of randomness becomes the bread and butter of the practicing applied mathematician, then Macrodynamic adhockery, however elegant, becomes a truly 'empty economic box'.

In this Recursive Universe we inhabit, it will be foolish to ignore the 13th step, and dangerous to forget that it was a path from Cantor's tentative first steps that led to the present stage. Research in Artificial Intelligence is a culmination, for the present, of that weird and wonderful path - a path that could have been generated even in the social sciences. And the really important lesson in that 13 stage trek is that almost every step was motivated by pure theoretical speculation.

REFERENCES

Arrow, K.J., 1986, "Rationality of Self and Others in an Economic System", *Journal of Business,* Vol. 59, No. 4, pt. 2, October, pp.s385-s400.

Arrow, K.J. and L. Hurwics, 1977, *Studies in Resource Allocation Process*, Cambridge University Press, Cambridge.

Carnap, R., 1935, *Philosophy and Logical Syntax*, Kegan Paul, London.

Hirsch, M.W., 1984, "The Dynamical Systems Approach to Differential Equations", *Bulletin (New Series) of the American Mathematical Society*, Vol. 11, No. 1-2, July, pp. 1-64.

Lucas, R.E. Jr., 1986, "Adaptive Behaviour and Economic Theory, in *Journal of Business*, Vol. 59, No. 4. pt. 2, October, pp. S401-S426.

Popper, K.R. and J.C. Eccles, 1977, *The Self and its Brain: An Argument for Interactionism*, Routledge and Kegan Paul, London.

Velupillai, K., 1987, "The Complexity of the 'Auctioneer'" (in preparation)

Åke E. Andersson, David F. Batten, Charlie Karlsson (eds.) Knowledge + Industrial Org. Springer-Verlag NY 1989.

CHAPTER 4

Knowledge, Nodes and Networks: An Analytical Perspective

David F. Batten, Kiyoshi Kobayashi and Åke E. Andersson

4.1 INTRODUCTION

This paper treats knowledge stocks as endogenous public goods. Rather than being used up in the process of production, knowledge is expanded and enhanced by way of exchange processes on a network consisting of nodes and links in geographical space. The nodes take the form of human settlements such as villages, towns or metropolitan regions, and the links between nodes consist of transportation routes and communication channels which facilitate knowledge acquisition and knowledge expansion.

Our focus is on the interdependencies between knowledge and its network characteristics in the vastly simplified form of interactive knowledge-handling workers (hereafter called K-workers) located within each node. Previous models stress distinctions between knowledge capital and material capital as basic inputs to the production process (see, for example, Andersson 1981). Our analysis is complementary to these previous studies (see Figure 4.1) in so much as it continues the embryonic tradition of treating technological and economic variables endogenously and interdependently. A novel feature introduced here is the possibility of a bilevel structure for the knowledge exchange network, i.e. consisting of transportation and telecommunications networks in some hierarchical relationship.

4.1.1 Elements of Our Analytical Approach

Four particular strands of scientific though play an important part in our analytical approach:

(1) We assume that the production of knowledge is separable from the conventional production technology. This is akin to the notion of Hicks neutrality.

(2) We regard the proportion of K-workers within each node as a fundamental attribute of its economy, its labour heterogeneity, and its ability to acquire and to exchange knowledge. This view of the labour force builds upon earlier work, notably by Machlup (1962); Anderstig and Hårsman (1986).

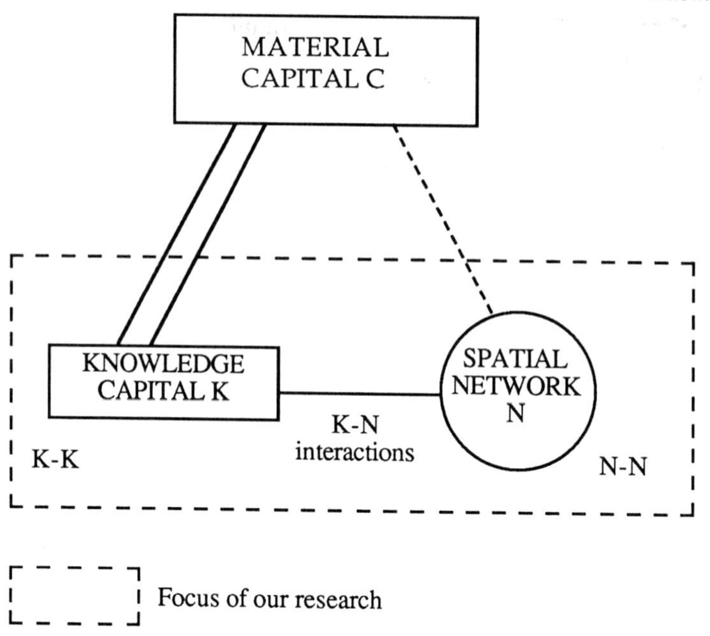

Figure 4.1 A triangle of economic interdependencies

(3) We posit accessibility to knowledge networks from the specific viewpoint of contact frequencies using transport/communication links which facilitate face-to-face knowledge exchange processes.

(4) We utilize methods of nonlinear systems analysis familiar to students of synergetics and bifurcation theory, drawing upon tools developed by Thom (1975) and Haken (1980).

Part of the original inspiration for our approach may be found in the pioneering research of Alistair Mees (1975), who demonstrated that even a gradual improvement in transportation or communication possibilities could result in an abrupt change to the structure of a human settlement system.

4.1.2 Outline of the Paper

In the following section, we develop an analytical foundation for the economic analysis of knowledge production and knowledge exchange. To cater for the public goods character of knowledge, the assumption is made that the production of knowledge is separable from the conventional production technology of the firm. A production function for knowledge stocks in each node is derived in terms of the size of its K-workforce and the aggregate level of accessibility to knowledge stocks in all nodes by means of both transportation routes and communication channels. The optimal R&D policy for each firm (node) in a perfectly competitive market is formulated as a profit-maximization problem in which the size of its K-workforce is included as an endogenous variable. Allowing for the non-cooperative decisions of reactive firms results in a Nash equilibrium solution to the general spatial equilibrium problem.

Section 3 contains the full derivation of the spatial equilibrium model of knowledge distribution, based on a qualitative definition of the knowledge production function (i.e. differences in the productivity of K-workers among nodes). The existence of a fixed point is demonstrated prior to consideration of a simple numerical example. This case of two nodes only is sufficient to illustrate that the system may possess multiple equilibria (in this case, two solutions) depending upon the values of the parameters which determine each nodes's accessibility function. Some of the bifurcation properties of the system are explored in greater detail to emphasize the full character of the model.

The final section of the paper contains some thoughts about additional refinements to the model which may be of interest, and concludes by stressing that the notion of *interactivity* across knowledge networks requires further exploration as a complementary measure to the conventional productivity measures generally associated with competing firms.

4.2 ECONOMIC ANALYSIS OF KNOWLEDGE PRODUCTION AND KNOWLEDGE EXCHANGE

4.2.1 Knowledge and the Production Function

Economic theory currently embraces a lengthy tradition wherein knowledge stocks and the associated flow variable, R&D activity, are regarded as factors exogenous to the production function. Given the active policy debate among the OECD nations on the role of R&D in the economic system, it seems more fruitful to examine the question of how knowledge-based inputs interact with conventional inputs in the production process. Some progress has been made in this direction by subdividing investment resources into those which are knowledge-based (R&D) and those which are material-based (tangible capital). Andersson (1981) has demonstrated that in such an interdependent technology-economy system, the stocks of knowledge and tangible capital should expand at the same rate. Lakshmanan (1986) points to the crucial role of (human and physical) infrastructure in the process of techno-economic transformation.

Perhaps a more pragmatic means of distinguishing knowledge-based inputs from conventional factor inputs is to decompose labour inputs into knowledge-handling occupations and other occupations. Machlup's (1962) pioneering statistical work on the US economy follows this course. More recently, Andersson et al. (1987) adopted both occupational and educational decompositions to demonstrate that labour should not be treated as a homogenous production factor in models of regional economic growth. An occupational classification of labour inputs in the production function parallels the disaggregation of interindustry inputs found in input-output formulations of the production problem. Such an approach may therefore provide us with a knowledge-oriented interpretation of the interregional differences in production technologies, which appear to be important for the theory of regional growth (see, e.g. Lande, 1978).

Formally, labour heterogeneity could be accommodated within the conventional production function framework where a vector of inputs is transformed into a vector Q of outputs (during a given period of time) as follows:

$$Q \leq f(K,L,G)$$

where K is the amount of capital, L the amount of non-knowledge-handling labour, and G the amount of knowledge-handling labour. One of the difficulties with this formulation is that knowledge acquisition and diffusion largely proceed independently of specific advances in applied technical knowhow (or process R&D). Most stocks of knowledge have a public goods character, since they contribute as a factor input without being used up by the production process. Although a few firms (the innovators) engage in their own private R&D activity, eventually this knowledge diffuses more widely so that many other firms (the imitators) have access to it.

To cater for this public goods character of knowledge, we shall redefine each firm's production function making the assumption that the production of knowledge is strongly separable from the conventional production technology. Thus we have

$$f(K,L,G) = g(G)h(K,L) \tag{1}$$

where g(G) denotes the knowledge production function and h(K,L) denotes the conventional production technology. Our assumption of strong separability implies that the production of knowledge (or knowhow) derived from process R&D shifts the frontier of the real production function f upwards with respect to the frontier of the conventional production function h. This is akin to the notion of Hicks neutrality.

Assuming that all firms (nodes) are price-takers competing only by way of differences in process R&D within an otherwise perfectly competitive marketplace, the optimization problem of the firm is to choose the best levels of K, L and G so as to

$$\underset{(K,L,G)}{\text{Maximize}} \{pf(K,L,G)-\omega K-\theta L-\rho G\} \tag{2}$$

where p is the f.o.b. price of the firm's product, ω is the rent on capital, θ is the wage rate of non-knowledge handling labour, and ρ is the wage rate of K-workers.

4.2.2 Accessibility to Knowledge

Stocks of knowledge are acquired and utilized in specific locations. R&D activities encourage flows of knowledge between these locations by way of exchange networks, which allow each R&D or decision unit to interact with other units within and outside the firm. We may identify two network classes. Telecommunications networks have been instrumental in improving our accessibility to information. Transportation networks facilitate face-to-face contacts which are necessary in order to expand our own stocks of knowledge. We shall assume that the location-specific production of knowledge involves transmission and exchanges of knowledge between K-workers by way of transportation and communication networks (see Figure 4.2):

For our purpose, knowledge exchange networks (hereafter called K-networks) are defined in terms of nodes and links across geographical space. The nodes take the form of human settlements such as villages, towns or metropolitan regions in which R&D units operate, and the links between nodes consist of those transportation routes and communication channels which facilitate knowledge acquisition and knowledge expansion. Possibilities for the latter within each node i are assumed to depend on the size of the node's K-workforce, G_i, and their relative level of *accessibility* to K-networks.

Accessibility to knowledge is determined by various frictional effects arising from geographical, social, political, educational or psychological "distances" between K-workers or K-centres. Firms employing K-workers who possess rather similar cultural backgrounds, educational skills or social interests may be considerably closer in this respect, so much so that they may interact more often with one another. Similarity of products may also be a key determinant of distance. But the impact of geographical distance is also important since like R&D units which are located within the same node are more likely to have frequent contact than like R&D units far apart. In this sense, accessibility measures can be looked upon as the spatial counterpart of discounting procedures over time.

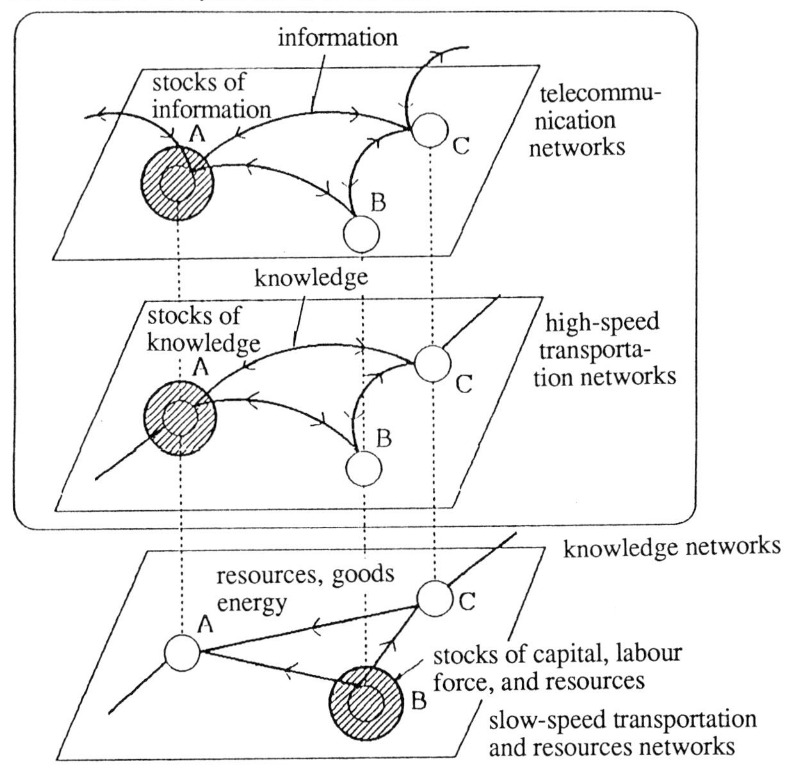

Figure 4.2 Knowledge exchange networks

The amount of knowledge which can be transferred from one unit to another is bounded from above by the size of each unit's K-workforce. We assume that unit j's knowledge stock, which we denote as G_j,, becomes perfectly accessible to unit i if, and only if, the "contact" distance from i to j, $d_{ij} = 0$. The standard assumption that accessibility decreases monotonically with increasing distance applies, so that accessibility to the stock of knowledge in an infinitely distant unit will be regarded as zero.

A simple function which fulfills these two boundary conditions is the exponential decay function:

$$A_{ij} = \sigma \exp(-\beta d_{ij}) G_j^{\gamma_1} \tag{3}$$

where A_{ij} denotes the accessibility of unit i to the knowledge stock of unit j, σ and γ_1, are parameters, and β is a system-wide parameter describing the technological efficiency of knowledge exchanges across the complete network. The accessibility of unit i to all other units - including its own public R&D units (e.g. universities) - may be given as

$$A_i = W_i + \sum_j A_{ij} = W_i + \sum_j \sigma f_{ij} G_j^{\gamma_1} \tag{4}$$

whre W_i is the scale of node i's public R&D units and $f_{ij} = \exp(-\beta d_{ij})$.

Accessibility measures such as (4) determine each unit's knowledge exchange potential. The A_i-values are therefore appropriate as additional arguments in the knowledge production function, g(G), such that

$$g(G) = g(G_i, A_i). \tag{5}$$

The above formulation will be elaborated upon in Section 3. Each firm chooses the optimal level of output and the required mix of inputs so as to maximize equation (2) given the distribution of K-workers in all other nodes, viz. $G_1, G_2, \ldots G_{1-i}, G_{i+1}, \ldots G_n$, and their own node's scale of public R&D. Here we assume that the spatial allocation of public R&D ($W_1, W_2 \ldots, W_i, \ldots W_n$) is exogenously determined by Governments.

4.2.3 The Optimal R&D Policy

Rewriting equation (2) using equation (1) and (5), we can specify the firm's profit-maximizing problems as

$$\underset{(K_i, L_i, G_i)}{\text{Maximize}} \ \{p_i \, g(G) h(K_i, L_i) - \omega_i K_i - \theta_i L_i - \rho G_i\} \tag{6}$$

where the values of ω_i and θ_i are specified for each firm i, but where ρ is assumed to be uniform across the whole network of firms. This distinction emphasizes that the wage levels of K-workers are assumed to be uniform rather than location-specific.

A typical form taken by the production function h is the Cobb-Douglas function:

$$h(K_i, L_i) = aK_i^b L_i^c. \tag{7}$$

If we assume for the moment that the size of firm i's K-workforce is fixed at $\hat{G}_i$ and that the output level is fixed at $\hat{Q}_i$, the above optimization problem simplifies to the following cost-minimization problem:

$$\underset{(K_i, L_i)}{\text{Minimize}} \ \{\omega_i K_i + \theta_i L_i + \rho G_i\}$$

subject to (8)

$$\hat{Q}_i = g(\hat{G}) h(K_i, L_i) \ .$$

Then the cost function becomes

$$C(Q_i, \hat{G}_i) = \{g(\hat{G})\}^{-1/s} T_i Q^{1/s} - \rho \hat{G}_i \tag{9}$$

where s = b + c and T_i is a constant given by

$$T_i = a^{-1/s}\{(b/c)^{c/s} + (b/c)^{-b/s}\} \omega_i^{b/s} \theta_i^{c/s} \tag{10}$$

Now allow θ_i to vary with G_i fixed at $\hat{G}_i$.Our original profit function becomes

$$\underset{(Q_i)}{\text{Maximize}} \ \{p_i Q_i - C(Q_i; \hat{G}_i)\}. \tag{11}$$

Assuming decreasing returs to scale, i.e. $b + c < 1$, there exists an optimal solution to (11) for arbitrary values of $\hat{G}_i$. Let Q^* be the optimal solution. Substituting Q_i^* into (11), we obtain the optimal profit function:

$$\Pi(\hat{G}) = \Psi_i[g(\hat{G})]^{\eta} - \rho\hat{G}_i, \tag{12}$$

where η and Ψ_i are constants given by

$$\eta = 1/(1-s)$$

$$\Psi_i = \Upsilon_i^{-\eta s}\{p_i(p_i s)^{\eta s} - (p_i s)^{\eta}\}.$$

According to Hotelling's lemma, by differentiating the profit function with respect to the factor prices, we obtain the factor demand functions:

$$K_i = \zeta(b\theta_i/c\ \omega_i)^{c/s}\{g(\hat{G})\}^{\eta}$$

$$L_i = \zeta(b\theta_i/c\ \omega_i)^{-b/s}\{g(\hat{G})\}^{\eta} \tag{13}$$

where $\zeta = a^{-1/s}(p_i s/T_i)^{\eta}$.

Now let the G_i become variables. Using (12), equation (11) becomes

$$\text{Maximize } \{\Psi_i[g(G)]^{\eta} - \rho G_i\} \tag{14}$$

The first-order optimality condition of eq. (14) is

$$\rho = \eta\Psi_i[g(G)]^{\eta-1}\Omega_{G_i}(G) \tag{15}$$

where $\Omega_{G_i}(G) = \partial g(G)/\partial G_i$. The second order optimality condition is given by

$$\Omega_{G_i}(G)\{\eta-1)g(G)^{-1}\Omega_{G_i}(G) + \partial\Omega_{G_i}(G)/\partial_{G_i}\} \leq 0 \tag{16}$$

for $G_i \geq 0$.

After rearranging (16), we obtain the following factor demand equation for K-workers in node i:

$$G^* = \Lambda_{G_i}(G^*,\rho) \tag{17}$$

The optimal value G_i^* follows directly from equation (16). By substituting this back into equation (13), we may also obtain the optimal levels of the conventional factor inputs, namely K_i^* and L_i^*.

4.2.4 Nash Equilibrium

In the above, we have restricted our discussion to a single firm (in node i) without considering the decisions or reactions of others. It is likely that the R&D decisions of each firm will influence the corresponding decisions of other firms. Such interactions over space may occur in a non-cooperative manner, in which case the resulting spatial equilibrium will correspond to a Nash equilibrium. A Nash equilibrium solution satisfies all firms' optimality conditions simultaneously. Such an equilibrium is given by

$$G_i^* = \Delta_{G_i}(G^*, \rho) \quad i = 1, \ldots\ldots, \eta \tag{18}$$

where $G^* = (G_1^*, \ldots\ldots, G_n^*)$, and may be formulated as a fixed point problem.

4.3 SPATIAL EQUILIBRIUM MODEL OF KNOWLEDGE DISTRIBUTION

4.3.1 General Formulation

Following on from (5), we shall assume that the knowledge production function takes the following form:

$$g(G) = \mu(G_i M_i)^\tau \tag{19}$$

where μ and τ are parameters and M_i is a qualitative index which measures the average interactivity of K-workers in i. The values of M_i are determined endogenously and depend on the contact frequency of K-workers in node i. Assuming decreasing returns to scale in the production of knowledge, i.e. $\tau < 1$, we have that

$$M_1 = \upsilon \log V_i \tag{20}$$

$$V_i = X_i / G_i \tag{21}$$

$$X_i = G_j^{\gamma_2} \exp(\propto A_1) \tag{22}$$

where V_i measures the average level of knowledge exchange in node i, X_i is the frequency of face-to-face communication, A_i is the accessibility to knowledge stocks [as defined in equation (4)], $\upsilon(>0)$, $\propto(>0)$ and $\gamma_2 (0<\gamma_2<1)$ are parameters. The logic of equations (20)-(22) is that the frequency of face-to-face contacts among K-workers depends partly on their number and partly on their accessibility to knowledge stocks. Each K-worker's level of interactivity may be measured in terms of this contact frequency which approximates the rate of knowledge exchange between K-workers.

According to (14), each firm's level of R&D is chosen to maximize its own profit function, i.e.

$$\underset{(G)_i}{\text{Maximize}} \ \{\Phi_i [G_i \log\{G_j^{\gamma_2 - 1} \exp(\propto A_i)\}]^{\tau\eta} - \rho G_i\} \tag{23}$$

where $\Phi_i = \Psi_i \mu^\eta \upsilon^{\tau\eta}$.

For simplicity, we shall assume that $\tau\eta = 1$. The first-order optimality condition for (23) is given by

$$\alpha(W_i + \Sigma_j \sigma f_{ij}' G_j^{\gamma_1}) - (1-\gamma_2)(1nG_i + 1) = \zeta / \Phi_i \tag{24}$$

where $f_{ij}' = f_{ij}(i \neq j)$, $f_{ii}' = (1+\gamma_1)f_{ii}$.

From (24), we obtain the following factor demand function:

$$G_i = \exp\{\lambda[\omega_i + \alpha\ (W_i + \Sigma_j\ \sigma f_{ij}' G_j^{\gamma_1})]\} \tag{25}$$

where $\lambda = 1/(1-\gamma_2)$ and $\omega_i = -(\zeta/\Phi_i) - 1 + \gamma_2$.

Our spatial equilibrium model is therefore specified as

$$G = F(G) \tag{26}$$

where F is a vector function, each element of which is given by (25). The choice of a different form of knowledge production function (19) would lead, of course, to a different spatial equilibrium model.

4.3.2 The Existence of a Fixed Point

Our function F is defined on the set C, C = $(0,\infty)$, which is not clearly compact. Thus the ordinary fixed point theorem cannot be utilized to ascertain whether there exist fixed points in (26). We must therefore derive the necessary conditions which guarantee the existence of such fixed points.

Assume that the wage rate of K-workers is endogenously determined in the labour market. For the moment, let us also assume that there exists a fixed point in (26) and denote the corresponding spatial equilibrium by $G^*(\rho)$, (i=1,...,n). Let G be the total size of the K-workforce. Now assume that there exists an equilibrium wage ζ^* which satisfies

$$\bar{G} = \sum_i G_i^*(\rho^*)$$

and let $\omega_i^* = -(\rho^*/\Phi_i) - 1 + \gamma_2$. Then for an equilibrium state (G^*, ω^*), the following holds:

$$G_i^* = \chi_i^* \exp\{\alpha\lambda\ (W_i + \sum_j \sigma f_{ij}' G_j^{*\gamma_1})\} \qquad (i=1,\ldots,n) \tag{28}$$

where $\chi_i^* = \exp(\lambda\omega_i^*)$. Summing up both sides of (28) with respect to i, we get

$$\bar{G} = \sum_i G_i^*(\rho^*)$$

$$= \sum_i \chi_i^* \exp\{\alpha\lambda\ (W_i + \sum_j \sigma f_{ij}' G_j^{*\gamma_2})\}. \tag{29}$$

At the Nash equilibrium, G^* and χ^* satisfy the following logit-like equation:

$$G_i^* = \frac{\bar{G}\,\chi_i \exp\{\alpha\lambda(W_i+\Sigma_j\sigma f_{ij}'G_j^{*\gamma_1})}{\Sigma_k\chi_k \exp\{\alpha\lambda(W_k+\Sigma_j\sigma f_{kj}'G_j^{*\gamma_1})} \tag{30}$$

On the contrary, let us assume that the value of χ^* in eq. (29) is fixed to a certain value. Obviously, eq. (29) is a continuous map defined on the compact set $R = \{G_j : \Sigma_j G_j = \bar{G}\}$. According to Brouwer's fixed point theorem, there exist fixed points in equation (29). Now define a parameter χ and denote the spatial equilibrium (fixed point) as a function of this parameter: $G_i(\chi)$. If there exists a χ^* which satisfies eq. (29), it is guaranteed that $G_i(\chi^*)$ satisfies eq. (28). Thus, the condition that there exist χ^*, $G_i(\chi^*)$ satisfying eq. (29) for a given $\bar{G}$ is necessary for the existence of the equilibrium wage satisfying (27).

It is straightforward to ascertain whether a fixed point exists in (28), since Φ_i is a function of a single variable ρ. Where the value of Φ_i happens to be uniform over the whole network, our spatial model simplifies to the following logit-like equation:

$$G_i^* = \frac{\bar{G} \exp\{\alpha\lambda(W_i+\Sigma_j\sigma f_{ij}'G_j^{*\gamma_1})}{\Sigma_k \exp\{\alpha\lambda(W_k+\Sigma_j\sigma f_{kj}'G_j^{*\gamma_1})} \tag{31}$$

In practice, equation (30) is quite convenient for illustrating the interesting properties of this spatial equilibrium model.

4.3.3 The Case of Two Nodes

In order to probe some of the basic properties of the above model, it is only necessary to consider a simplified example involving two nodes (cities). The contact geography for this situation is depicted in Figure 4.3. Defining $N = G^*_1/\bar{G}$ as the share of K-workers located within node 1, it follows that node 2's share of all K-workers is 1-N. Then equation (30) simplifies to the following:

$$N = 1/\{1+\exp[C_2(N)-C_1(N)]\} \tag{32}$$

where

$$C_1(N) = W_1+\sigma\{(1+\gamma)f_{11}N^{\gamma}+f_{12}(1-N)^{\gamma}\} \tag{33}$$

$$C_2(N) = W_2+\sigma\{f_{21}N^{\gamma}+(1+\gamma)f_{22}(1-N)^{\gamma}\}$$

and $\gamma = \gamma_1$. The contact distance d_{ij} is a composite measure of the various frictional influences affecting the intra- or inter-nodal exchange of knowledge [We further assume that $d_{ij} > d_{ii} \geq 0$ and $d_{ji} > d_{ii} \geq 0$].

It is noteworthy that the equilibrium structure defined by (32)-(33) may have zero, one or two solutions depending upon the values of certain parameters. For any given value of β, the inverse function of the bifurcation diagram, $N = (W,\beta)$ is given by

$$W = AN^{\gamma} - B(1-N)^{\gamma} + \ln\{(1-N)/N\} \tag{34}$$

where $W = W_2 - W_1$, $A = \sigma\{(1+\gamma)f_{11}-f_{21}\}$ and $B = \sigma\{(1+\gamma)f_{22}-f_{12}\}$.

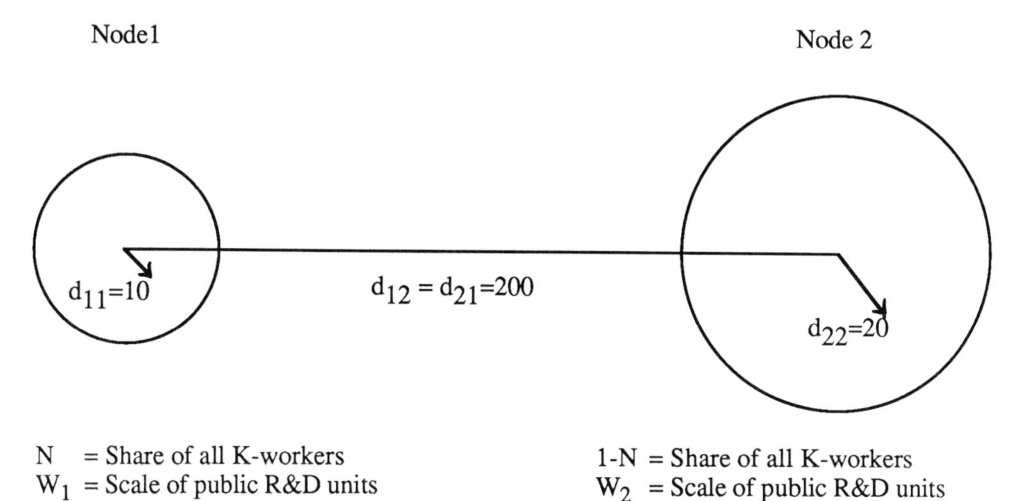

Figure 4.3 A bilateral numerical example

A bifurcation point is then a solution to the differential equation $dW/dN = 0$ (alternatively, $dN/dW = \infty$). Differentiating (34) and equating to zero we obtain

$$h(N) - r(N,\gamma) = 0 \tag{35}$$

for which $h(N) \triangleq (N^2-N)^{-1}$, $r(N,\gamma) = -A\gamma N^{\gamma-1} + B\gamma(1-N)^{\gamma-1}$ and $0 \leq N \leq 1$.

Figure 4.4 depicts the function h(N) and the family of functions r(N,γ) which result for given values of $\beta = 0.03$ and $\sigma = 3.0$. Note that no solutions exist when $\gamma \leq 0.4$, since the two functions do not intersect. At a value of γ just below 0.5, the two curves are tangential depicting a unique equilibrium. Higher values of γ yield two solutions corresponding to the points of intersection of the curves.

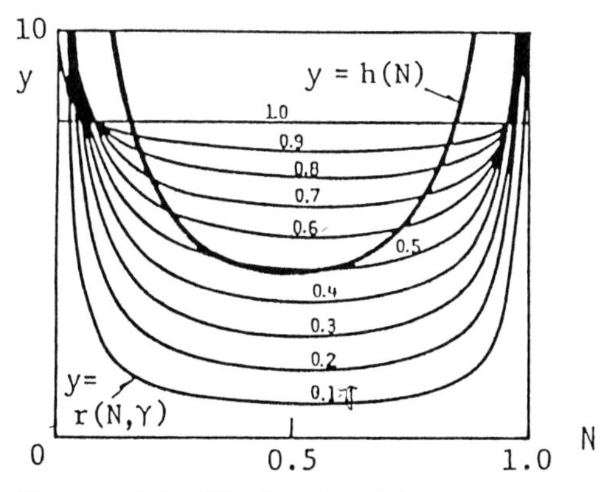

Figure 4.4 The function h(N) and a family of r(N,γ) functions ($\beta = 0.03$, $\sigma = 3.0$, γ varied)

The corresponding family of bifurcation possibilities is displayed in Figure 4.5. For given values of $\beta = 0.03$, $\sigma = 3.0$ and $\gamma = 1.0$, the outermost curve depicts a fold catastrophe possessing bifurcation points at B and D (the turning points where $dW/dN = 0$). Similar types of fold catastrophe result for values of γ in the range $0.5 \leq \gamma \leq 1$, but revert to ogives when $\gamma < 0.5$.

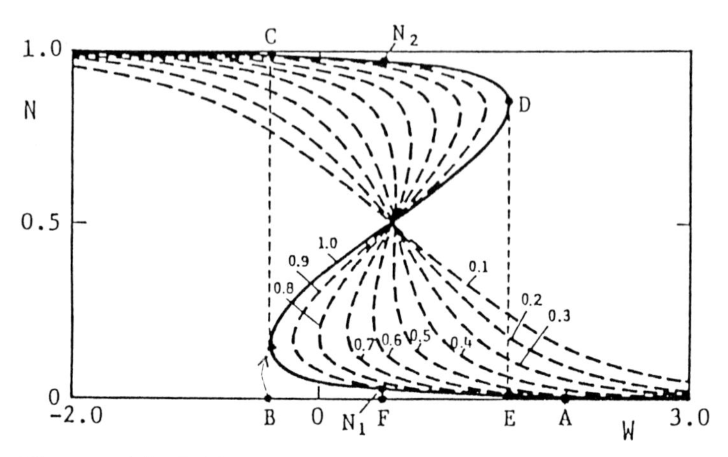

Figure 4.5 Fold catastrophes in the (W,N) plane
($\beta = 0.03$, $\sigma = 3.0$, γ varied)

Consider the implications of this fold catastrophe, which corresponds to one of Rene Thom's elementary family for which many behavioural contexts have been found [see, for example, Poston and Stewart (1978)] including binary choice problems in urban modelling [by Wilson (1976)]. Adopting the outermost curve (γ=1.0) for demonstration purposes, let us assume that point A defines our initial state. This corresponds to a greater share of K-workers within node 2 (since $W>0$ and $N<0.5$). Suppose that the attractiveness of node 1 for K-workers increased through the establishment of a new university or public research unit. This steady decrease in W follows the given trajectory passing through points E and F, causing a small increase in the share of K-workers choosing node 1. If the critical point B happens to be reached, the value of N rises abruptly (to the point C). The equilibrium state of the knowledge-handling labour market undergoes a structural transition of a profound nature, with the vast majority of K-workers now preferring node 1.

A key feature of this nonlinear behaviour is its cyclic nature. For example, the attractiveness of node 1 may decline over time in terms of R&D capacity. In this case the equilibrium state of the market shifts gradually to the right of point C. A second critical point at D may be encountered, in which case node 1 loses the major share of K-workers just as abruptly as the earlier gain (falling from point D to point E). This hysteresis effect emphasizes the need for prolonged investment on R&D capacity expansion in order to attract and retain a high proportion of K-workers.

In Figures 4.4 and 4.5, γ was varied in order to demonstrate the model's sensitivity to the level of accessibility to K-workers (in all nodes) enjoyed by each node. Other parametric variations of interest include σ, β, d_{11} and d_{12}, since each of these also typically affect each node's level of accessibility to knowledge stocks, but in contrasting ways. For example, Figure 4.6 depicts the bifurcation possibilities as σ varies (for given values of $\beta = 0.03$ and $\gamma = 0.9$). Higher values of σ correspond to a broadening of the bifurcation base so that the attractiveness of node 1 for K-workers has a much higher threshold value (in other words, W must decrease further before the possibility of bifurcation arises). The opposite is the case for different values of d_{12} (with $\beta = 0.03$, $\gamma = 0.6$ and $\sigma = 3.0$), where Figure 4.7 confirms that the bifurcation base is quite narrow. Figures 4.8 and 4.9 present similar results for the cases where d_{11} and β vary, respectively.

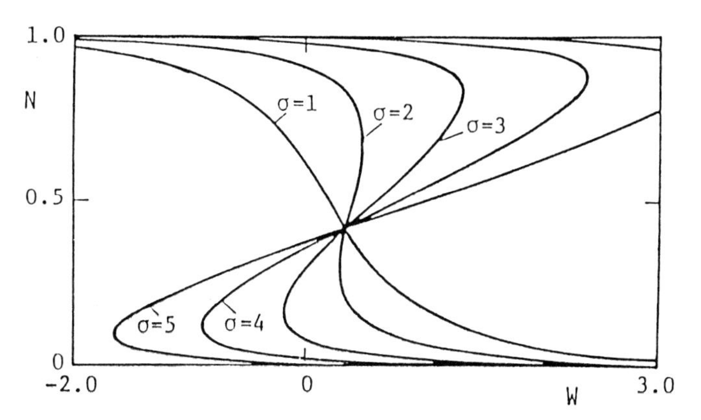

Figure 4.6 Fold catastrophes in the (W,N) plane ($\beta = 0.03$, $\gamma = 0.9$, σ varied)

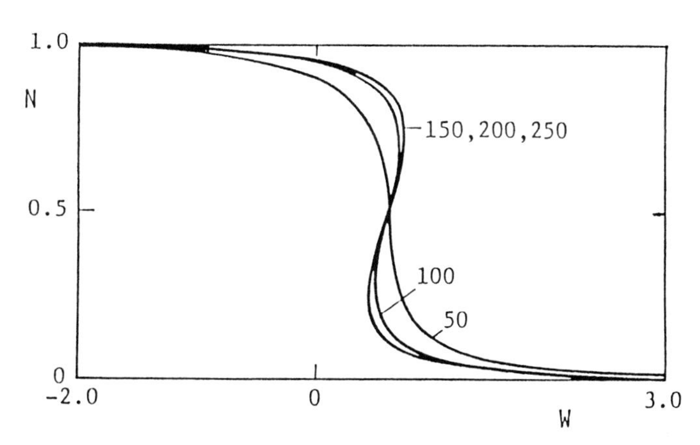

Figure 4.7 Fold catastrophes in the (W,N) plane ($\beta = 0.03$, $\gamma = 0.6$, $\sigma = 3.0$, d_{12} varied)

Figure 4.9 is quite revealing since the parameter β corresponds to an overall network technology coefficient. In the evolutionary situation where β is decreasing gradually as the overall efficiency of network exchanges improve, it is apparent that the system passes through a bifurcation zone ($0.002<\beta<0.04$ approximately). The shape of this zone in, say, the (β-W) plane is asymmetrical and may not correspond to any of Thom's elementary catastrophes. It therefore warrants further examination.

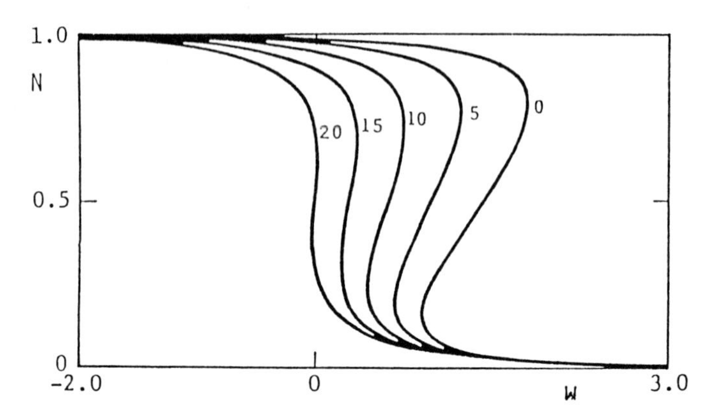

Figure 4.8 Fold catastrophes in the (W,N) plane
($\beta = 0.03$, $\gamma = 0.6$, $\sigma = 3.0$, d_{11} varied)

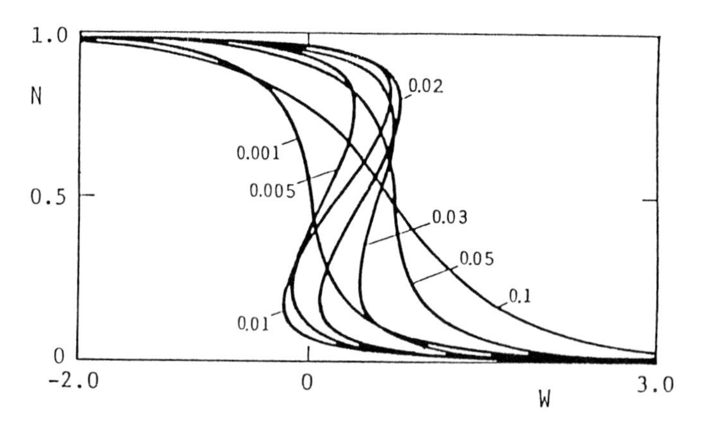

Figure 4.9 Fold catastrophes in the (W,N) plane
($\gamma = 0.6$, $\sigma = 3.0$, β varied)

4.4 CONCLUDING REMARKS

Some of the ramifications of the numerical results sketched above warrant brief explanation. The existence or absence of a bifurcation zone within an actual network of

K-workers is in itself a major research issue. In a qualitative sense, the implication is that those nodes which are able to develop and maintain a "research-rich" development path by employing a higher-than-average share of productive K-workers are most likely to enjoy technological leadership and higher-than-average economic growth. Furthermore, retention of a higher-than-average proportion of the K-workforce will depend on continuing investments to expand the capacity of local R&D units. The possibilities of discontinuous change merely highlight the risks associated with "do-nothing" policies and inappropriate priorities.

The notion of "interactivity" across knowledge networks may be an important one in this context. Measures of nodal interactivity may be defined in terms of the frequency and productive efficiency of all levels of interaction per unit of time. For example, if transport and communication are the two primary networks for knowledge exchange, a measure of interactivity would combine the average distance travelled per K-worker per year with the average number of messages (sent and received) annually by the same K-worker. Such a composite measure might be quantified in monetary units (e.g. as a share of the cost-budget) or in time units (e.g. as a share of the time-budget). Although it awaits further substantiation, in the transition towards a knowledge-based economy there is evidence to suggest that various measures of interactivity might play a complementary role to the productivity measures traditionally reserved for the firm.

The analytical discussion is far from complete. In addition to our desire to explore more deeply the multiple equilibrium states of the more general (n-nodes) model, the important issue of multi-layered networks possessing different rates of exchange has not been addressed. We have concentrated out attention in this paper on a single inter-dependency, namely the knowledge-network (K-N) interaction depicted in Figure 4.1. A more comprehensive approach would need to address a hierarchy of network categories - such as financial, information, commercial and cultural networks. We plan to examine various aspects of multiple network (N-N) interactions in future papers.

Further items of interest which have not yet been considered include:

(a) alternative formulations to represent the synergistic effects of multi-layered networks;
(b) the self-organizing character of the dynamic model of knowledge distribution;
(c) nonlinear accessibility functions which allow for such phenomena as economies of scale or decreasing knowledge returns from congested networks;
(d) the possible use of fractal theory to analyze variations in the density and quality of networks;
(e) appropriate investment functions to represent the response patterns to changing relative demands for nodal and link infrastructure which facilitate knowledge expansion.

REFERENCES

Andersson, Å.E., 1981, "Structural Change and Economic Development", *Regional Science and Urban Economics* 11, 351-361.

Andersson, Å.E., C. Anderstig and B. Hårsman, 1987, "Knowledge and Communications Infrastructure and Regional Economic Change", *Working Paper from CERUM* 1987:25, University of Umeå.

Anderstig, C. and B. Hårsman, 1986, "On Occupation Structure and Location Pattern in the Stockholm Region", *Regional Science and Urban Economics* 16, 97-122.

Haken, H., 1980, *Dynamics of Synergetic Systems*, Springer Verlag, Berlin.

Lakshmanan, T.R., 1988, "Infrastructure and Economic Transformation", in Å.E. Andersson, D.F. Batten, B. Johansson and P. Nijkamp (eds.), *Advances in Spatial Theory and Dynamics*, North Holland, Amsterdam.
Lande, P.S., 1978, "The Interregional Comparison of Production Functions", *Regional Science and Urban Economics* 8, 339-353.
Machlup, F., 1962, *The Production and Distribution of Knowledge in the United States*, Princeton University Press, Princeton, N.J.
Mees, A., 1975, "The Revival of Cities in Medieval Europe", *Regional Science and Urban Economics* 5, 403-425.
Poston, T. and I. Stewart, *Catastrophe Theory and its Applications*, Pitman, London.
Thom, R., *Structural Stability and Morphogenesis*, Benjamin, Reading, Mass.

CHAPTER 5

Innovation, Diffusion and Regions

Manfred M. Fischer

5.1 INTRODUCTION

Several explanations have been offered for the slow down of the world economy since the early 1970s and the international recessions of the mid 1970s and early 1980s: the destruction of the international monetary system when the Bretton-Woods regime of fixed exchange rates was replaced by generally fluctuating exchange rates in 1973, the oil-price crises of 1973/74 and 1979/80 which clearly reduced domestic real disposable incomes, rising inflationary pressures, stagnating markets for many major product groups etc. Even if all these factors have certainly played a more or less considerable role, they seem to have acted mainly to accelerate and exacerbate previously established trends.

Indeed, there is strong evidence that the roots of the current economic and employment crisis go back further than 1973 and refer to fundamental changes that have taken place in the structure of industry and technology (Rothwell and Zegveld, 1981). This structuralist explanation of the crisis has received increasing popularity in recent years and has been put forward within the framework of a renewed interest in long wave research aiming at explaining structural changes in long-term economic development (see, e.g., Mensch, 1985; Freeman, Clark and Soete, 1982).

The growing awareness of the role of technological innovations as a driving force in structural changes has stimulated policy makers in several countries (especially in France and Japan) to design and implement policy measures aimed at regulating and stimulating the creation of new ideas and the diffusion of knowledge. As is often the case, such policy initiatives have been generally more an object of faith rather than an understanding of the complexity of the innovation-diffusion process (Rothwell and Zegveld, 1981).

Evidently a fuller understanding of the conditions under which technical advance takes place is warranted. Technical change refers to all the changes in technology and techniques which lead to new products, new processes and new methods in industrial and distributional organisation and covers all the activities related to the innovation process, but also those related to the transfer and diffusion of knowledge. Research on technological change deals explicitly, or at least implicitly, with the questions why innovations occur, where innovations take place and how innovations diffuse in time and space. Detailed knowledge of these questions is still rather fragmentary. The paper attempts to summarize some of the major findings and to point to some deficiencies and gaps in knowledge. Particular emphasis is laid on spatial aspects of the innovation-diffusion process.

Section 2 starts with a brief characterisation of the innovation process, considers Research and Development (R&D) - a fundamental component of the innovation process - in more detail, describes industry- and size-specific variations in R&D, moreover, attempts to address the question why R&D effort is located where it is. Section 3 moves the focus from R&D to the diffusion of technological innovations and briefly characterizes four major conceptual developments in diffusion research. Most research on innovation diffusion relies on the adoption perspective, viewing the diffusion in terms of the adoption behaviour of firms. The major purpose of Section 4 is to identify the key elements which affect the rate and degree of adoption and to develop an understanding of the temporal and spatial nature of the diffusion process. Some major problems arising in diffusion research are briefly discussed in Section 5. The paper concludes with a discussion of some policy implications.

5.2 R&D, INNOVATION AND SPACE

The term innovation commonly refers to the initial commercial introduction of a new product, the first utilization of a new process or an organisational technique, i.e., to the result of an innovation process (see Freemann, 1974). The innovation process covers a succession of operations (i.e. the transition from the idea to the materialization) and may be considered as a logically sequential, though not necessarily continuous or linear process which can be disaggregated into three functionally separate, but interacting stages: first, recognition and idea conception; second, research and development (R&D), and third, innovation (see Figure 5.1). Interactions and feedbacks are inherent characteristics of the innovation process. The various R&D-functions are not only linked with other functions inside a business company, but are also connected to external developments in the scientific and technological infrastructure and the market.

Of course, activities carried out in all these stages are influenced, by the locational and the wider environment in which the innovation organisation operates and the perceived market needs. Such external conditions include inter alia world, national and regional economic conditions, environmental and economic regulations, political climate, cultural aspects etc. and the whole range of public measures designed to facilitate the technological transformation process within the firm (see Rothwell, 1983). Innovation-relevant ideas may result from the technology and its development (technology push), the market (demand pull) or from a linking of both, i.e. an increasing recognition and clarification of technological possibilities and assessments of relevant market needs.

5.2.1 Three Major Categories of R&D

R&D activity is a fundamental component of the innovation process. It aims at expanding and applying the stock of knowledge to commercial needs and encompasses work of different kinds. The distinction between categories of work is often hazy. But to generalize, it seems to be useful to distinguish between three broad types of activities (see Figure 5.1): Basic research; applied research; and experimental development. These categories can be associated with particular task environments. the most distinguishable attributes of these environments are inter alia the relative presence of commercial objectives, the operational time horizon, the degree of uncertainty associated with the particular R&D activity and barriers to entry (Howells, 1984), which are used in Table 5.1 for characterizing the corresponding task environments.

Basic research has strong ties with pure science and refers to original investigations for the advancement of scientific knowledge without any particular commercial application in view. Of course, basic research is a very costly and risky long-term exercise with unpredictable commercial benefits. Thus, it is not surprising that basic research is primarily undertaken in research units of higher education and in governmental research establishments rather than in the industrial firm even if greater efforts in fundamental research may be observed in industrial R&E-laboratories very recently. In Great Britain for example,

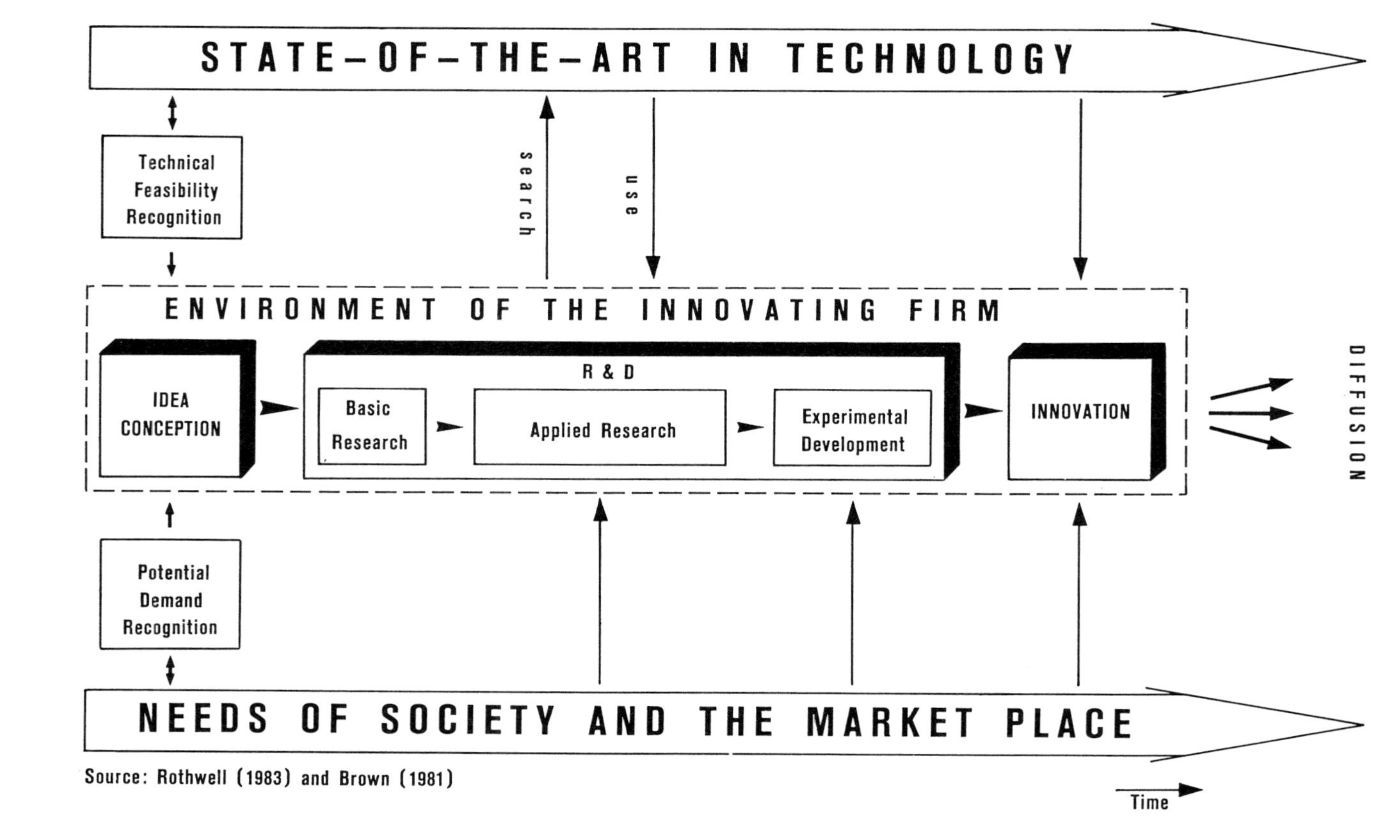

Source: Rothwell (1983) and Brown (1981)

Figure 5.1 An interactive model of the innovation process

higher education undertakes about 45 per cent of the total expenditure on basic research, governmental research establishments about 35 per cent and industry the remaining 20 per cent (Kennedy and Thirlwall, 1972).

Table 5.1 Task environment characteristics of the three categories of R&D

Category of R&D	Presence of Commercial Objectives	Operational Time Horizon	Degree of Uncertainty	Barriers to Entry
Basic Research	Low	Long Run	High	High
Applied Research	High	Medium Run	Moderate	Medium
Development Activity	Medium/High	Short Run	Low	Low

Source: Howells (1984)

The greatest number of industrial R&D efforts are directed towards applied research and experimental development, with an emphasis on development (including the design, production and testing of prototype and pilot developments). Both types of activities are important for a manufacturing firm to maintain or enhance its commercial position. Applied research may be defined in this context as that work which is undertaken with commercial objectives in view, in terms of new or improved products, processes or devices.

Basic innovations such as developments in microelectronics, have major direct and indirect impacts upon national and regional economies and form the basis for a continuous flow of incremental and improvement innovations which are based on user feedbacks. Relatively little research has been done to analyse the aggregate impacts of innovations on regional economies. Nevertheless, it might be suggested that old industrial areas and peripheral regions create much fewer major innovations than core regions do. Technical progress in such regions comes more from the diffusion of innovation (especially of process technologies) than from the creation of new products and processes.

The question, whether basic inventions are mainly the creation of individuals working independently or the result of large organised R&D departments, is discussed controversy in the literature. Following Kennedy and Thirlwall (1972) there are good reasons to assume that, on the one hand, independent inventors account for a significant, but decreasing percentage of major inventions and that, on the other hand,the large firm R&D laboratories seem to be only a relatively minor source of major inventions, though a major source of improvement inventions.

The impact of R&D activity on new knowledge in general and on inventive output in particular is extremely difficult to measure. there is no unique measure of inventive output. Quite often the number of patents is used as a crude indicator of advances in knowledge. But this proxy is not without problems because the quality of patented inventions varies markedly. A high proportion of all patents have no commercial value and some incorporate very little new knowledge at all. Moreover, inventions still require development work. Development work essentially translates inventions into production and marketing. Even the technical success of an innovation does not guarantee its commercial success.

Clearly, the opportunity of a firm to innovate successfully depends on the flow of new knowledge or resources to put new knowledge to work and, thus, on good internal and external communication links between the functionally separate departments within the firm on the one hand and between the firm and outside sources of science and technology as well as between the firm and customers on the other. It is, however, not only the opportunity to innovate, but also the capacity to innovate which is important. And the capacity to innovate depends not only on the flow of information, but also on the availability of financial resources to undertake innovation.

Commercially successful innovations require efficient development and especially the elimination of technical defects in the prototype and, in particular, before commercial launch. Successful firms, on average, out-perform failures in all the areas of competence encompassed by the innovation process and pay more attention to marketing and sales in order to understand user needs better.It is basically the balance of functions of R&D, production and marketing, in quantitative and qualitative terms, which is important for success (Rothwell, 1980).

The time interval between an invention and its eventual utilisation as an innovation is often rather long and costly, too (see Thomas and Le Heron, 1975). Empirical evidence suggests that the average lag between invention and innovation is approximately 10-15 years. Of course, there are inter-industry variations. For example, mechanical innovations require only rather short intervals and electronic innovations rather long ones (Mansfield, 1968). R&D expenses, on average, will be distributed approximately in the ratios of 1:6:27 on basic research, applied research (including inventive activity) and development, until a product is brought to the point of continuous production (Buswell, 1983).

The issue, whether R&D can be said to lead to innovation, is a problem which has long been under scrutiny, and points to the problem of how to measure innovation performance. There is no universally accepted method to solve this measurement problem. In general, statistics derived from the input side (such as R&D and patent statistics) rather than the output side, are used and regarded as an indication of the intensity of the search for innovations (Thomas and Le Heron, 1975). Counts of successful innovations would be certainly more satisfactory measures (see, e.g., Oakey et al., 1980, 1982). There are, however, considerable problems in data collecting, in measuring incremental versus radical innovations and especially in defining satisfactorily the population of innovations (Pavitt and Soete, 1980).

5.2.2 Industry- and Size-Specific Variations in R&D

Innovation research over the last decade has clearly demonstrated that some industries show a greater propensity for innovation via R&D, as reflected in R&D investment, than others. R&D investment measured in terms of expenditure or employment varies considerably from industry to industry. The more science-based industries of electrical/electronics, office equipment and computers, aerospace and chemicals show relatively high ratios of professional manpower in relation to total employment or R&D expenditures to net output in several countries, while relatively little R&D activity is generated in such traditional sectors as textiles and leather goods.

The inter-industry variations in R&D intensity in Great Britain have been the subject of considerable research undertaken by the Science Policy Research Unit at the Sussex University (see, e.g., Pavitt, 1980). Some general conclusions which may be drawn from these research efforts can be summarized as follows (Pavitt, 1980). First, the nature of innovations varies widely among sectors. In mechanical engineering, for example, the main focus has been product innovation. In iron- and steel-based industries the major innovations have been process innovations even if they have been induced or resulted in some changes in the quality of the product. Second, in nearly all sectors in-house R&D capacity has become a prerequisite for successful innovation. The slow pace of innovation in the British steel industry, however, illustrates the fact that innovation will not follow automatically if volume of investment is right. Third, much R&D effort in Great Britain (also in France and the USA) has been devoted to defence and related activities (aerospace) and nuclear energy, to the neglect of other sectors of the economy. But in contrast to the USA, the spin-off of defence into new industrial technologies and products has been rather low in Great Britain.

Expenditures in R&D are not only concentrated in the more science-based industries, but also in larger enterprises. Larger firms dominate in all research-based sectors. In Belgium, for example, larger firms (with more than 500 employees) accounted for 70 per cent of all R&D spending in the 1970s (Sweeney, 1987) and in Sweden even for 94 per cent (Andersson, 1985). In general, smaller single-site enterprises tend to have only a very limited in-house R&D capability. Especially those carrying out subcontract work seem to be dependent upon customer specification rather than original research, even if it happens that technology-based small enterprises which have a particular expertise sustained by intensive R&D in a very narrow field outperform large firms in innovation (e.g. in medical instrumentation), especially in new areas of economic activity. Nevertheless, there is no doubt that there is a positive relation between the size of firms (measured in terms of total employment) and research intensity, although not in a simple linear correlation as data for the USA indicate (see Malecki, 1979b).

5.2.3 Spatial Aspects of the Innovation Process

The importance of R&D to both national and regional economic development (measured in terms of production and employment) is widely recognized (see e.g., Mansfield, 1968). Nevertheless, spatial aspects of technological change in general and the question why R&D effort is located where it is, in particular, have received only little attention, - thus, empirical knowledge is relatively scarce despite the fact that a number of economic geographers and regional scientists have addressed the question of R&D location at a conceptional, theoretical level.

The location of industry-specific R&D facilities in the USA has been documented for the mid 1970s by Malecki (1979a,b). His findings provide evidence that R&D is largely an urban-based activity. About 60 per cent of the R&D departments were concentrated in the ten largest SMSAs. In California, where a very large amount of R&D activity is located, R&D is concentrated almost exclusively in the three largest urban areas, Los Angeles-Long Beach-Anaheim, San Diego and San Francisco-Oakland-San Jose. Accessibility to major research-oriented universities and federal government facilities, to (inter)national repositories of information such as libraries and patent offices, the availability of highly qualified scientific and technological labour force, and close proximity to the headquarters of firms are found to be important location factors.

Naturally, there are inter-industry variations in the degree of R&D concentration. Food, miscellaneous manufacturing, chemicals and machinery have a large fraction of their R&D departments in non-metropolitan regions and are, thus, more evenly distributed over the regions. In contrast, instruments, drugs, aerospace, car electrical and oil industries show a strong preference for metropolitan locations, though the latter four have increased their R&D engagement in non-metropolitan areas during the 1970s. Changes in the location of R&D facilities of the 300 largest corporations over the period of 1965-77 indicate a general trend towards a moderate decentralisation in the location of R&D activity in the USA (Malecki, 1979b).

New locations, including metropolitan and non-metropolitan areas, have developed in the South Eastern USA, the South and West. This trend is clearly associated with other shifts in the Frostbelt-Sunbelt dichotomy of recent American regional economic development (see, e.g., Fischer and Nijkamp, 1987). The decline of the large-city locations in the North East has been at least partially offset by nearby surrounding small- and medium sized city locations. These areas seem to have increased their innovative potential with regard to large firms, to the disadvantage of the larger urban areas, especially in the Boston and the Philadelphia area (Malecki, 1979b).

Several of Malecki's findings coincide with research results obtained in Europe. For France, for example, Brocard (1981) reports on a strong bias of the R&D activity towards the metropolitan region of Paris, despite a recent decline of about 10 per cent in the concentration of R&D employment and a growth in the more peripheral regions, especially in the Cote d'Azur-Provence. In the UK a preference for smaller towns and cities can be observed in the metropolitan region of London and in the regions based on the provincial conurbations (see Goddard and Thwaites, 1986). The overall picture shows a strong concentration of R&D in the South East, reinforced by the concentration of corporate headquarters.

Quite recently, attempts have been made to analyse the location of large firm R&D within the corporate context, especially by British and American geographers and regional scientists. The studies - especially by Howells (1984) and Gibbs and Thwaites (1985) - have provided useful and interesting insights into the organisation of R&D within certain sectors of industry and the factors influencing its location. The work by Gibbs and Thwaites (1985) for example, is based on four UK sectors of industry: Pharmaceuticals; metal working machine tools; scientific and industrial instruments; radio and electronic components.

The findings obtained in these studies suggest the following major conclusions. The location of the three major types of R&D activity (see Table 5.1) is largely determined by the location of higher-order corporate functions rather than by external (environmental) factors. Good communication links between R&D and other corporate activities are of central importance (Howells, 1984). Long-term basic research is usually concentrated in central research units benefitting from economies of scale and is located in close proximity to the decision making functions. The location of applied research units in a multi-divisional corporation tends to be more likely associated with the individual product divisions. Each product-line division tends to have its own R&D department. Only more important activities seem to be centralised. Short-term and, in particular, minor production-oriented development work is much more widely spread among the locations of the product-line divisions than basic and applied research, and is linked to the location of production. This certainly enables production problems to be solved more quickly and efficiently. Of course, the level and quality of R&D work at production sites depends on the technological context of the product range. A productive unit producing standardised and mature products is less likely to need in-house higher level R&D facilities than a plant with technology-based products. Consequently, both type and level of R&D activity within a given region largely depend upon the other corporate functions which are located there (Gibbs and Thwaites, 1985).

5.3 DIFFUSION OF INNOVATIONS: SOME ALTERNATIVE PERSPECTIVES

The diffusion of a new product or process is the last stage in the sequence of the innovation diffusion process. Without doubt the innovation stages have, to date, attracted far more academic attention than diffusion (see, for example, Kennedy and Thirlwall, 1971, for a fairly comprehensive survey). A deeper understanding of the diffusion process may be potentially relevant to the study of industrial economics in general and to the implications for the technical progressiveness of high concentration and large firm size in particular.

Research on the diffusion of innovations has traditionally taken two directions. The first focuses on consumer innovations, i.e. innovations which are adopted and used by households. This research direction has been and still is a major concern in geography. The second research tradition has been developed in economics and focuses upon technological or firm innovations which are adopted and used by firms themselves. The majority of this work is concerned with the spread of process innovations across industrial firms. Spatial aspects of the diffusion process, however, are by and large ignored, in contrast to the first research direction.

Process innovations diffuse and are adopted by firms on the open market as purchased machines. Product innovations are more confidential, both in conception and construction, since they are the means by which advantage is gained by the innovating firm over competition. Consequently, product innovations are markedly less mobile in diffusion terms and may diffuse only in a limited sense through multisite corporations or licensing arrangements (Oakey et al., 1982).

In attempting to integrate elements of both the economic and geographic research traditions, Brown (1981) has identified four major approaches to the study of innovation diffusion:

• the adoption approach,
• the market and infrastructure approach,
• the economic history approach and
• the development approach.

The *first approach* focuses on the process by which adoption occurs and, thus, on the demand side of the diffusion process. This perspective is indicative for must of the research of diffusion of technological innovations and will be characterised in greater detail later on.

The second *approach*, the market and infrastructure approach, emphasises the role of supply factors in the diffusion process and removes attention from the characteristics of adopters to those of organisations, such as firms and government agencies, which market innovations (see Brown, 1981; Malecki, 1977). The market structure of the supplying industry, the objectives of the suppliers in general and price setting and output policy in particular, promotional communications and market selection, the degree of learning economies and the extent of any financial constraints are considered to have an important influence on the diffusion path. Stoneman (1983) presents various diffusion models which allow for interaction between the supplying industry and the using industry and which may be considered as prominent examples of this approach to innovation diffusion.

The *economic history perspective* on innovation diffusion emphasizes the dynamic evolving nature of innovations. A critical element of a successful innovation is its continual technological improvement and adjustment to an increasing variety of uses, leading to adoption by an increasingly wider range of firms. This continuity aspect of technological innovations is addressed by the economic history approach in an economic frame of reference where the form and function of the innovation and the environment into which it might be adopted are modified throughout the life of the innovation and these changes affect the pace of diffusion (Brown, 1981).

The *development perspective* focuses on the interrelationship between diffusion and development in general and the impact of innovation diffusion on economic development and social change in particular. This approach suggested by Brown (1981) is still not developed very well, but might aim at analysing the role of technological innovations within the hierarchy of economic systems at different spatial scales on the one hand and at examining the consequences of innovation diffusion in specific socio-economic and spatial contexts on the other.

Most research on innovation diffusion relies on the adoption perspective. This perspective views the diffusion of innovations in terms of the adoption behaviour of the firms deciding to use the innovation in question. Despite the relative paucity of empirical studies, there is a general consensus that the characteristics of the innovation, firm characteristics, industry characteristics such as the intraindustry competitive structure and institutional factors such as laws and regulations are of crucial importance in the adoption decision (see Kennedy and Thirlwall, 1977; Nabseth and Ray, 1974; Thomas and Le Heron, 1975; Malecki, 1977; Brown, 1981).

The innovation characteristics identified as being important are the cost of adoption and the relative profitability or economic advantage of the innovation. Adoption cost includes the initial investment in implementing the innovation, operating expenses and the risk of adopting an unproved innovation (Malecki, 1977). Lack of financial resources might delay the adoption of an innovation at least in an early stage or lead to an adoption in a limited manner only, especially in the case of innovations which are most difficult or costly to learn to use, install and integrate, owing to their technical complexity. In general, low cost innovations tend to diffuse more rapidly than high cost ones. Moreover, it is widely expected that economically more advantageous innovations will diffuse more rapidly. But the measurement of economic advantages which make the new process profitable relative to alternative more conventional technologies usually involves serious difficulties, mainly because profitability changes over time for a given new process and for a given firm (Nabseth and Ray, 1974). This occurs because of the time pattern changes over the expected life of the innovation in investment requirements, output levels, product prices, input factor qualities and prices, marketing and distribution.

Firm characteristics are also considered to be important variables influencing adoptive behaviour. The influence of firm size is based on the rationale that larger firms can more easily accept the adoption risks, have easier access to capital and the expert device needed to evaluate the innovation and to implement it within the firm (see Nabseth and Ray, 1974; Malecki, 1977; Thwaites et al., 1982). The adoption of a more complex innovation requires not only transfer of information about the innovation in the form of blue-prints, drawings and other data, but also technical devices to modify the innovation and sometimes even further innovation before a given innovation can be introduced successfully.

Mansfield et al. (1977) discovered that larger firms were generally early adopters of new technologies. Not all case studies, however, confirm this relationship. Nabseth and Ray (1974), e.g., report that smaller firms took the lead in certain fields, for instance, in oxygen steel and continuous casting. The findings lead to the conclusion that, first, firm size operates more as a threshold rather than as a scale and, second, the threshold size of firm tends to vary with different technologies (see Nabseth and Ray, 1974).

Another important firm characteristic is the attitude of its management to the adoption of innovation (see Nabseth and Ray, 1974; Thomas and Le Heron, 1975; Malecki, 1977). Aggressive firms with an incentive to grow and improve their competitive position tend to display a much greater propensity to adopt innovations than passive or defensive firms. This feature may be more prevalent in small- and medium-sized firms whose managements tend to be more flexible and willing to adjust to new technologies and market possibilities more rapidly than large, organisationally and technologically more rigid corporations. Such receptivity to innovation might offset the several advantages which directly vary with firm size, especially for lower cost and risky innovations which exhibit a lower firms size threshold for adoption. The concept of aggressiveness and innovativeness is, however, not easy to quantify and has remained largely a qualitative explanation of residuals from adoption patterns (see, e.g., Mansfield, 1968; Nabseth and Ray, 1974).

5.4 TIME AND SPACE PATTERNS OF DIFFUSION

The majority of the work on diffusion focuses on the adoption perspective and is concerned with the spread of new processes across firms. The studies concentrate on individual innovations diffusing in single industries and attempt to explain differences between firms in the time taken to adopt. A common place of much of this research is the heavy reliance that is placed on the epidemic model approach (see, e.g., Mansfield, 1968; Romeo, 1975; Stoneman, 1983). The basic tenet of this conceptualization of the spread of innovation is that the adoption of an innovation is seen to be primarily the outcome of a communication process. The approach assumes equal probability of adoption within the a priori defined population of potential adopters and, moreover, homogeneity of the space. These assumptions lead to the well-known sigmoid path of development of the diffusion rate showing a slow take-off, an intermediate stage of varying rapid diffusion and a final stage of retardation in which diffusion asymptotically approaches a saturation ceiling.

The logistic curve, the most commonly used representation of the class of sigmoid curves may be described (see Stoneman, 1983) as

$$p(t) = k/(1 + \exp(-G(t)) \tag{1}$$

where t denotes time, $p(t)$ the proportion of adopters who have adopted the innovation in question at time t, k the asymptotic ceiling value of adoption (saturation) or equilibrium level of adoption and $G(t)$ some - usually linear - function of time so that

$$d\,p(t)/dt = g(t)\,p(t)\,(1-p(t)/k) \tag{2}$$

with

$$G(t) = \int_0^t g(\tau)\,d\tau \tag{3}$$

with $g(t)$ denoting the diffusion speed. According to (1)-(3) adoption by potential adopters increases at an increasing rate during the first half of the diffusion process, reaches an inflection point (in the case where $g(t)$ is constant: $k/2$), then increases at a decreasing rate until the potential of potential adopters is exhausted at the asymptotic ceiling k.

In the simplest version of the epidemic model approach, $G(t)$ degenerates to $a+bt$ where 'a' denotes the intercept parameter representing the starting point of adoption and 'b' the rate of increase of adoption. Taking the linear transform of this logistic function $\log(p(t))/(k-p(t)) = a+bt$ the model parameters a, b and k can be directly estimated by least squares. Even if information about the innovation is essential, it cannot be used alone to account for the diffusion pattern of technological innovations. Thus, the more general version of the logistics has clearly to be preferred where in addition to informational aspects other factors influencing the adoption decision can easily be taken into account and enter into the function $g(t)$.

Within an intra-firm diffusion context internal characteristics of firms such as the organisational structure and resource endowment may be inserted as determinants of diffusion, together with communication related variables. The importance of the sectoral, technical-economic environment of a firm is emphasized in the case of intra-sectoral diffusion. Industry and firm characteristics such as the degree of market concentration, the size structure of firms, the profitability of an innovation related to its market potential, the extent of technological experience and access to capital are frequently used to explain the intra-sectoral diffusion process, in addition to informational variables.

Of course, there are different speeds of diffusion and also irregularities in this general time pattern, from one innovation to another and from one industry to another. Nevertheless, there is considerable agreement that the cumulative pattern of adoption of a new technique tends to conform rather closely to a S-shaped time distribution when the number of adopters is measured over time. Most studies on innovation diffusion have emphasized the sectoral and temporal dimensions, but largely neglected any explicit consideration of the spatial dimension of the diffusion process. Research into the spatial dimension of the innovation diffusion process has been mainly left to geographers primarily concerned with consumer rather than technological innovations.

Two spatial regularities of innovation diffusion need to be considered briefly here. First, *the neighborhood effect*, which is well documented in the case of consumer innovations (see Hägerstrand, 1967), emphasizes the relationship between distance and the sequence of innovation adoption (usually modelled in form of a distance decay function) where distance may be measured in physical, socio-cultural or technical-economic terms. The spatial pattern is thought to start with a diffusion source and to spread outward over time in a contagious fashion. This effect, it has been suggested, plays a role in spatially agglomerated industrial complexes and in urban areas (Pred, 1966; Thomas and Le Heron, 1975), in particular, in cases where bandwagon effects predominate and where information occurs principally through contacts among neighbours.

The second, *the spatial regularity of innovation diffusion* which has received much attention, emphasizes the urban-size hierarchy as the prime determinant of the diffusion process. According to this approach adoption principally takes place in the largest cities of a settlement system first and then the diffusion occurs horizontally between cities of the same size and vertically downwards through the settlement hierarchy. The main causes of hierarchical diffusion patterns are considered to be, on the one hand threshold problems in the market potential or the resources necessary for the adoption of an innovation and, on the other, the hierarchical structure of important communication channels (Ewers and Wettmann, 1980). Operational hierarchic diffusion models are largely based on either Christallerian- and Löschian-type central place theory or the rank-size rule. Frequently, hierarchical and contagious diffusion patterns are incorporated in the models (see, e.g., Pederson, 1970).

The explanations of the diffusion process of technological innovations offered in such studies are, however, not satisfactory for several reasons. First, Pred (1975) has provided convincing evidence that diffusion - no matter whether intra- or inter-organisationally - need not be merely comprised of larger city to smaller city sequences, but may also include spread from smaller cities to larger ones. Moreover, Pred (1975) points to the fact that an adoption decision made in a headquarter city does not necessarily imply innovation implementation in that city. Second, contagious and hierarchical diffusion models basically deal with the question how adoption develops in time and space, but provide no answer as to why a particular diffusion pattern arises (see Ewers and Wettmann, 1980).

Few studies of innovation diffusion have focused upon regional differentials in the diffusion process or have attempted to explain the causes of such disparities. One notable exception is the work undertaken at the Centre for Urban and Regional Development Studies at the University of Newcastle upon Tyne (see, e.g., Thwaites et al., 1982) which relates the adoption of technological innovations to a number of spatial and aspatial contextual variables. There are good reasons to assume that the explanations of regional differentials in the innovation diffusion process lie not only in regional differences in industrial structure or in the size structure, but are also related with the local industrial milieu and the corporate context of industrial organization. The geographical path of diffusion is likely to be different in single-site and multi-site companies. Studies on the influence of local environmental conditions and of industrial organisation on the adoption behaviour of firms certainly would help to improve our knowledge about the nature and speed of diffusion processes.

5. SOME PROBLEMS IN MEASURING DIFFUSION RATES

One of the major problems one faces in diffusion analysis refers to the way in which the diffusion of innovation at any particular time is measured. The most commonly used measures are the proportion of firms or plants, using the innovation, within the a priori defined population of potential adopters. The first of such measures diffusion among firms, and the second, diffusion among and within firms. A shortcoming of these measures is clearly the fact that they fail to indicate whether the adoption represents only limited developmental applications or pervasive commitments (Gold, 1981).

The major problem consists in defining the denominator in the fraction of the diffusion measures, i.e. the population of potential adopters. This problem has at least two roots. First, there is the general problem of identifying the population of potential adopters at a certain point in time. This is rather difficult and creates severe difficulties, especially in cases of process innovations only suitable for applications to a certain type of production within industry.

A second major difficulty arises because technological innovations continuously undergo numerous kinds of changes. Improvements may involve increases in reliability, in operating quality and efficiency etc. and, thus, increase the range and benefits of applications of earlier forms of the innovation. Significant improvements over time, however, may lead to an expanding array of potential adopters. This points to the need to relax the assumption of a static potential adopter population. One strategy to tackle this widely neglected problem is to replace the denominator by successively adjusted estimates of the pool of potential adopters corresponding to technological changes in the innovation over time (see Gold, 1981). For a discussion of further problems and shortcomings see Gold (1981), Nijkamp and Rietveld (1987).

Finally, it is worthwhile to mention that several important aspects of technological diffusion are largely neglected by diffusion research relying on the adoption perspective, aspects which may, however, provide useful insights into the matter for policy making. One of these aspects involves measurement of the effects of increasing diffusion of an innovation not only on its adopters, but also on non-adopters, on the suppliers and the customers of adopters, on the relevant factor and product markets, on the growth and competitive strength of the industry as well as on the development of regional economies in which adopters are located.

5.6 CONCLUDING COMMENTS: SOME POLICY IMPLICATIONS

In recent years it has become increasingly clear that regions, least of all the problem regions, can no longer rely on an external of solution for their regional employment problems. Local authorities in several countries are often now promoting the retention of local industry and employment, quite in contrast to the situation in the 1960s and early 1970s. Consequently, regional policies increasingly emphasize the need for indigenous regeneration of problem areas. One approach to achieve this is by stimulating innovation. This reorientation in regional policy is also reflected in the growing attention geography and regional science are now paying to the relation between innovation, technology and regional development.

In this context the innovation potential of a region plays a central role as a decisive bottleneck factor. A wide range of factors are considered to make up the innovation potential of a region. The most important ones seem to be the following (see Sweeney, 1987):

- the sectoral and technological mix of the industry which forms a major component of the diversity leading to a region's higher creativity and innovativeness (especially higher level technological skills and products strongly influence the technological orientation of a region and its ability to achieve self-generated growth);
- the strength in the indigeneous engineering sector with a sufficient number of medium to large-sized firms which seems to be essential for the technical culture providing the environment for innovation and rapid diffusion of innovation;
- the autonomy of decision making in the industries and industrial infrastructure;
- the strength of the quarternary information sector consisting of public and of private research institutes, information centres, financial institutions, advertising, public and

private bureaucracies etc. which make up a large segment of the innovation potential of a region.

Naturally, different regions are characterized by quite different innovation potentials. Core central regions, as defined by Malecki (1981), show very favourable conditions for technological change. Core central regions are large conurbations in which the headquarters of large high-technology industries with the centralised information-intensive functions of R&D, corporate planning and decision making are located and in which there are, moreover, information-intensive activities in consultancy and information centres, financial institutions and government services. Clearly, such regions have a high stock of knowledge, which is continuously being enlarged, and a high awareness of and access to the best current technical practice where the larger firms being technically progressive as major innovators (see also Sweeney, 1987).

Problem regions are characterized by a weak innovation potential. Different types of such regions are lacking in different ways. The industrially declining regions such as those based on coal and steel do not have a mix of sectors, technologies and occupations. Their former prosperity was built around specialisation in one or two sectors. As the regions matured in production and marketing, concentration took place in a few large firms. The industries became incapable of renewing their technologies and of creating new economic activities, leading to a low entrepreneurial vitality and, thus, to a declining small firm sector. The regions suffer from a petrification of their socio-economic system. Not only the firms, but also regional and local government institutions, banking, labour organisations etc., are predominantly large and inflexible. Moreover, most of the public and private organisations which serve the industry have largely adjusted their activities to the specific needs of large industrial firms (see Ewers and Wettmann, 1980; Sweeney, 1987). For such regions it is very difficult to design and implement an adequate innovation strategy, especially it would be necessary - as Ewers and Wettmann (19800) emphasize - to destroy well-established organisational networks, cartels of vested rights, mental barriers etc. first, before a promising regeneration strategy can be started.

While innovation policies in old industrial areas confront the major problems of hostile environments, this is less valid in the case of regional policies aimed at peripheral and rural regions, the second type of problem regions. The problem in these regions is quite different. They are largely characterized by branch plants of multi-plant companies which have no autonomy in production technology and product development and are, thus, inhibited in technological innovation (see Malecki, 1981). There is only little diffusion of the best practice of such branch plants into the regional economy because their information-intensive functions are rather limited and highly production-oriented. The lack of advanced in-house technological research in such plants also reduces their possible role as incubator plants for new firms (Thwaites, 1978).

A regional innovation strategy in this case has clearly to be directed at the small- and medium-sized firms as catalysators to induce technological change. But Rothwell's (1982) suggestion to establish an indigenous innovation potential via the formation of new technology-based firms seems to be too optimistic. Certainly not all regions have the capacity to pursue such a kind of techno-economic regeneration (see Buswell, 1983). More realistically, regional policy has to rely on already existing small- and medium-sized firms and to provide assistance in terms of information supply, consultation, financial and training services, and to encourage interfirm cooperation which may help these firms to become early adopters of innovations. In the long term the creation of a suitable socio-economic environment and infrastructure seems to be decisive to the development of the intraregional R&D potential which - no doubt - is desirable because of the high-multiplier activity of R&D. But this is not an easy task, even in the long run.

REFERENCES

Andersson, A.E., 1985, *Creativity and Regional Development*, WP-85-14, IIASA, Laxenburg.

Brocard, M., 1981, "Amenagement du territoire et developpement regional: Le cas de la recherche scientifique", *L'Espace Geographique* 6:61-73.

Brown, L.A., 1981, *Innovation Diffusion. A New Perspective*, Methuen, London and New York.

Buswell, R.J., 1983, "Research and Development and Regional Development", in Gillespie, A., *Technological Change and Regional Development*, 9-22, Pion, London.

Davies, S., 1979, *The Diffusion of Process Innovations*, Cambridge University Press, Cambridge et al.

Ewers, H.-J. and R.W. Wettman, 1980, "Innovation-Oriented Regional Policy", *Regional Studies* 14:161-179.

Fischer, M.M. and P. Nijkamp, 1987, "Current Trends in Regional Labour Markets", in Gordon, I., *Unemployment, Regions and Labour Markets: Reactions to Recession*, 15-43, Pion, London.

Freeman, C.,, 1974 *The Economics of Industrial Innovation*, Penguin Books, Harmondsworth.

Freeman, C., J. Clark and L. Soete, 1982, *Unemployment and Technical Innovation. A Study of Long Waves and Economic Development*, Frances Pinter, London.

Gibbs, D.C. and A.T. Thwaites, 1985, *The Location and Potential Mobility of Research and Development Activity: A Regional Perspective*, Paper presented at the 25th European Congress of the Regional Science Association, Budapest.

Gillespie, A., ed., 1983, *Technological Change and Regional Development*, Pion, London.

Goddard, J.B. and A.T. Thwaites, 1986, "New Technology and Regional Development Policy" in Nijkamp, P., *Technological Change, Employment and Spatial Dynamics*, 91-114, Springer, Berlin et al.

Gold, B., 1981, "Technological Diffusion in Industry: Research Needs and Shortcomings", *The Journal of Industrial Economics* 29:247-269.

Hägerstrand, T., 1967, *Innovation Diffusion as a Spatial Process*, University of Chicago Press, Chicago.

Howells, J.R.L., 1984, "The Location of Research and Development: Some Observations and Evidence from Britain", *Regional Studies* 18:13-29.

Johansson, B., and C. Karlsson, 1986, "Industrial Applications of Information Technology: Speed of Introduction and Labour for the Competence", in Nijkamp, P., *Technological Change, Employment and Spatial Dynamics*, 401-428, Springer, Berlin et al.

Kennedy, C. and A.P. Thirlwall, 1972, "Surveys in Applied Economics: Technical Progress", *The Economic Journal*, 82:11-72.

Malecki, E.J., 1977, "Firms and Innovation Diffusion: Examples from Banking", *Environment and Planning* A9:1291-1305.

Malecki, E.J., 1979a, "Agglomeration and Intra-Firm Location in the United States", *Tijdschrift voor Economische en Sociale Geografie* 70:322-332.

Malecki, E.J., 1979b, "Location Trends in R&D by Large U.S. Corporations, 1965-1977", *Economic Geography* 55:309-323.

Malecki, E.J., 1980, "Corporate Organization of R and D and the Location of Technological Activities", *Regional Studies* 14:219-234.

Malecki, E.J., 1981, "Science, Technology and Regional Economic Development: Review and Prospects", *Research Policy* 10:312-334.

Mansfield, E., 1968, *Industrial Research and Technological Innovation. An Econometric Analysis*, Norton, New York.

Mansfield, E., J. Rapoport, A. Romeo, E. Villani, S. Wagner and F. Husic, 1977, *The Production and Application of New Industrial Technology*, Norton, New York.
Mensch, G.O., 1975, *Stalemate in Technology. Innovations Overcome the Depression*, Ballinger, Cambridge, Ma.
Nabseth, L. and G.F. Ray, 1974, *The Diffusion of New Industrial Processes. An International Study*, Cambridge University Press, London.
Nijkamp, P. ed., 1986, *Technological Change, Employment and Spatial Dynamics*, Springer, Berlin et al.
Nijkamp, P. and P. Rietveld, 1987, "Technological Development and Regional Labour Markets" in Fischer M.M and P. Nijkamp, *Regional Labour Markets: Analytical Contributions and Cross-National Comparisons*, 117-138, North-Holland, Amster–dam.
Oakey, R.P., A.T. Thwaites and P.A. Nash, 1980, "The Regional Distribution of Innovative Manufacturing Establishments in Britain", *Regional Studies* 14:235-253.
Oakey, R.P., A.T. Thwaites and P.A. Nash, 1982, "Technological Change and Regional Development: Some Evidence on Regional Variations in Product and Process Innovation", *Environment and Planning* A 14:1073-1086.
Pavitt, K,. 1980, *Technical Innovation and British Economic Performance*, MacMillan, London.
Pavitt, K. and L. Soete, 1980, "Innovative Activities and Export Shares: Some Comparisons" in Pavitt, K., *Technical Innovation and British Economic Performance*, 38-66, MacMillan, London and Basingstoke.
Pedersen, P.O., 1970, "Innovations Diffusion Within and Between National Urban Systems", *Geographical Analysis* 2:203-254.
Pred., A.R., 1966, *The Spatial Dynamics of U.S. Urban-Industrial Growth 1800-1914*, The M.I.T. Press, Cambridge, Ma.
Pred, A.R., 1975, "Diffusion, Organisational Spatial Structure, and City-System Development", *Economic Geography* 51:252-268.
Romeo, A.A., 1975, "Interindustry and Interfirm Differences in the Rate of Diffusion of an Innovation", *Review of Economics and Statistics* 57:311-319.
Rothwell, R., 1980, "Policies in Industry", in Pavitt, K., *Technical Innovation and British Economic Performance*, 299-309, MacMillan, London and Basingstoke.
Rothwell, R., 1982, "The Role of Technology in Industrial Change: Implications for Regional Policy", *Regional Studies* 16:361-369.
Rothwell, R. and W. Zegveld, 1981, *Industrial Innovation and Public Policy*, Frances Pinter, London.
Stonemann, P., 1983, *The Economic Analysis of Technological Change*, Oxford University Press, New York.
Sweeney, G.P., 1987, *Innovation, Entrepreneurs and Regional Development*, Frances Pinter, London.
Thomas, M.D. and R.B. Le Heron, 1975, "Perspectives on Technological Change and the Process of Diffusion in the Manufacturing Sector", *Economic Geography* 51:231-251.
Thwaites, A.T., 1978, "Technological Change, Mobile Plants and Regional Development", *Regional Studies* 12:445-461.
Thwaites, A.T., 1982, "Some Evidence of Regional Variations in the Introduction and Diffusion of Industrial Products and Processes Within British Manufacturing Industry", *Regional Studies* 16:371-381.
Thwaites, A.T., A. Edwards and D.C. Gibbs, 1982, *Interregional Diffusion of Production Innovations in Great Britain. Volume I*, Centre for Urban and Regional Development Studies, University of Newcastle upon Tyne.

CHAPTER 6

Technological and Institutional Innovations in the Service Sector

T.R. Lakshmanan

6.1 INTRODUCTION

A structural transformation has been underway over several decades in the industrialized affluent economies. The provision of services has replaced the production of goods as the predominant economic activity. The emerging picture of the economies is not so much one of factories as of a society engaged in the production and consumption of services - a society of learning, of travelling, of legal and business consultancies, of dining, of dental visits and of data transmission.

While there is a growing literature on the nature and dimensions of this transition, much of the discussion is in terms of final output of services, of labor intensity, of lower productivity and of "post-industrial" services. What is missing in this literature, barring a few exceptions (Fuchs, 1968; Gershuny, 1978; Stanback et al., 1981) is a systematic inquiry into the vitality and dynamism of the service sector.

Two aspects of this dynamism are relevant. First is the vital role that services play in enhancing the productivity and growth of primary and industrial sectors by reducing transaction costs and adjustment costs. The incidence of transaction costs, that represent the costs of bringing supply and demand sides of a market together (e.g. uncertainty, asymmetrical distribution of information, risk, problems of coordination, control and the incentive structure), give rise to specialized dynamic producer services. Adjustment costs incurred in increasing the supply of resources that have long gestation periods (e.g. educated labour, sophisticated capital goods, organizational resources, etc.) are being addressed in the service sector.

A second and even less analysed aspect is the dynamism in the service sector deriving from the application of technical and organizational innvations. The emergence of robust technologies in the computing and telecommunications industries has lead to predictions of their diffusion (Lakshmanan, 1986; Barras, 1986; Gillespie and Hepworth, 1986); but there is not yet a systematic analysis of the process of diffusion of such technologies into user service sectors. Similarly, innovations in work organization in personal services in response to labor cost pressures have been noted (Gershuny, 1979). Yet there is little, if any discussion of the fundamental characteristic of the service sector - the participation and modification of the consumer in service production - and the consequent importance of various organizational changes to facilitate this "coproduction" and further evolution of services. Moreover, the few discussions of technical innovations are not linked to those of organizational innovations in the service sector. We suggest that these links are particularly significant in the transformation of the service sectors. Indeed, given the special characteristics of the service sector industries, which may well influence what Nelson and Winter (1982) call "technological trajectories" and "selection environments", the interplay between technical and institutional innovations determine the further evolution of the service sector. The purpose of this paper is to explore the process in which institutional innovations induced by special characteristics of the service sector interact with emerging proposed technologies in a dynamic fashion.

Technological and institutional innovations, encompass a highly complex and broad range of activities. Further, not only do these technologies change over time, but there are often several technologies that coexist in any society at one time. This heterogeneity suggests that efforts to treat innovations in the service sector in a highly aggregated manner are suspect. The service sector is too aggregate a notion in order to comprehend the complexities of technology and their interrelations with institutions and other components of the social system. Consequently, in section II of this paper we disaggregate the service sector into functional categories, each of which provide 'time, space and form utility' and cause a change either in or on behalf of the consumer. Further, we identify their attributes (e.g. informational requirements or asymmetries, production technologies, etc.) that render them hospitable to certain technologies or that may induce certain types of organizational changes.

In order to characterize the transformation of the service sector more satisfactorily we proceed in sections III and IV of this paper to review two major theoretical strands of innovation literature. Technical change, considered first, may be viewed at various levels of aggregation and for different time periods. Neoclassical and evolutionary theories (Nelson and Winter, 1982; David, 1975) tend to analyze technical change at the level of the firm and industry, while broad historical syntheses are offered by Marx and Schumpeter. We focus here on theoretical notions relating to the process by which technical innovations get diffused through the economy. Drawing upon the work of Rosenberg (1976), Nelson and Winter (1982), and Barras (1986) we sketch the broad outlines of a process of innovation spread in the service sector.

Next we highlight the growing literature on induced innovation of institutions. To gain a deep rich understanding of institutional innovation processes, one must step outside disciplinary boundaries and engage in intellectual trespassing. Three classes of issues - *why* there is institutional innovation, *how* the process of innovation evolves in a prior institutional context, and *where* does innovation take place more often - can be pursued with the help of ideas from economics, public administration and geography. Thus the idea that institutional structures of a society could be understood and broadly derived from a material base continues to fascinate researchers in many traditions, such as the neoclassical (economic), Marxian (radical) and structural-functional (sociological and anthropoligical (Field, 1981). Public administration analysts (Rogers and Kim, 1985; Mohr, 1978) focus on how innovations are initiated, reinvented, structured and implemented in the context of a prior institutional structure. The third issue of *where* relates to the structure of the environment in which institutional innovations occur. Geographers have emphasized how processes of innovation, adoption and diffusion are facilitated by urbanization and by growing speed, density and capacities of communications and transportation networks and the knowledge accumulation. Moreover, we know that barriers to institutional innovations in the form of social rigidities that Mancur Olson (1982) identifies, are often less formidable in specific cities and communications, so that many service sector related institutional innovations occur in localities and spread in a 'bottom-up' fashion, rather than 'top down' from the nation.

Next, the paper interprets the ongoing transformation in the service sector in the light of the ideas gleaned from the innovation literature and identifies emerging institutional styles and forms, modifications of existing ones and blends of old and new forms. We focus in particular on one institutional form - coproduction - induced by the participation of the consumer in the service production. We analyse in this case the potential for feedback between and reinforcement of technical and institutional innovations in a sequence that improves initially the efficiency of existing services by work process changes, then proceeds to quality augmenting process innovations, and then on to development of new services that represent a blend of technical and organizational innovations.

It appears that there is a major need for new theoretical models appropriate for the study of the transformation of the service sector. Not only is there a need to include consumer as a factor in the production function, to consider labor-embodied technical change but also to link interactively physical and institutional technologies in the analysis of change.

6.2 THE SERVICE TRANSFORMATION: SOME CHARACTERISTICS

6.2.1 A Capsule Profile of the Service Sector

The structural shift towards services in the industrialized countries has been widely documented (Inman, 1985; Summers, 1985; Daniels, 1985; Stanback et al., 1981). Whether measured by output or labor shares, services are an important part of total economic activity. Employment in service industries and white collar occupations has greatly outpaced the growth of employment in goods production in the U.S. (Stanback et al., 1981; Leveson, 1985).

Three hypotheses have been put forward for the relative growth of service employment in the economy (Fuchs, 1968; Gershuny, 1978). First, since services have an income elasticity demand higher than 1, as real incomes per capita climb, real services per capita grow faster than the proportional rise in income; thus services account for an increasing proportion of income and national employment as well. Second, as automation and increasing division of labor rise with economic growth, services once produced in the household or the firm can be contracted out more efficiently to specialists outside the organizations (often with higher service quality and/or lower average costs). Thus services such as housekeeping, day care, food production, data processing, financial services, legal and advertising services are provided in new establishments that are measured now as service activity. Third, given the slower relative growth of labor productivity in services, as the economy expands, the share of services in total employment increases.

The broad range of economic activities in the service sector have been classified in numerous ways each focusing on some aspects of service production, somewhat independent of data source requirements (the exception is Riddle, 1986). Table 6.1 displays 4 classifications based on societal function of the service and one on productivity potential of the service. Since none of these is satisfactory for our purpose of identifying receptivity to technical change, we return to this issue later in this section.

The major growth segments of the service sector are:

1. *Human Capital Services*: Largely non profit services: e.g. education and health - reflecting the greater need for investments in human capital in an increasingly sophisticated, internationalized arena of production of goods and services.
2. *Producer Services*: These result from increasing division of labor; the growth in size and importance of large corporations; to the rising importance of planning, development and complex managerial functions; to the growth of markets; and to changes in the organizational and institutional arrangements of the private sector, and
3. *The Producer-Service-Like Functions in the Public Sector*: These functions reflect the need to protect consumer interests and worker rights, to further equal opportunity; to regulate markets as necessary, to promote national economic interests in a rapidly expanding international system of production, service delivery and exchange (Stanback et al., 1981).

Table 6.1 Alternate classification of services

A. Function based

1. Foote and Hatt (1953)

   Tertiary (restaurants, hotels, repair & maintenance, laundry)
   Quartenary (transportation, communication, business services)
   Quinary (health, education, recreation)

2. Katouzian (1970)

   Complementary Services (retail & wholesale, finance, transportation)
   New Services (health, education, entertainment)
   Old Services (domestic)

3. Browning and Singelmann (1975)

   Distributive Services (transportation, retail, communication)
   Producer Services (professional, business)
   Personal Services (leisure, domestic, hotels)
   Social Services (defence, health, education)

4. Riddle (1986)

   Infrastructure Services (transportation, communication)
   Trade Services (wholesale and retail)
   Business services (finance, professional)
   Community Services (public administration, social/personal services)

B. Productivity based

5. Baumol (1985)

   Stagnant Personal Services (non-routine, labor-intensive)
   Progressive Impersonal Services (technology intensive, no contact with client)
   Asymptotically Stagnant Impersonal Services (an amalgam of the above two)

The significant growth of these classes of services does not represent a major movement to freestanding services purchased on the market as much as a joint provision of goods and services. Rising real income and increasing diversity of consumer demand have led to greater product differentiation that is responded to by joint provision of goods and services. The consumption of services turns out to be strongly complementary to consumption of goods (Gershuny, 1978; Hirschborn, 1981; Stanback et al., 1981; Daniels, 1985).

6.2.2 Service Sector Technologies

The enormous diversity of activities encompassed in the service sector has often led to loose definitions and ascription of attributes to the sector that are both misleading and analytically unhelpful. Typical of such characteristics attributed to the sector as a whole are its low rates of productivity growth, its labor intensity, low capital intensity and its major role in the slowdown in productivity in the overall economy in the last 15 years.

Since these characterizations are not accurate[1], we engage in a brief definitional digression.

The first attribute of service activities is that they involve a change in (or for) the consumers or the goods they possess. Three types of such changes in the asset structure of the consumer during service production can be noted (Garn et al., 1976). In one type, there is a change in the tangible asset structure of the consumer and provider of services (e.g. retail services, repair and maintenance services for durable goods, economic development activities). In a second type of service activities, there are changes in the human capital of the consumers (e.g. education and training; counselling and health care services). The third type of activities include primary brokerage activities (e.g. referral programs and real estate sales) where the output appears as a change in the personal assets of clients who are providers and clients who are consumers of the service. However, a few services involve changes neither in the consumers or in their goods. For example Rental services allow persons to make use of goods without acquiring then - a change in accessibility created for the consumer. Thus the change process involved is *in* or *for* the consumer and confers on him or her time, place and form utility (Riddle, 1985).

The second general attribute of service activity is that consumers are intrinsic participants in the production process. This distinguishes the production of goods from that of services. In goods production, resources - labour, capital, energy and materials - are combined (Figure 6.1).

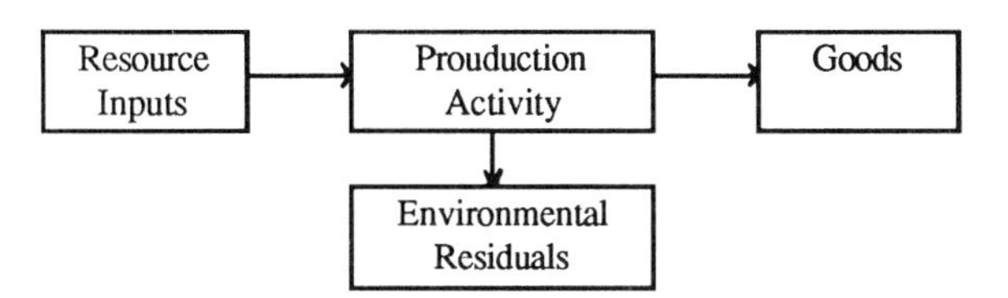

Figure 6.1 Production of goods

A production unit is subject to a variety of opportunities and constraints on its actions. However, the transaction between the producer and consumer will *not* directly affect the level of output. The asset positions of the seller and buyer are altered during sales but the output is not affected by the attitude or behavior of the seller or buyer. This enables the analyst to view goods producing units as self-contained units for analysis of efficiency.

This is not the case with production of services (Figure 6.2), where the consumer of services is part of the process of service production. The final output is jointly produced. Becker (1965) conceptualised market goods as intermediate goods that individuals combine with their time and skills to produce final products or commodities (e.g. meals, entertainment, recreation, etc.) they consume. Becker's notions have been extended by social indicator researchers (Garn et al., 1976) so as to view consumers as producers in the realm of service production. For example, the provision of quality health services requires the participation of the patient in the form of producing accurate medical histories, cooperating in the treatment etc. In other words, the efficiency of the service

[1] An analysis of 145 industry groups in terms of capital stock per work hour shows that the service industry groups formed nearly one-half of the 30 groups in the top 2 deciles and the bottom 3 deciles of industry ranking by capital intensity did not contain any service industries (Kutscher and Mark, 1983). Further the range of productivity growth noted in the service sector is not significantly different from the range among goods producing sectors (Kutscher and Mark, 1983). The growth of services can not possibly be the cause of the recent slowdown in the larger economy (Fuchs, 1985).

production cannot be solely attributed to the activities of the service provider. The consumer's activities bear upon the efficiency and effectiveness of the services.

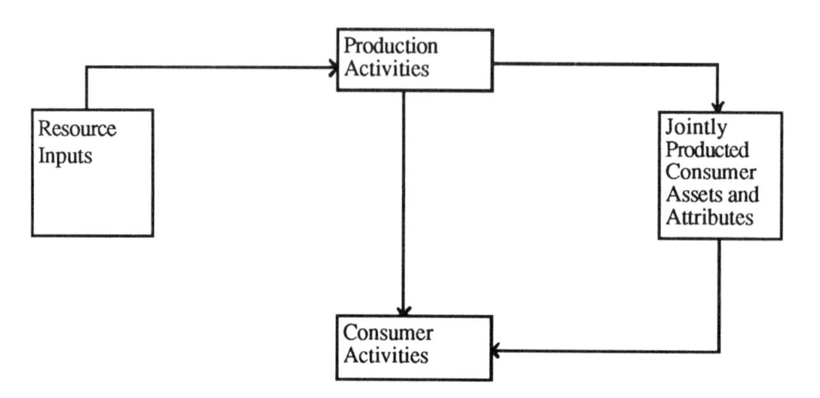

Figure 6.2 The production of service output. (After Garn et al., 1976)

We briefly characterize the nature of service provider - consumer interface along two dimensions. First, in terms of the four *functions* the service client can perform (Normann, 1984):

- *Specification* of the service: e.g. by participation in the diagnosis of the problem via providing health data; specifying items in a menu of services that are desired as in a restaurant.
- *Self Service*: where the consumer is used as a primary input without interaction with the producer (e.g. retailing, automatic banking etc.). The idea is to reduce the labor costs of service.
- *Coproduction*: where the client carries out part of the production (e.g. client's staff doing data collection and some analysis for a market analysis; residents participating in housing services delivery; the student contribution to education, etc.).
- *Development* of the service system in a professional organization: (e.g. in auditing, consulting).

A second way to characterize the producer-client interface is via the labor-capital dimension (Figure 6.3). While service producers position themselves in one of the four cells initially, they may change their technologies over time. Thus a producer engaged in traditional labor intensive conventional training (cell 1) may shift over time into computer aided training which involves the client using producer's specialized capital (cell 2).

Given these patterns of producer-consumer interface and our interest in service technologies, we proceed to a classification of service activities that may be appropriate for analysing technical change. This classification builds on the ideas of Baumol (1985) and Mills (1986) and is based on the nature and intensity of producer-consumer contact and the types of skills and knowledge brought to the service production by clients. The kinds of encounters between producers and clients will depend upon the degree of environmental uncertainty (reflecting the complexity of the information exchanged).

We adapt an existing taxonomy and recognize three types (Mills, 1986):

- Service Dispensing Activities,
- Task-interactive Services, and
- Personal Interactive Services.

Client / Producer	Labor	Capital
Labor	Conventional training welfare agency 1	Car repair; computer software 2
Capital	Computer-aided training; Automatic bankteller 2	Automatic carwash 4

Figure 6.3 Different types of producer client interfacing

(i) *Service Dispensing Activities*:

These activities occur in environments with low uncertainty. The customer needs and production technologies are well known. The service provider is less involved with production and more in dispensing them, There is a tendency to standardize activities and set up large scale operations, thereby increasing productivity. The consumer may be incorporated in the self service mode and the information exchanged is simple in nature. This group includes (a) impersonal services such as telecommunications, (b) distributive services using self service e.g. retail, (c) personal services amenable to "industrialization" (Levitt, 1976) e.g. fast food restaurants. Customer involvement is minimal and is often completely removed in some predictable environments by employing appropriate equipment (ATMs in banks).

(ii) *Task Interactive Services*:

Moderate to high degree of uncertainty characterize these activities. The technologies to solve these task oriented problems is known, but large amounts of information and effort are expended before the customer's needs are known. Further the uncertainty exists because of the uniqueness of the clients' needs. Examples are accounting, engineering and architectural firms, advertising, financial, legal and marketing services.

While the organization may possess relevant specific knowledge, the vast information requirements provide an opportunity for introducing the new information technology. While the earlier efforts in this area have been focused on data processing activities (that provided efficiency) subsequently on-line systems of computerised services have been adopted.

(iii) *Personal Interactive Services*:

The environments of these services are both dynamic and complex. The consumers are unaware or imprecise about their problems or solutions. thus the service producer and consumer jointly strive to diagnose the problem. The body of knowledge that exists is such that the cause-effect relationship between what the service producer does and the outcome of the action is not sound. Examples of such services are health care, psychological consulting, welfare agencies, etc. The characteristics of services as outlined here poses special problems for the market provision of services. As Holmstrom (1985) suggests, the quality of services provision is difficult to assess. Whereas the service provider may know product quality, the buyer often does not. A doctor will clearly know much more about the probabilities of success of a medical procedure than his or her patient. So does a lawyer often understand more often basic merits of a case than her or his client. This "asymmetry" of information between sellers and buyers of service poses severe problems for market provision (Holmstrom, 1985). Two kinds of problems arise: "adverse selection" and "moral hazard".

Adverse selection occurs when the consumer is unable to observe and monitor the activities of the service producer. The patient may be unable to determine whether the procedures and tests the doctor implemented are appropriate. There is potential for adverse selection when the service is oversupplied. A law firm may not utilise the billed time most productively for the client.

Moral Hazard occurs when the quantity and quality of the service rendered is difficult to verify. In the case of unsatisfactory outcomes of "failures" the recipient of the services can not distinguish perfectly between incompetence and bad luck and this has had incentive effects on the service provider. Such problems of monitoring and assessing the performance of such service providers often create a variety of new services - informative advertising and certifications, warranties and contingent contracts, seller reputation etc. More generally there are two mechanisms for addressing the quality problem resulting from adverse selections and moral hazard; monitoring and bonding activities: in the private sector, the importance of reputation of the service producer becomes important.

6.3 TECHNOLOGICAL INNOVATION AND DIFFUSION IN THE SERVICE SECTOR

Technical change is a very complex social process, that has dimensions that do not stay within the confines of any single academic discipline. It covers a continuum of inventive efforts, stretching from initial conceptualization to establishment of technical feasibility (invention) to commercial feasibility (innovation) and finally to the subsequent diffusion into the larger economy.

Theoretical inquiry into this process takes two major approaches (Elster, 1985). First, technical change may be viewed as a rational choice of the best innovation among a set of feasible changes. A second approach is to view technical change as a process of trial and error, as the accumulation of small, individually minor, improvements of the production process. Neoclassical and Marxian theories of technical change at the level of the firm adopt the rational actor approach although they impute different goals to the entrepreneur - project maximization in neoclassical theory and innovation as power in the class struggle according to Marxists. Schumpeter is perhaps the most influential single writer in this grand tradition of explaining technical change and its consequences. Rather than using the straight jacket of profit maximization, he focused on creativity and disequilibrium and the dynamic character of capitalism. All these theories focus on the earlier part of the technical change process - from the conceptualization of the idea to the innovation.

On the other hand, the second approach to the analysis of cumulative change processes is evident in the work of economic historians and "evolutionary" theories of technical change (Usher, 1954; Rosenberg, 1986; David, 1975; Nelson and Winter, 1982). This approach conceptualizes the technology diffusion process historically and focuses on the factors that affect the pattern and spread of diffusion of innovation into the larger economy. Given our interest in the adoption of technical change in service industries, we are less interested in the grand questions of technical change but more with the issues that come to grips with technical change and adoption such as: What circumstances promote successful accumulation of knowledge? What conditions promote rapid diffusion? We briefly highlight here the literature on technology diffusion and technology trajectories.

The earlier literature on technical change emphasized the role of demand side considerations on the rate and direction of technological advance. Rosenberg (1976) and others have pointed out the importance of supply side considerations. Even big technological breakthrough inventions have much more gently declining slopes of cost reduction than is indicated in many studies and thus longer gestation periods for innovation to be generally adopted. If a technology innovation is to widely diffused, several contributory factors must operate. First, there is a period of critical "secondary inventions" and design modification that improve the quality and efficiency and thus the adoption chances. Second, there is the role of development of human skills and capital goods capacity upon which the rise of the new technique depends in order to be effectively exploited. In the earlier stages, IBM and other high technology firms have had to develop not only the sophisticated machinery and the trained labor they need to produce and sell but also to educate the consumers. At a later stage of development of the technology, there is a significantly larger level of appropriate resources in the market at large for new human skill development.

Third, the diffusion of an innovation is facilitated by complementarity in productive activity between different techniques. The growing productivity of an economy is the complex outcome of large numbers of interlocking, mutually reinforcing technologies, the individual components of which are not that significant. Thus an interrelated clustering of innovations will stimulate rapid diffusion (Rosenberg, 1976). Nelson and Winter (1982) develop first the concept of technological trajectories or directions of advance common to a wide range of technologies. Second, they propose the concept of a "selection environment" which influences the path of productivity growth. Can we speculate about these technological trajectories and selection environments in the case of the service sector?

First, it appears that in what we have called the service dispensing activities, (which are close in their technologies to goods production), the considerable potential for standardizing operations suggest the two "natural trajectories" of Nelson and Winter (1982), namely progressive exploitation of latent scale economies, and increasing mechanization of labor inputs. This trend is noticeable in the distributive services, telecommunication etc.

In some task-interactive and personal-interactive services, where the contact with the consumer may range from moderate to high, the 'technical trajectory' may lie in reduction of costs of communication. In this regard, initial efforts have been primarily towards data acquisition economies. In this phase, information acquisition and processing operations are being carried out more efficiently with the new computing equipment. In the next phase, increased amounts of low cost information are used to improve the quality of the services. In the next decade the potential move is likely to be towards Integrated Network Services (INS). INS may not only be cost saving but also an opportunity to develop new information service products (Lakshmanan, 1986).

In the case of services where the producer is in the non-market sector, the direction of technical change is influenced by consideration of public interest - expressed in areas such as environmental policy. government regulations play a role in such cases.

For services where the participation of the consumer is crucial, the introduction of technical change is influenced by the functional nature of the producer - consumer interface. Thus the technical innovations that would be appropriate will depend upon whether the producer-consumer interface is for specification of the service, for self service, for coproduction of development of the service delivery system. We will return to this issue in the next section.

6.4 INSTITUTIONAL INNOVATIONS IN THE SERVICE SECTOR

Institutions are the rules of a society or of organizations that provide a framework within which people interact. They establish the conventions that have emerged in a society regarding the behavior of individuals and groups relative to their own behavior and that of others. In order that institutions can perform their role of guiding reasonable expectations in interpersonal and group dealings, stability of institutions is valued. However, institutions change over time in order to realize latent gains by overcoming disequilibria caused by changes in product demand, resource endowments and technical change.

We briefly review the literature on institutional innovation from three perspectives. Why does institutional innovation come about? How does the institutional innovation process operate? In what types of locations are such innovations favored?

6.4.1 The Rationale for Institutional Innovations

The notion that when the economic base of a society is changed, innovations are induced in the institutional structure is subscribed to by a variety of scholars-Marxist ("with the change in the economic foundations the entire immense superstructure is more or less rapidly transformed"), structural-functional (sociology and anthropology) and neo-classical.

The neo-classical theory of induced innovation suggests that such innovation occur in order to realize gains from transaction costs, reduction of risk, economies of scale, internalization of externalities and redistribution of income (Davis and North, 1971; North, 1982; Hayami and Ruttan, 1985; Field, 1981). In the course of technical change specialization and division of labor increase and so do a variety of transaction costs. These costs include:

- Transportation and communication costs
- Measurement costs (to determine the quality, quantity and dimensions of the service exchanged)
- Insurance costs (hedge against unforeseen circumstances) and
- Enforcement costs

Corresponding to each type of transaction costs (Williamsson, 1980), a producer service activity develops to lower such costs.Overtime, specialization in that producer service leads to changes in industrial organization (externalization of such services). Such developments in turn, through cost reductions, lead to growing demand for such services. The government also provides some of the producer-service like functions - e.g. those designed to regulate competition, protect employees working conditions, the environment, etc.

Technical change leads also to changes in relative prices, thereby putting pressure on lowering labor costs. In many private services, this objective to reduce labor costs had led to innovation in the organization of the personal services. Specifically, a market service is organized into a 'do-it-yourself' service - through a combination of market capital goods (TV, car, etc.), infrastructure (telecommunications, transportation, etc.), software (programming) and customers' unpaid time (Gershuny, 1978).

Public services, which are provided to deal with cases where privately expressed choices will not lead to ideal outcomes or cases where the objective is to redistribute towards target groups. In such cases, the thrust towards cost reduction is being met by innovations such as "privatization" of service delivery and the use of user charges for many services provided by the public sector.

In some cases, the reorganization of older systems of service lead to better quality of service often at comparable if not lower costs of delivery. Examples of institutional innovation in formerly fossilized or regulated contexts are the health maintenance organizations, Federal Express and nursing home chains.

For a growing number of services, which may be described as "positional services" (Hirsch, 1976; Scitovsky, 1987), the innovations emphasize uniqueness rather than efficiency e.g. Euro class of SAS, Design Labels etc. Here quality at high prices is provided in a strategy of playing market niches.

6.4.2 The Process of Innovation in an Organizational Context

Since institutional innovations emerge and succeed in the context of earlier organizations, public administration theorists (Rogers and Kim, 1985; Downs and Mohr, 1976) have attempted to characterize the process of adoption, modification and implementation of institutional innovations.

Two main subprocesses are involved: initiation and implementation. Initiation consists of all the information gathering, conceptualizing, and planning for the adoption of an innovation in a larger organizational context. During implementation, the innovation is modified, adapted, and changed as it is embedded into existing structures and interconnected to other parts in the work flow. The point to note here is that during adoption the innovation is often "*reinvented*".

We note this process of "reinvention" because it is in this stage that the potential for interplay of technical and institutional innovations is exploited synergestically - a topic to which we will return later.

6.4.3 Locales for Institutional Innovation

While institutional innovations can take place in many locales, it results from a cumulative and reinforcing process of change. Such processes are disequilibrium in nature and the speed of adoption of innovations depend upon the structure of the environment in which they occur. In urban areas, characterized by growing fast, dense and sophisticated networks of transportation and communication networks, and accumulations of knowledge and expertise, the opportunities for interaction, learning, and adopting are much improved. Hence the origin of many service business organizational innovations in urban areas (Stanback et al., 1981; Pred, 1977).

Clearly the spatial organization of the economy has important implications. It is not surprising that the dynamism of the service sector has been reflected in the recent renewing vitality of the larger urban centers.

However, the ability of any organizational unit to innovate and adapt is constrained often by the prior organizational structure. Mancur Olson (1981) has drawn attention to the social rigidities in older democratic societies. These rigidities appear in the form of organizations at the national level that reflect the power of interest groups that are unwilling to permit changes in the existing distribution of resources. These are often however, "gaps" or "slacks" in these structures at different localities. Taking advantage of such slack, innovations appear in specific localities and spread in a "bottom up" fashion. If the "slack" is combined with specific economic inducements to organizational reform innovations emerge. For instance, under pressure from tax payer revolts such as Proposition 13 (California), some localities have entered into joint equity relationships with private developers in return for authorizing large development projects. The emergence of institutional innovations in a society is often a "bottom up" process.

6.5 THE INTERPLAY OF TECHNICAL AND INSTITUTIONAL INNOVATIONS IN SERVICES

Even from this all too brief review, it must be evident that technical and institutional innovations have been analysed independent of one another. Independent and separate roles have been assigned to technical and organizational innovations (e.g. North offered the opinion that organizational innovations contributed to the same degree as technological innovations to the development of the Western World during 1550-1830). In reality both types of innovations exert their effects upon productivity be reinforcing one another and accelerating the process of technology diffusion.

The evolution of urban infrastructure services (water supply, sewer system and transportation) in the U.S. in the 19th century illustrates this interplay very well (Tarr, 1984). The patterns and pace of development of infrastructure reflects the interplay of mutually reinforcing organizational innovations in the private, voluntary and public sectors and technical developments in materials, metallurgical and chemical industries. The earliest innovations were by the public sector to reduce risk, provide investment and an institutional framework for further private activity[2]. The latter consisted of structural changes in city government in the form of new charters or authorized revisions of old charters (Tarr, 1984).

The rapid development of water supply and sewer systems in the late nineteenth century depended upon key innovations in the private, voluntary and government sectors. The growth of investment banking provided the vast amounts of infrastructure capital through a national bond market. Voluntary Associations such as professional organizations (e.g. American Water Works Organization, American Public Health Association, American Society of Civil Engineers) provided a venue for both lobbying for infrastructure investments and for increasing the knowledge and supply of skilled personnel. On the political side the replacement of commercial and upper class groups by the localised and patronage - spoils - equity oriented machine politics increased infrastructure services. Other innovations included special districts for the pursuit of infrastructure. Parallel to these organizational innovations there were significant complementary technical innovations: improved methods of manufacturing cast iron pipe and coating interiors for pressure maintenance; newer paving and construction materials. Once these technical innovations were adopted in a few places, their diffusion throughout the urban network was made possible by the next round of private organizational innovations: packaging and franchising these water supply technologies by the early innovators (Tarr, 1984). Figure 6.4 provides a general equilibrium view of the development and diffusion of urban infrastructure services.

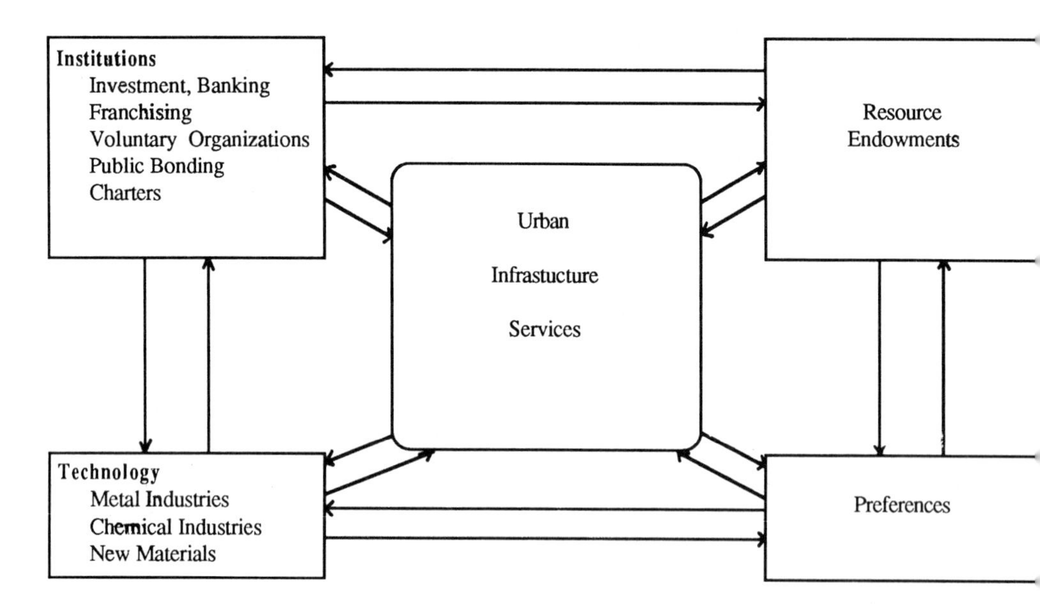

Figure 6.4 The development of 19th century urban infrastructure: A general equilibrium view

[2]Capital for transport services was raised by sales of state and local bonds largely to foreigners (Tarr, 1984) in early 19th century.

Table 6.2 Evolving technologies in the service sector

Type of Service	Key Characteristics	Technical Innovations	Institutional Innovations	Synergistic Develop– ments
1. Service Dispensers (e.g. retail wholesale, telecommu- nication, fast food, etc	Consumer Conctact minimal; production technologies known, amenable to scale	Automation of many proces- ses, ATM etc. High volume machine tech- nologies	Self service - standardi- zed service	Federal Express
2. Task inter- active servi- ces e.g. accounting, legal, finan- cial	Moderate to high customer contact; unique customer needs; high information subject to different inter- pretations. Client goals known but outcome of solutions uncertain	Telecommuni- cation for ef- ficiency; Quality of service, on- line informa- tion systems	User of con- sumers for specification of output; Forms of co- production	Information network services (INS): New service products
3. Personal interactive services e.g. health care, welfare agencies	Dynamic uncertain environ- ments; client goals im- precise; cause-effect relationships between solu- tions and outcomes un- certain. Adverse selection; Moral hazard	Machine tech- nology growth rapid (e.g. health) - on line in- formation system	Coproduction Bonding Monotoring	Distributed coproduc- tion Innovation in service out- put

6.5.2 The Evolution of Coproduction: A Contemporary Example

A brief and selective introduction to the diffusion of technologies in the contemporary service sector appears in Table 6.1. We return to the threefold classification of services. For each, we highlight the characteristics that encourage certain types of technical and institutional changes. For example, in some personal interactive services (e.g. Health Services) technical change appears in two forms: (a) advances in instrumentation, biotechnology, diagnostic technology, etc. (b) Informations systems. In order to carry out older manual operations more efficiently, the scale economies of mainframe computers were utilised, substituting away from clerical labor in the process. In the next stage, with the arrival of micro and minicomputers and on-line data systems, there were efforts to improve the quality of a variety of administrative and outreach operations.

Meanwhile, given the characteristics of the personal interactive services, a variety of institutional innovations have emerged. We briefly comment on one - coproduction. Co-production entails the participation of the consumers in the production of services. Examples are the utilisation of residents in the delivery of housing services, paramedical services, training, recreational services (Chatterjee, 1986). In other words, clients become "members" of the service producer organizations during the service delivery and are crucial to he effectiveness of the service. Though there is a range of client behavior that facilitate production, certain factors are common to these behavior. As active participants, consumers *have to acquire knowledge, skills, disposition and material resources that will enable them to performs as effective participants.* As Chatterjee (1986) suggests, the consumers need to play a crucial role even in the specification of service, in addition to production. Thus, where coproduction is operative, the producer cannot seal off the production technology (as in service dispenser agency). That technology is at least partially open to the consumer.

How does one augment the productivity and quality of the service output whose technology is open, and interactive with the consumer? Business theorists (Mills, 1986; Normann, 1984) emphasize a client "socialisation process" from the service producer's perspective. The more general point about service productivity is that there should be an *effective transfer and exchange of knowledge, skills and disposition between the producer and consumer.* At this development is likely to increase communication and information costs, efficient coproduction would require more efficient exchange of knowledge, skills and information.

It is in this context, the development of an emerging technology is significant. This is the Information Network Services (INS). INS may be viewed as a communications network over which digital information - data, voice, video and text - can be transmitted, displayed and processed. Stimulated by the array of satellite, optic fiber, microwave and switching technology, INS is an economic reality, as a global intercity development, though widely used by only a few large private corporations and public sector units. Lakshmanan (1986) has argued that in order to realize its full potential, two other components of INS need to be developed. These components are *intracity* and *intra-building*. A policy view of INS as an infrastructure service will facilitate rapid development of the intracity and building components of INS (Lakshmanan, 1986).

The relevance of INS to our discussion is that with its fuller diffusions, synergistic development possibilities arise in the service sector. It is possible to combine coproduction (institutional innovation) with (the technology of) INS to develop a new service activity. We call this *Distributed Coproduction.*

To illustrate, one may conceptualize a number of old age home care centers in a metropolitan area all linked together by INS with work stations in each center. From a central educational and diagnostic facility, it is possible to convey information, health care instructions, diagnostics a variety of treatment procedures *interactively* with para-professional or trained community workers in each center. Such a system is likely to promote old age care at considerable economies.

They key point is that the analysis of the service sector requires models of change that are consistent with the special characteristics of the sector. Further, in the evolution of services, technical change and institutional change are important and have to be viewed together in a dynamic interactive manner.

REFERENCES

Bailey, F.G. 1973, "Debate, Comprise and Change", in F.G. Bailey, ed., *Debate & Comprise*, Rowman and Littlefield, Totowa, New Jersey, 309-328

Barras, Richard, "Towards a Theory of Innovation in Services", *Research Policy* 15,4, 161-173

Baumol, William, 1967, "The Macroeconomics of Unbalanced Growth", *American Economic Review* 57, 425-426

Baumol, William, 1985, "Productivity Policy and the Service Sector" in R.P. Inman, ed., *Managing the Service economy: Prospects and Problems*, Cambridge University Press, London, 301-317

Becker, Gary, 1965, "A Theory of Allocation of Time", *Economic Journal* 75, 493-517

Beyers, William B., M.J. Alvine and E.G. Johnsen, 1985, "The Service Sector: A Growing Force in the Regional Export Base", *Economic Development Commentary*, Fall, 3-7
Browning, H.C. and J. Singlemann, 1975, *The Emergence of a Service Society*, National Technical Information Service, Springfield, VA
Chatterjee, Lata, 1986, "Coproduction and Low Income Strategies", paper presented at the International Housing Research Conference, Gävle, Sweden, June 10-13. Boston University Center for Energy and Environmental Studies Research Paper No. 4
Daniels, P.W., 1985, *Service Industries: A Geographical Appraisal*, Methuen, London
Chase, Richard B. and David A. Tansik, 1983, "The Customer Concact Model for Organizational Design", *Management Science* 29, 1037-50
David, P., 1975, *Technical Choice, Innovation and Economic Growth*, Cambridge University Press, Cambridge, Mass.
Davis, Lance E. and Douglass C. North, 1971, *Institutional Change and American Economics Growth*, Cambridge University Press, Cambridge, U.K.
Deutsch, Karl W., 1985, "On Theory and Research in Innovation" in Richard L. Merrit and Anna J. Merrit, eds., *Innovations in the Public Sector*, London Sage Publications, 17-35
Downs, G.W. and L.A. Mohr, 1976, "Conceptual Issues in the Study of Innovation", *Administrative Science Quarterly*, 700-714
Elster, Jon, 1985, *Explaining Technical Change*, Cambridge University Press, Cambridge, U.K:
Field, Alexander J., 1981, "The Problem with Neoclassical Institutional Economics: A Critique with Special Reference to the North-Thomas Model of Pre 1500 Europe", *Explorations in Economic History* 18, 74-98
Foote, N.N. and P-K- Holt, 1953, "Social Mobility and Economic Advancement", *American Economic Reveiw* 43, 364-378
Fuchs, Victor, 1986, *The Services Economy*, National Bureau of Economic Research and Columbia University Press, New York
Garn, Harvey A., 1973, "Public Services on the Assembly Line", *Evaluation*, vol. 1, No. 2
Garn, Harvey, A., 1975, "Program Evaluation and Policy Anaiysis of Community Development Corporations", in *The Social Economy of Cities*, vol. 9, Urban Affairs Annual Reviews, Sage Publications Inc.
Garn, Harvey A., M.J. Plex, M. Springer and J.B. Taylor, 1976, *Models for Indicator Development*, Urban Institute, Paper 1206-1217, Washington D.C.
Gemmell, Norman, 1982, "Economic Development and Structural Change: The Service Sector", *The Journal of Develepent Studies* 19, 37-66
Gershuny, J.I., 1978, *After Industrial Society: The Emerging Self Service Economy*, MacMillan, London
Gershuny, J.I. and R.E. Pahlm, 1979, "Towards an Alternative Society", Paper read at Social Science Research Council Workshop on Unemployment and Employment, Londo, September 28
Gillespie, A.E. and M.E. Hepworth, 1986, "Telecommunications and Regional Development in the Information Society", New Castle Studies of the Information Economy, Working Paper No. 1, University of New Castle upon Tyne, October
Gramlich, Edward M., "Government Services" in Robert Inman, ed., *Managing The Service Economy*, Cambridge University Press, London, 273-289
Hatch, Stephen and Ian Macroft, 1979, "The Relative Costs of Services Provided by Voluntary and Statutory organization", *Public Administration* 57, 397-455
Hayami, Yujiro and Vernon Ruttan, 1985, *Agricultural Development*, The John Hopkins University Press, Baltimore
Hirsch, Fred, 1976, *Social Limits to Growth*, Harvard University Press, Cambridge, Mass.

Hirschman, A., 1958, *The Strategy of Economic Development*, Yale University Press, New Haven

Hirschhorn, L., 1985, "Information Technology and the New Service Game", in Manual Castels, ed., *High Technology, Space and Society*, Sage Publications, Beverly hills, 179-190

Holmstrom, Bengt, 1985, "The Provision of Services in a Market Economy" in Robert P. Inman, ed., *Managing the Service Economy: Prospects and Problems*, Cambridge University Press, London, 183-213

Katouzian, M.A., 1970, "The Development of the Service Sector: A New Approach", *Oxford Economic Papers* 22, 362-382

Kikuchi, Massao and Yujiro Hayami, 1980, "Inducements to Institutional Innovation in an Agrarian Community", *Economic Development and Cultural Change*, October 21-36

Kutscher, R.E. and Jerome A. Mark, 1983, "The Service Producing Sector: Some Common Perception Reviewed", *Monthly Labor Review*, April 21-24

Lakshmanan, T.R. and A. Elhance, 1983, "Public Expenditure and the Performance of Regional Production Enterprises: Analytical and Policy Issues" in Lata Chatterjee and Peter Nijkamp, eds., *Urban and Regional Policy Analysis in Developing Countries*, Gower, London

Lakshmanan, T.R. and Lata Chatterjee, 1986, "Technical Change and Metropolitan Evolution", *Regional Science and Urban Economics*, February

Lakshmanan, T.R., 1986, "Telecommunications Infrastructure and the Service Sector Transformation", paper presented at the North American meeting of the Regional Science Association, Columbus, Ohio, November

Lancaster, Kelvin J., 1971, *Consumer Demand: A New Approach*, Columbia University Press, New York

Leff, Nathaniel H., 1984, "Externalities, Information Costs, and Social Benefit - Cost Analysis for Economic Development: An Example from Telecommunications", *Economic Development and Cultural Change*, Vol. 32 #2, 255-275

Leveson, Irving, 1985, "Services in the U.S. Economy", in Robert P. Inman, ed., *Managing the Service Economy: Prospects and Problems*, Cambridge University Press, London, 89-102

Levitt, Theodore, 1976, "The Industrialization of Service", *Harvard Business Review*, September

Linder, Steffan, 1970, *The Harried Leisure Class*, Columbia University Press, New York

Lynn, Leonard, 1984, "Japan Adopts a New Technology: The Roles of Government, Trading Firms and Suppliers", *Columbia Journal of World Business*, Winter, 39-45

Mark, Jerome A., 1982, "Measuring Productivity in Service Industries", *Monthly Labor Review*, June 3-8

Mensch, Gerhard O., 1985, "Trends and Perspectives in Innovation Policies" in Richard L. and Anne J. Merrit, eds., *Innovation in the Public Sector*, Sage Publications, London, 253-268

Merrit, Richard L. and Anne J. , eds., *Innovation in the Public Sector*, Sage Publications, London, 9-16

Mills, Peter K, 1986, *Managing Service Industries: Organizational Practices in a Post Industrial Economy*, Ballinger, Cambridge, Mass.

Nelson, Richard R., 1986, "Institutions supporting Technical Advance in Industry", *American Economic Review* 76:2, 186-189

Nelson, Richard R. and Sidney G. Winter, 1982, *An Evolutionary Theory of Economic Change*, Harvard University Press, Cambridge, Mass.

North, Douglass C. and R.P. Thomas, 1970, "An Economic Theory of the Growth of the Western World", *The Economic History Review* 23:1, 1-17

North, Douglass C., 1981, *Structure and Change in Economic History*, W.W. Norton Co., New York

North, Douglass C. and John J. Wallis, 1982, "American Governmental Expenditures: A Historical Perspective", *American Economic Review*, May, 336-40

Normann, richard, 1984, *Service Management*, John Wiley & Sons, New York

Olson, Mancur, 1982, *The Rise and Decline of Nations*, Yale University Press, New Haven

Porat, Marc V., 1982, "Information, Communication and Division of Labor" in M. Jussawalle and D.M. Lamberton, eds., *Communications, Economics, and Development*, Pergamon, London, 75-81

Riddle, Dorothy I., 1986, *Service-Led Growth*, Praeger, New York

Roehrich, robert L., "The Relationship Between Technological and Business Innovation", *The Journal of Business Strategy*, 5:2, Fall, 60-73

Rogers, E.M. and Joving I.M. Kim, 1985,"Diffusion of Innovations in Public Organizations" in Richard L. and Anne J. Merrit, eds., *Innovation in the Public Sector*, Sage Publications, London, 85-108

Room, Graham, 1986, *Cross-National Innovation in Social Policy*, St. Martins Press, New York

Rosenberg, Nathan, 1976, *Perspective on Technology*, Cambridge University Press, Cambridge, U.K.

Scitovsky, Taber, 1987, "Growth in the Affluent Society", *Lloyds Bank Review*, Januari 1-14

Stanback, Thomas M. Jr., P.J. Bearse, T.J. Noyelle and Robert A. Karasek, 1981, *Services: The New Economy*, Allanheld, Osmun, Totowa, N.J.

Stanback, Thomas M. and T.J. Noyelle, 1982, *Cities in Transition*, Atlanheld, Osmun & Co., Totowa, N.J.

Summers, Robert, 1985, "Services in the International Economy" in R.K. Inman, *Managing the Service Economy: Prospects and Problems*, Cambridge University Press, London, 27-48

Usher, A.P., 1964, *A History of Mechanical Inventions*, Harvard University Press, Cambridge, Mass.

CHAPTER 7

Research and Development, Corporate Organisation and Industrial Location: Prospects for Regional Development

Jeremy Howells

7.1 INTRODUCTION

Increasing attention by policy makers has been focussed on research and technology as a means to stimulate and expand national and regional economies. Thus although the association between research and development (R&D) and economic growth is difficult to specifically determine, nevertheless research and technological innovation plays an important part in the overall growth and development of modern industrial economies. Recent work, for example, on a broad macro-scale by the OECD (1986a; 1986b), has confirmed the contribution of research and technology in the growth of national economies (Pavitt, 1982). They indicate that high R&D intensive industries grew the fastest within the OECD manufacturing system during the 1970-1980 period as measured in terms of production, domestic demand, employment, imports and exports (OECD 1986b, 3-8). Similarly on a micro, firm scale it has been shown that R&D can lead to an improved market and competitive position, generating future growth and profit, and this can be essential for long term company survival and growth (see, for example, Comanor, 1965; Mansfield, 1968; Shen, 1969; Bailey, 1972; Branch, 1974; Schwartzman, 1976; Nolan et al., 1980; Ong and Pearson, 1982; Odagiri, 1985; Odagiri and Iwata, 1986).

However although a broad association between research and technological innovation and economic growth has been established, together with increasing evidence to show that research and innovation schemes have been successful in their economic objectives (Kunze, 1982; Gronhaug and Fredriksen, 1982, 1984; Rothwell, 1982, 1986a, 1986b), there are a number of key issues which are still of great concern to both academics and policy makers alike. They revolve around the continued problems of: isolating more specifically the key links and parameters between R&D and industrial performance and growth; the actual functioning, organisation and location of the research process; the increasing spatial fluidity of research and technological development (RTD; i.e. benefits of research undertaken in one locality may be felt in another); and the development of appropriate RTD schemes and initiatives for particular countries and regions.

A number of these policy problems involve spatial issues and indeed an increasingly important policy area is the uneven spatial distribution of research and technology and its associated implications for regional economic growth and development. This paper will attempt to provide some insights into industrial R&D organisation and location, using the pharmaceutical industry as a basis, and will then go on to link this with some of the problems associated with developing an RTD strategy for regional economies. The final section will draw upon some of the work undertaken by the Centre for Urban and Regional Development Studies for the European Commission in formulating and establishing an RTD strategy for the less favoured regions of Europe, known as the STRIDE (Science and Technology for Regional Innovation and Development in Europe) programme.

7.2 INDUSTRIAL R&D LOCATION AND CORPORATE ORGANISATION

Organised research activity undertaken by industry and, to a lesser extent, government is now the main generator of technological progress in mature industrial economies (Little, 1977, 87-91; Rothwell and Zegveld, 1982, 46-48; Malecki, 1981, 21). This section proposes to reveal some of the underlying factors and processes associated with the location of industrial R&D and the way it functions geographically, in relation to how research activity is structured and located within a corporate (large firm) context. This is important because large firms still play a dominant role in R&D and technological innovation in industrial economies (Rothwell and Zegveld, 1982, 63-66; OECD, 1982 a, 62). Thus, for example (leaving aside the continuing debate over the relationship between innovativeness and firm size) the sheer dominance of industrial research by large firms is indicated by the fact that in the UK the top 100 firms accounted for 90% of total industry R&D employment, 91% of intramural R&D expenditure and received 99% of all government funding going to industry in 1981 (Business Statistics Office, 1985, 74).

With industrial research having become an institutionalised and integrated part of corporate activity (Kay, 1979, 69; Taylor, 1977, 332) it is likely that behavioural and organisational factors will have an important impact on the location of industrial innovation. If we accept that the organisational structure of industrial firms has a corporate activity (Wood, 1978, 147-149) it is not unreasonable to suggest the the corporate organisation of a firm will influence how R&D activity within the company is structured and, in turn, located. Since it is hypothesised that 'internal' factors, relating to the behaviour and structure of the firm, now play an important part in influencing the overall distribution of research activity, a review of the functioning and structure of R&D within the context of corporate activity is now presented.

Two factors can be seen as providing an overall delimitation of the functioning and structure of R&D within a company: the status and the scale of research within the organisation and secondly, the actual structure of the firm itself. The status of research varies between firms and this will obviously depend on the type of technological environment which confronts a particular firm. Thus for companies in industrial sectors with a rapidly changing and uncertain technological environment greater stress is likely to be placed on industrial research in overall corporate policy, which will be exemplified by high R&D investment ratios (Ansoff and Stewart, 1967, 78-79).

The more complex structure and functioning of R&D within large corporations has obviously been influenced by the evolution of the overall corporate structure of industrial enterprises (Howells, 1984a, 21). This was associated with the trend away from single product, unitory enterprises, with their functional structure based upon specialised departments, and into multi-divisional (M-form) organisations (Chandler, 1962, 53) composed of a series of semi-autonomous divisions, responsible for administering a particular product (or geographical) market sector. The evolution of many large industrial enterprises into M-form type of structures had important implications for research in that R&D was no longer necessarily a corporate-wide task. Thus even if a firm did maintain a centralised corporate establishment it would find it increasingly necessary to have specialised research departments to cope with the particular needs of the different product divisions. This structural change in many large enterprises, associated with the process of diversification, has meant that the possible permutations of R&D structure and functioning, and consequently locational pattern, within these firm could be much greater and more complex.

In terms of the nature and objective of research, industrial R&D can be divided into three broad categories (National Science Foundation, 1973, 19; OECD, 1981, 53-6): basic, applied and development. Basic research can be defined as being an original investigation in scientific knowledge having no specific commercial objectives although it may eventually prove to have future commercial potential. Applied research has been seen as an investigative process directed towards the discovery of new scientific knowledge having specific commercial objectives with respect to products and processes. Lastly, development work was defined as a non-routine technical activity concerned with translating resource information obtained by the organisation into products or processes. The distinctions between the three categories can be extremely blurred (Kay, 1979, 11), but these three R&D types can generally be associated with particular 'task' environment (Table 7.1) which have been discussed in detail elsewhere (Howells, 1984a, 21; Howells, 1984b).

Table 7.1 Task environment characteristics of the main categories of R&D

Type of Research	Degree of Orient-ation	Presence of Commercial Objectives	Payback Criterion	Operational Time Horizon	Degree of Uncert-ainty	Barriers to Entry
Basic research	Minimal	Low	Long term	Long run	High	High
Applied research	Medium/ High	High	Medium term	Merium rum	Moderate	Medium
Develop-ment work	Medium/ High	Medium/ High	Short term	Short run	Low	Low

The particular R&D structure adopted by a company will therefore depend on the type of mixture of research carried out and the relative operating merits of the various alternative structures considered. In terms of the spatial structure of R&D this is related to whether the research is to be essentially concentrated or decentralised in its operation. Linked with this spatial element are the scale economies associated with research establishment size and the coupling (communication activities) arrangements between R&D and other corporate activities.

Indeed the main advantages of research concentration as against decentralisation revolve around the benefits of scale economies and research efficiencies (Scherer, 1970, 353) associated with large-scale research laboratories, with the additional gain of improved internal communication between research groups. Research is also less likely to become involved in trying to solve short term operational problems (Twiss, 1974, 227). In order to effect just such research concentration advantages, in particular scale economies, companies making acquisitions often rationalise their own and newly acquired research departments (Massey and Meegan, 1979, 192-3). Against this, a centralised research establishment tends to be at a disadvantage when it comes to coupling arrangements with other corporate functions, in particular with production, and in terms of operational efficiency and administrative control. Thus most companies recognise that there is an optimum size of research centre in terms of general efficiency (NEDO, 1972b, 29).

The main benefits arising from decentralised research facilities result from improved coupling with other organisational functions. Decentralised R&D, on a divisional basis, allows improved communication links with production and marketing, effectively overcoming the problem of 'programme dislocation' when a project is transferred from R&D to production (Twiss, 1974, 222). Improved research orientation and better recognition of the enterprise's operating environment is also more likely to occur via close locational contact. Lack of suitable specialist staff and equipment (NEDO, 1972a, 53) can however present a problem in certain small decentralised laboratories. In addition, laboratory costs pro rata are also usually higher (Olin, 1973, 130) and research viability often becomes critical because of size constraints, although (NEDO, 1984, 14-17). This has been confirmed in a recent study of pharmaceutical research which has also suggested that the minimum viable size of a research laboratory has risen dramatically over recent years (NEDO, 1987, 18-20). Another disadvantage is that research may become too involved with trying to solve short run, technical problems at the expense of more fundamental, longer term research programmes.

For the large multi-product companies, however, which now dominate most mature industrial economies, industrial research is undertaken both on a centralised and decentralised basis (Twiss, 1974, 227). Thus longer term, basic research is carried out in central research laboratories, whilst applied research is undertaken at a divisional level and short term development work takes place within each product division in smaller development laboratories attached to production units. The role of these latter small technical and development units is largely one of facilitating the transfer and application of new innovations to the manufacturing plants. The dispersed nature of such R&D activity (important in numerical terms if not in relation to the overall share of research activity) has been indicated by a number of studies (Thwaites, Oakey and Nash, 1981, 18, see also Crum and Gudgin 1977, 118; Lall, 1979, 325) and they have shown that certainly not all research activity is concentrated in large centralised research facilities.

The hierarchy of R&D activity presented above can, in turn, be related to the hierarchical nature of corporate structure as a whole (Simon, 1965, 98-99; Thorngren, 1970, 416). The different roles and operational requirements of the three main types of R&D are therefore responsible for producing this spatial hierarchy of research activity, moving from a centralised to a decentralised framework.

In the pharmaceutical industry a 'mixed' R&D structure (Olin, 1973, 131) is common in large multi-divisional companies, where the pharmaceuticals division is only one of several product sectors in which the company is involved. However, since pharmaceutical R&D represents a highly distinct technical environment the overwhelming majority of pharmaceutical research is carried out at the divisional level. Indeed within this divisional structure research is further organised on a centralised/decentralised basis with most basic and applied research being undertaken at a centralised research laboratory whilst development work is carried out at smaller units attached to production units. Research associated with a particular product group may therefore be virtually autonomous from the rest of R&D within the company (NEDO, 1972b, 26). As a consequence the centralised corporate R&D laboratory may only be used for certain research requirements, such as toxicological testing. Indeed these specialist functions may themselves be based in a separate laboratory.

An important conclusion from this discussion of the organisation of R&D in large multi-product companies concerns the importance of communication links and coupling arrangements within the corporate structure. This is indicated by studies undertaken by Thorngren (1970) and Goddard and Morris (1976). Thorngren (1970, 418-9) in his model of organisation/environment relations, has identified three fundamental types of contact patterns related to different time horizons: orientation, planned and programmed processes. Orientation processes are those contacts which involve the long term scanning of the organisations' knowledge and task environment in order to identify future possibilities and alternatives for the firm. Planning processes are principally concerned with the development of specific alternatives which have been identified through orientation processes. Lastly, programmed processes involve the large volume of routine communication concerned with the day-to-day control and management of operational activity (the production process is usually included in this functional category). Programmed processes occur on a regular, and largely predictable, basis so that standardised decision procedures can be applied to them. It has been shown by Goddard and Morris (1976, 49) that planned functions pre-eminently involve international contacts (94.7% of all contacts compared with 42.7% and 53.2% for orientation and programmed processes). Of pertinence here is that Thorngren (1970, 419) and Goddard and Morris (1976, 46) both classified research, with its primary function directed towards the development of specific products or projects for the intermediate future, as being a planned function.

The importance for innovation success of good communication channels between different sections of a company's R&D network, and between research and other corporate functions, has also been stressed elsewhere (CSII, 1971, 19). In part the problems and cost of communication increase simply with greater size. However, when an organisation is geographically dispersed there are even greater barriers to the diffusion of knowledge and flow of technology. A study by Hough (1972, 2-4; see also Tomlin, 1981, 21), for example, has emphasised the impact of distance on the transfer of technical information between research laboratories and sales/marketing units within the same company. Work by Hall and Ritchie (1975) however questions the validity of examining the impact of distance on communication links without acknowledging the fact that the structure of an organisation may have been developed to take distance into account when the decision to locate various functions over space was made.

The importance of the internal structure of the company, in particular, good communication links between R&D and other corporate activities as an influence in the location of research units was confirmed in a survey of pharmaceutical research establishments in the UK (Howells, 1984a, 24). The survey revealed that the two most important factors deciding the location of R&D were 'internal' factors: namely proximity to the company's and/or division's head office and proximity to the main production unit. This supports the hypothesis that the dominating influence in the research location is the siting of research relative to other corporate operations, in particular head office (strategic) and production (programmed) activities. In turn, it suggests that the importance of good communication channels if there is to be any success in developing innovations may be a correct assumption (Carter and Williams, 1957; Myers and Marquis, 1969; CSII, 1971; Langrish et al., 1972; Schock, 1974; Rothwell, 1976). It appears therefore that *internal* corporate considerations are more important than external, environmental factors in influencing the pattern of corporate R&D location. Proximity to headquarters units and manufacturing plants appear to dominate the R&D location decision. External factors, in particular proximity to other research establishments and general accessibility, were much lower in terms of importance.

A final issue in relation to the organisation and location of R&D (Estrin, 1986; Hepworth, 1986) is the impact of the growth and development of computer-communication networks. Computer-communication networks on an inter-organisational level has been shown to increase external R&D contacts (Estrin, 1986, 16-18), which is in turn likely to have implications for internal R&D communication flows. However, as yet the spatial impact of such computer-communication networks has on the location of industrial research still has to be evaluated, although the importance of face-to-face contacts will undoubtedly remain (Törnqvist, 1970, 27-30).

7.3 THE ORGANISATION OF INDUSTRIAL R&D: LOCATIONAL IMPLICATIONS

The discussion in the previous section has centred on the organisation and location of R&D on a corporate, firm based perspective. However, obviously the way that R&D is structured on a corporate basis is going to provide a wider regional significance for R&D and consequently economic growth and job generation. Thus the type and function of industrial R&D activities listed in Table 7.1 can be linked to wider inter-, and intra-, organisational patterns and location (Figure 7.1). As we progress from basic to applied and development work in research, we move from a pattern broadly of concentration to one of dispersal, and this has important regional ramifications. More specifically on an inter-organisational perspective (Figure 7.1) basic research by the private sector is largely restricted to large corporations due to cost considerations, the high element of risk involved and the long term nature and pay-off of the research (Table 7.1). By contrast development and applied research is more widely dispersed amongst firms where entry barriers to this type of R&D are much lower and where the benefits are more immediate and short term. However, even with development work the sheer cost of certain development programmes (for example, with the development and testing of new drugs or the development of aero engines), will restrict such activity to large firms only.

On a intra-organisational basis, for large firms undertaking the range of R&D activities, again basic research will be concentrated in a small number of research establishments located at, or near to, corporate or divisional headquarters. Applied research will be found more widely scattered amongst a range of R&D establishments, from applied research activities located in central research laboratories to smaller R&D units attached to factories. Finally development work involved with, for example, testing or the setting up of pilot plants will be the most dispersed R&D operation and will be primarily attached to manufacturing plants (often in conjunction with other servicing operations, such as quality control and routine testing).

Finally these different R&D functions will be concentrated or dispersed more specifically on a locational basis. Basic research, and to a lesser extent applied research, will be more spatially concentrated in core metropolitan regions. This is because basic and applied research establishments tend to be sited near to the key strategic, orientation functions of the company (the corporate and divisional headquarters and key 'lead' manufacturing plants) but also such units will benefit from being located in information-rich environments which are more important to more externally oriented, research environment scanning operations found in basic and, to a lesser extent, applied research laboratories. By contrast development activity, associated most closely with on-site factory R&D work, will be more widely scattered on a regional basis. However even this type of R&D is likely to favour more central regions on the basis of recent evidence relating to overall R&D location on both an intra- and inter-organisational level. Thus Howells (1984b, 262-6) found that, in the case of pharmaceutical research activity, the likelihood of a factory having an R&D unit attached to it was increased the nearer the plan was to its corporate, or relevant divisional, headquarters. Given the spatial concentration of headquarters functions in core metropolitan regions, there will therefore be a higher on-site R&D level for such areas. This is supported on an inter-organisational basis by two studies covering a range of UK industries which indicated that plants with the highest proportion of on-site R&D were located in the core regions of Britain (Thwaites, Oakey and Nash, 1981, 18-23; Thwaites, Gibbs and Edwards, 1982, 38-41).

This broad framework of R&D structure can be seen in Figure 7.2 representing a model of R&D functional organisation in a large multinational, multi-divisional firm. The model indicates how as one moves towards the development end of the research process, the R&D operation becomes more firmly embedded in the manufacturing process. For large, research-intensive companies operating sizeable operations abroad, a similar but smaller and more embryonic research structure may be established in foreign countries. Thus

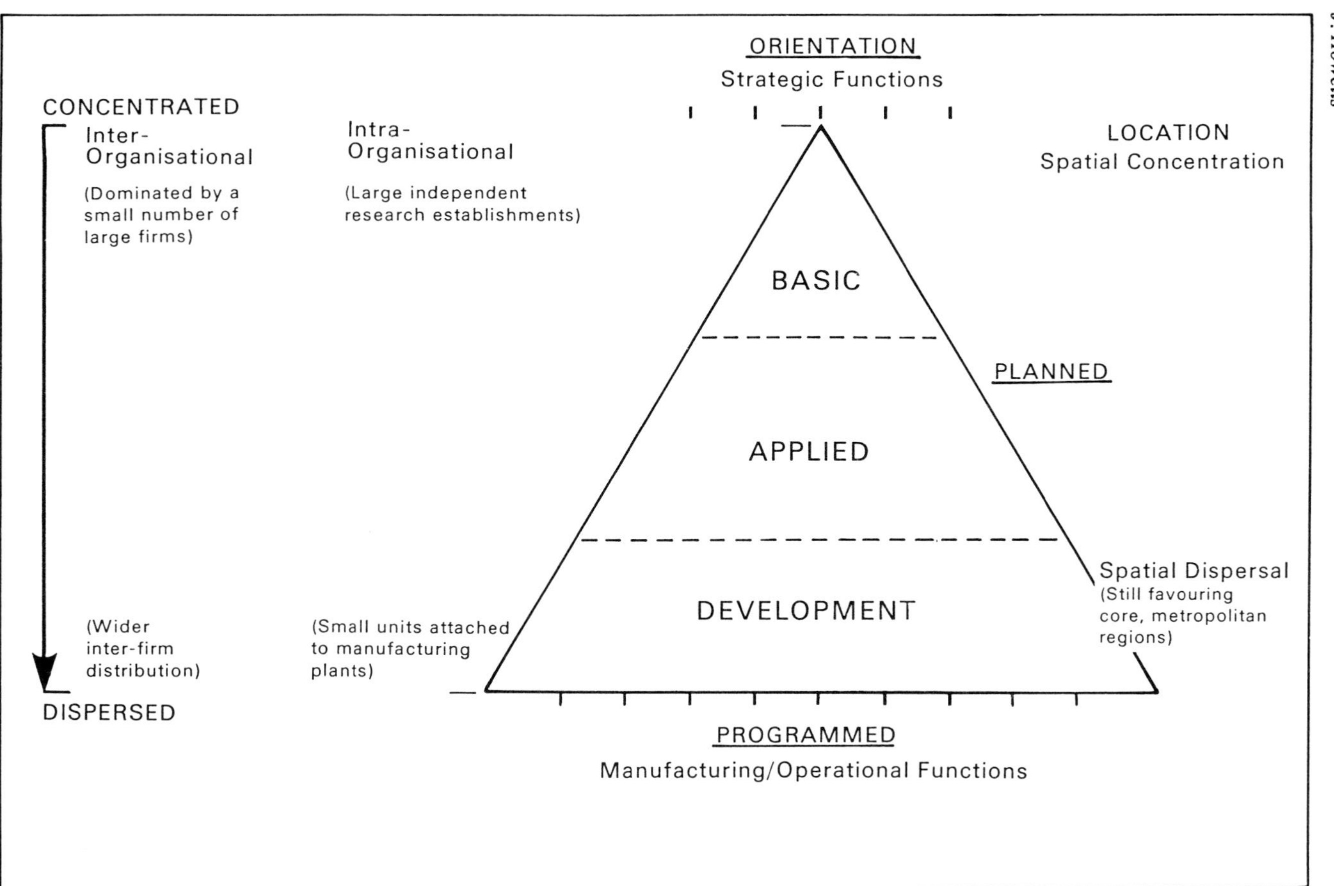

Figure 7.1 R%D functional organisation and location

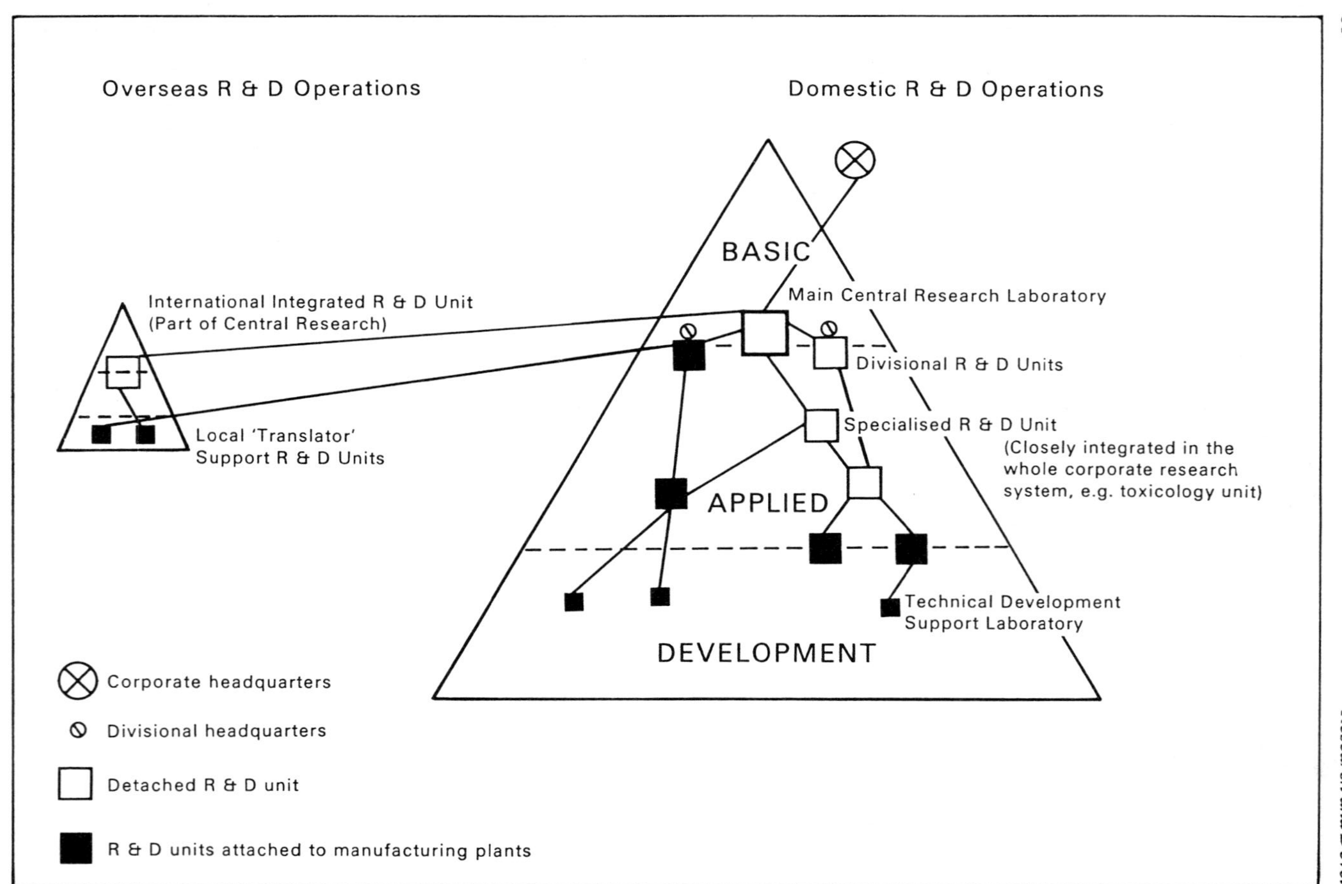

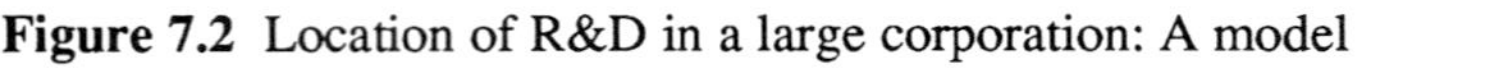
Figure 7.2 Location of R&D in a large corporation: A model

there may be a large international research establishment linked directly to the group R&D centre in the domestic country, as well as more applied/development research laboratories. These latter units will be associated with the transfer and modification of innovations developed outside the host country for the domestic market (for example, modifications associated with the regulatory environment or local market conditions; see Hood and Young, 1981; Charles, 1987).

Obviously these two models represent simplified and stylised representation of R&D functional organisation and location. As noted in the previous section different corporate structures will profoundly influence and alter the structure and location of R&D. For example, a large industrial conglomerate with operations centred on product division basis may not have a centralised, group R&D establishment.

A further issue is that although large firms dominate the industrial R&D process in advanced industrial economies, small and medium sized enterprises (SMEs) can make significant contributions, at least in certain industrial sectors, to the innovative capacity of a nation or region. Indeed recent U.K. research (Pavitt, Robson and Townsend, 1987, 313) suggests a U-shaped relationship between innovative activity and firm size with both large and small firms showing greatest innovative activity, whilst medium-sized firms are shown to be increasingly lagging behind in innovation generation. Again however, as has been indicated earlier in relation to on-site R&D location, evidence suggests that research and innovation intensive new and small firms favour core region, metropolitan locations (Thwaites, Oakey and Nash, 1981, 50-3; Mason, 1985; Wever, 1986, 60).

7.3.1 Research and Technological Development and Regional Economic Growth: Policy Issues and Problems

In terms of policies designed to foster RTD in regions the previous discussion indicates that important functional and organisational variations in R&D on a regional, core-periphery basis have to be acknowledged, as well as the more obvious problem of low levels of R&D capacity in peripheral regions. Of policy significance therefore is that less favoured regions not only possess relatively low levels of R&D but that it tends to be of a certain type and orientation. As such less favoured localities are likely to possess largely short term development (mainly process related) research activities which are unlikely to have a significant impact on their local or regional economy. In addition, as noted earlier, the SME base of less favoured regions will tend to have a low research innovation capacity. The discussion has also suggested that R&D undertaken by plants externally controlled by large firms will be linked into the internal R&D system of the company and have few external research and technical contacts with its local area.

The conditions outlined above in relation to the research base of peripheral, disadvantaged regions are associated with, and lead on to, a wider set of problems which have to be recognised when developing an RTD strategy for such localities (see also Howells and Charles, 1988, 1989). One issue in the development of research and technology in less favoured regions is the role of *inward investment and external control.* Clearly, for depressed regions which have little or no scientific or technological base the policy of attracting high technology ventures into the area is an attractive proposition not only in terms of its immediate impact, via job creation and prestige, but also, and perhaps more particularly, because their other options are highly constrained and very long term in nature. A key issue here is whether disadvantaged regions adopting this strategy will be 'locked into' a system of technology dependency or imperialism which destroys what local indigenous technology base there is and in turn becomes increasingly dependent on large foreign multinational companies which have little commitment to the region, particularly in terms of investing in research and development (Firn, 1975; Britton and Gilmour, 1978). The theme of technological dependency which is associated with inward investment and external control embraces a number of sub-issues. The first is whether externally-controlled branch plants of multi-plant companies actually integrate with their local economies, particularly in relation to technology. Another is the absorption, or 'creaming off', by externally controlled high technology branch plants, of scarce scientific and technical staff in less favoured regions. This makes recruitment of such key personnel even harder for smaller, indigenous based companies. More fundamental, though less tangible, is the lack of indigenous ownership, control and key decision-making over research and technologically oriented functions (Britton, 1985, 86).

Moreover, inward investment may not only take the form of greenfield developments, but also the acquisition of indigenous companies. This may lead to 'technological leakage' and the transfer of control of high technology away from the locality. This is evident on a national scale, where for example in the U.K. there have been a number of foreign takeovers of new and small U.K. high technology firms (Table 7.2). Such firms become satellite units within large corporate complexes, and the results of the research and development originating from the unit, may be applied and manufactured in establishments located in other regions or continents across the globe. Finally, inward investment may have little impact on traditional indigenous firms in less favoured regions which assume a satellite or subordinate role either in relation to more dynamic firms located outside the region or to externally controlled plants within it. Such firms make no effort to initiate technical or product change and typically they have lost all initiative in product design and possess no R&D facilities (Freeman, 1974, 274; see also Britton and Gilmour, 1978, 135). In the following paragraphs we examine these conditions in greater detail.

One of the key problems, in relation to technological innovation, for less favoured regions with a high degree of external control is that of the *'technological isolation'* (or 'independence effect') of branch plants (Howells, 1983, 153). This phenomenon is associated with branch plants which are not dependent on their local environment for sources of technical information and services, or on their own unit-based resources, but rather are provided with resources and information via corporate contacts and linkages (Parker, 1974, 206). Thus McDermott (1979, 300) in a study of Scottish electronics enterprises found over 80% of the externally controlled enterprises surveyed were dependent on corporate technical support and direction (see also Lever, 1974, 332). As such, branch plants, in terms of research and technological innovations, can only be considered as intrinsic elements within large, integrated corporate RTD systems. The expectation that high technology branch plants will rapidly integrate with the local and regional economy in terms of input linkages is therefore misguided. This is particularly true of high technology operations where they are likely to have much more extensive linkage patterns than less technology sophisticated, indigenous manufacturing establishments (Howells, 1984b, 175-6).

Another issue relating to the problem of developing RTD in peripheral, depressed regions is that of *'technological leakage'*. Technological leakage from regions can occur in a variety of ways ranging from: (i) out-migration of scarce scientific and technical personnel trained and educated in the regions; (ii) through Higher Education Institutes (HEIs) and Public Research Establishments (PREs) in the disadvantaged regions linking-up with outside firms to provide research and technical know-how; and (iii) by HEIs and private companies in peripheral regions allowing (via, for example, licensing and patenting agreements) companies located outside the region to manufacture and develop the products and processes they have invented. Finally it can involve, (iv) the acquisition of high technology firms, or even the outright closure of existing high technology branch establishments, which in both cases can lead to the transfer of research and technical know-how out of the region.

An important problem in the expansion and development of RTD in peripheral regions is the 'brain drain' of talented scientific and technical personnel to core regions or abroad to the U.S. or Japan. This problem has been noted not only in the U.K., on both a national (ACARD, 1986) and more specific sectoral basis (Pearson & Parsons, 1983), but has also been identified in a broad range of other European countries notably Ireland, Spain and Belgium. As such, in both industrial decline and structural development regions of the Community it is in many instances not the local education and training of scientific and technical personnel which is the problem but rather the subsequent out-migration of such highly qualified manpower. Thus although the long term effects are difficult to evaluate, the net export of scientific personnel away from peripheral represents a serious haemorrhaging of scientific and technical opportunities which seriously constrains the development of a strong, indigenous technology base in such localities.

Similarly there has been a leakage of technical information and innovation from HEIs and PREs in less favoured regions to industry located in the core regions. What appears to be a common phenomenon in peripheral, disadvantaged areas is that the public research and technology sector in terms of HEIs and PREs has often developed leading-edge technologies but local industry, primarily centred on low technology manufacture, is unable to utilise such developments. What happens therefore is that results, and more particularly the economic benefits of publicly funded scientific research in the less favoured regions, do not remain in the region but go to private companies located in the core regions or abroad who further develop and then manufacture the new products and processes that are generated. Table 7.3 provides an illustration of this type of leakage from the U.K., and indicates that many of the ides and innovations developed by HEIs and PREs in the less favoured regions of Britain are not even remaining within the U.K. or Europe but going elsewhere, most notably to the USA. This problem is particularly apparent in the U.K. where universities have traditionally sought and developed industry-academic links on a national rather than regional basis (Howells, 1986). In addition centrally-funded universities in the U.K. up until recently have not expected to specifically tailor the types of departments they support, and the research that they undertake, to the particular characteristics and needs of local industry. It also stems from a desire by universities in the U.K. and elsewhere in Europe to work with more prestigious, multinational companies which are most likely to have their research facilities located in the core regions rather than with local firms which are often little known and have fewer funds to develop and implement the products and processes being generated by academic research. Above all local industry in less favoured regions often has little or no interest in innovation and research, and therefore HEIs are naturally forced to seek collaboration and funding from outside the area. Indeed, simply in terms of the relative size and consequent diversity of the research and technology base of the core regions vis-a-vis the peripheral regions it is not surprising that innovations being developed by HEIs and PREs in these regions will lead to links with companies outside the locality rather than in it (over and above the issue of technological mismatch between the generators and potential users of technology in a particular region; see later).

A final element in the process of technological leakage is that of the acquisition, restructuring or closure of high technology establishments. The acquisition of high technology firms in the U.K. by foreign multinationals has already been noted (Table 7.2). Of more concern is the downgrading or sometimes outright closure of high technology establishments located in peripheral regions, with the removal, or selling off, of the high technology component and research base of these establishments to corporate sites, or companies, located outside the region. At its most extreme it can involve what has become known as 'intellectual asset stripping'. This was highlighted in the recent case of BTR's acquisition of Dunlop. In the break up and disposal of parts of Dunlop, BTR sold to Goodyear, the U.S. tyre company, a British research based tyre operation, Engineering Consultants Services (ECS), located in sites in the North West and Wales which had developed a world lead in computer tyre assembly. Subsequently Goodyear closed the British operations down and transferred the technology to Ohio and Luxenburg.

Table 7.2 Takeovers of U.K. high technology firms by foreign companies

U.K. high technology firm	Activity	Location	Foreign firm
Columbia Automation	Automatic test equipment	Berkshire, South East	Zehntel
Precima	Automated assembly equipment	Essex, South East	Emhart
Applied Research of Cambridge	Computer sided design (CAD)	Cambridge, East Anglia	McDonnell Douglas
Hoskyns	Computer Systems & consultancy	London, South East	Martin Marietta Data Systems
Shape Data	CAD	Cambridge, East Anglia	Evans and Sutherland
Acorn Computers	Computer manufacture	Cambridge, East Anglia	Olivetti
General Computer Systems	Computer maintenance	London, South East	Bell Canada
Immediate Business Systems	Computer manufacture	Buckinghamshire, South East	Allied Signal
Remek	Robotics/vision systems	Bedfordshire South East	Asea
JK Lasers	Lasers	Warwickshire East Midlands	Lumonics
Compeda	CAD marketing/ research	Cambridge, East Anglia	Prime Computers
Computer Inter-active Systems	CAD	Cambridge, East Anglia	Computerisation
Sevcon	Microprocessor controls	Tyne & Wear North	Tech Ops Inc
Aston Technology	Microcomputers/ software	WMCC, West Midlands	General Automation
Systime	Computer systems services	West Yorkshire, Yorkshire & Humberside	Central Data

Source: Updated from Howells and Charles (1989)

Activity technology firm	Nationality	Other Comments
Automatic test equipment	U.S.	
Electronics & engineering	U.S.	
Aerospace/ Computer Systems	U.S.	Sold by the U.K. government
Aerospace/ Computer systems	U.S.	
Computer simulation	U.S.	Founders moved out and formed new company Three-Space
Electronic business systems	Italy	Acorn in severe financial difficulty before Olivetti takeover
Telecommunications	Canada	Company renamed - Bell Technical Services
Computer/ Communications	U.S.	
Automation	Sweden	Remek acquired by VS Engineering which was then acquired by Asea
Lasers	Canada	
Computers	U.S.	Founded in 1977 by U.K. Government's National Research Development Corporation (now part of British Technology Group)
CAD/CAM	U.S.	Sold by U.K. Government in 1981
Electronics	U.S.	Sevcon was a division of Joyce Loebl; the division was acquired by Vickers then sold off to Tech Ops
Computer systems and software	U.S.	Largest company in Aston Science Park
Computer manufacture and systems	U.S.	

Table 7.3 'Technological leakage' from U.K. regions

Name of HEI/PRE	U.K. Region	Sector	Name of Foreign or U.K. 'Core' Region Manufacturer	Location
Bristol University	South West	Pharmaceuticals	Sandoz AG	Switzerland
University of Birmingham	West Midlands	Biotechnology (Diognostics)	Amersham International	South East, U.K.
Royal Signals and Radar Establishments, Ministry of Defence (+ University of Sussex)	West Midlands	Chemical processing	Balzers AG and Establishment, Pieme Angenieux SH (+ OCLI Optical Coatings Ltd	Liechtenstein St. Heard, France Dumfermline, Scotland)
University of Ulster	Northern Ireland	Electronic Components	Boeing	U.S.A.
University of Leeds	Yorkshire & Humbside	Polymers	Celanese Corporation (owned by Hoechst, W. Germany)	U.S.A.
Health and Safety Executive, Sheffield	Yorkshire & Humbside	Gas detectors	Mine Safety Appliance	U.S.A.
Wolfson Unit (Cambridge Industrial Unit) at Cambridge University	East Anglia	Computer graphics	Control Data	U.S.A.
Newcastle Polytechnic	Northern	Advanced surface engineering technologies	Multi Arc Inc. Siemens Interatom Sumituomo Nissan	U.S.A. FR Germany FR Germany Japan Japan

Source: Updated from Howells and Charles (1989) HEI = Higher Education Institute, PRE = Public Research Establishment

Another issue relating to the problems of developing RTD in less favoured regions on the *technological mismatch* between a region's research base and the research needs, and absorptive capacity, of local industry. This stems in part from the fact that often the most significant research capacity for less favoured regions lies in the public sector, centred on the HEIs and PREs. As such, there is a mismatch in many depressed regions between the potential *generators* of innovations, often in the public sector, and potential *users and developers* of such innovations located in private industry. Thus the HEIs and PREs are often involved in leading-edge research of considerable commercial value, which local industry with its limited development and design capacity finds difficult, if not impossible to utilize. The issue of technological mismatch in the regions appears to be a common problem throughout the Community and one which raises more fundamental issues in terms of developing a research and technology strategy for such regions. The issue of technological mismatch on a national scale has been touched upon by a recent OECD (1984) report on Greece which noted the high research and scientific potential of its universities, but the general weakness of industry in terms of research and its ability to use the research originating from the universities and public sector in Greece. Similarly in relation to the education and training of scientific and personnel there is a mismatch in the ability of local firms to effectively use their talents. In many European countries trained scientific and technical personnel leave the less favoured regions because firstly of simply the lack of innovative firms and secondly more widely most firms in peripheral, disadvantages regions have difficulties in incorporating these highly trained manpower because they are low technology operations.

7.4 CONCLUSION

Obviously the simple lack of research and technology base in a region represents a major problem in the development of an RTD strategy. Moreover, R&D should only be viewed as one element, although a major one, in the overall research and technology capacity of a region. However, this paper has also sought to indicate that is is not only the location and level of R&D and technological capacity that is important in the transmission of research capacity into more specific economic benefits for a particular locality but also its function and corporate context and its relationships with the wider technological/institutional environment. The attraction or indigenous growth of R&D capacity into a region is therefore not a sufficient policy objective for an RTD strategy if it is to lead to economic growth and development in the region. Above all, with the increasing fluidity of research and technological linkages both on an intra- and inter-corporate basis it is becoming increasingly common for the R&D effort generated in one locality to be commercially developed and exploited in another. In this context Rugman (1981) has argued for R&D incentives to be discriminatory in order to encourage the technological capacity and economic benefits of research activity to be retained in the home region or country.

Finally, it has already been noted that R&D is only one element in an regionally oriented RTD strategy, and R&D should also be viewed in a wider technological setting. Equally RTD strategies and mechanisms for localities must also be set in a wider economic, social and institutional framework, and that non-RTD policies (associated with 'parallel' schemes to encourage venture capital, or to improve the information service base of a local economy; Howells 1988) are often important precursors to the successful RTD initiatives in regional economies.

REFERENCES

Aaronvitch, S. and M. Sawyer, 1975, *Big Business: Theoretical and Empirical Aspects of Concentration and Mergers in the United Kingdom*, Macmillan. London.

ACARD, 1986, *Exploitable Areas of Science*, HMSO, London.

Ansoff, H.I. and J.M. Steward, 1967, "Strategies for a Technology-Based Business", *Harvard Business Review* 45, 71-83.

Baily, M., 1972, "Research and Development Costs and Returns", *Journal of Political Economy* 80, 70-85.

Bowie, R.M. 1963, "The Direction and Control of Research and Development", *Research Management* 6, 277-288.

Branch, B., 1974, "Research and Development Activity and Profitability", *Journal of Political Economy* 82, 999-1011.

Britton, J.N.H, 1985, "Research and Development in the Canadian Economy: Sectoral, Ownership, Locational and Policy Issues" in Thwaites, A.T, and R.P. Oakey, *Technological Change and Regional Economic Development*, 67-114, Frances Pinter, London,

Britton, J.N.H. and J.M. Gilmour, 1978, "The Weakest Link - A Technological Perspective on Canadian Industrial Development", *Background Study No. 43*, Science Council of Canada.

Business Statistics Office, 1985, "Industrial Research and Development Expenditure and Employment - 1981", *Business Monitor M014*, HMSO, London.

Carter, C.F. and B.R. Williams, 1957, *Industry and Technical Progress*, Oxford University Press, Oxford.

Centre for the Study of Industrial Innovation (CSII), 1971, *On the Shelf: A Survey of Industrial R and D Projects Abandoned for Non-Technical Reasons C.S.I.I.,* Science Policy Research Unit, University of Sussex.

Chandler, A.D., 1962, *Strategy and Structure*, Doubleday, New York.

Charles, D., 1987, "Technical Change and the Decentralized Corporation in the Electronics Industry" in Chapman, K., *Technical Change and Industrial Policy*, 176-198, Basil Blackwell, Oxford.

Comanor, W.S., 1965, "Research and Technical Change in the Pharmaceutical Industry", *Review of Economics and Statistics* 47, 182-190.

Crum, R.E. and G. Gudgin, 1977, "Non-Production Activities in U.K. Manufacturing Industry, *Regional Policy Series, 3-Collection Studies*, Commission of the European Communities, Brussels.

Estrin, D.L., 1986, *Access to Inter-Organisation Computer Networks*, (mimeo), Department of Electrical Engineering and Computer Science, Massachusetts Institute of Technology, Cambridge, Mass.

Firn, J., "External Control and Regional Development: The Case of Scotland, *Environment and Planning* 7, 393-414.

Freeman, C., 1974, *The Economics of Industrial Innovation*, Penguin, Harmonworth.

Goddard, J.B. and D. Morris, 1976, "The Communications factor in Office Decentralization", *Progress in Planning* 6, 1-180.

Gronhaug, K. and T. Fredriksen, 1982, "Government Innovation Support in Norway: Micro- and Macro-Level Effects, *Discussion Paper 2*, Institute of Industrial Economics, Bergen, Norway.

Gronhaug, K. and T. Fredriksen, 1984, "Stimulate or Substitute? Governmental Innovation Support Revisited", *Discussion Paper 10*, Institute of Industrial Economics, Bergen, Norway.

Hall, K.R. and E. Ritchie, 1975, "A Study of Communication Behaviour in an R and D laboratory", *R and D Management* 5, 243-245.

Hepworth, M.E., 1986, "The Geography of Technological Change in the Information Economy", *Regional Studies* 20, 407-424.

Hood, H. and S. Young, "US Multinational R&D: Corporate Strategies and Policy Implications for the UK", *Multinational Business* 2, 10-23.

Hough, E.A., 1972, "Communication of Technical Information between Overseas Markets and Head Office Laboratories", *R and D Management* 3, 1-5.

Howells, J., 1983, "Filter-Down Theory: Location and Technology in the UK Pharmaceutical Industry", *Environment and Planning* A15, 147-164.

Howells, J., 1984a, "The Location of Research and Development: Some Observations and Evidence from Britain", *Regional Studies* 18, 13-29.

Howells, J., 1984b, *Location, Technology and Filter-Down Theory: An Analysis of the United Kingdom Pharmaceutical Industry*, Unpublished Ph.D. thesis, University of Cambridge.

Howells, J., 1986, "Industry-Academic Links in Research and Innovation: A National and Regional Development Perspective", *Regional Studies* 20, 472-6.

Howells, J., 1988, *Economic, Technological and Locational Trends in European Services*. Avebury-Gower, Aldershot.

Howells, J. and D. Charles, 1988, "Research and Technological Development in the 'Less Favoured Regions of the European Community: A UK Dimension, in Dyson, K., *Local Authorities, and New Technologies: The European Dimension*, 24-48, Croom Helm, Beckenham.

Howells, J. and D. Charles, 1989 (forthcoming), "Research and Technological Development and Regional Policy: A European Perspective" in Gibbs, D., *Government Policy and Industrial Change*, Croom Helm, Beckenham.

Kay, N.-M., 1979, *The Innovating Firm*, Macmillan, London.

Kunze, U., 1982, "Evaluation of Funding R&D Personnel in Germany", *Paper presented to Six Countries Programme on Innovation Workshop on: Evaluating Innovation Policies and Instruments*, 22-3 November, Windsor, UK.

Lall, S., 1979, "The International Allocation of Research Activity by U.S. Multinationals", *Oxford Bulletin of Economics and Statistics* 41, 313-333.

Langrish, J., M. Gibbons, W.G. Evans and F.R. Jevons, 1972, *Wealth from Knowledge*, Macmillan, London.

Lever, W., 1974, "Manufacturing Linkages and the Search for Suppliers and Markets" in Hamilton, F.E.I., *Spatial Perspectives on Industrial Organisation and Decision-Making*, 309-333, Wiley, London.

Little, A.D. Ltd. 1977, *New Technology-Based Firms in the U.K. and the Federal Republic of West Germany*, Anglo-German Foundation/Wilton House Publications, London.

Malecki, E., 1982, "Federal R and D Spending in the United States of America: Some Impacts on Metropolitan Economies", *Regional Studies* 16, 19-35.

Mansfield, E., 1986, *The Economics of Technological Change*, Longmans, London.

Mason, C.M., 1985, "The Geography of 'Successful' Small Firms in the United Kingdom, *Environment and Planning A* 17, 1499-1513.

Massey, D.B. and R.A. Meegan, 1979, "The Geography of Industrial Reorganisation: The Spatial Effects of the Restructuring Corporation", *Progress in Planning* 10, 157-237.

McDermott, P.J., 1979, "Multinational Manufacturing Firms and Regional Development: External Control in the Scottish Electronics Industry, *Scottish Journal of Political Economy* 26, 287-306.

Meyer-Krahmer, F., G. Gielow and U. Kuntze, 1983, "Impacts of Government Incentives towards Industrial Innovation", *Research Policy* 12, 153-169.

Mueller, D.C and J.E. Tilton, 1969, R and D Costs as a Barrier to Entry. *Canadian Journal of Economics* 2, 570-579.

Myers, S. and D.G. Marquis, 1969, *Successful Industrial Innovation*, National Science Foundation, Washington DC.

National Economic Development Office, 1972a, *Focus on Pharmaceuticals*, HMSO Pharmaceuticals Working Party, Chemicals Economic Development Committee, London.
National Economic Development Office, 1972b, *Organising R and D*, HMSO Pharmaceuticals Working Party, Chemicals Economic Development Committee, London.
National Economic Development Office, 1984, *Chemicals R and D: Support for Research and Development in Small and Medium-Sized Firms in the UK Chemical Industry*, NEDO, Chemicals Economic Committee, London.
National Economic Development Office, 1987, *Pharmaceuticals: Focus on R & D*, NEDO, Pharmaceuticals Economic Development Committee, London.
National Science Foundation, 1973, "Research and Development in Industry 1971, *Surveys of Science Resources Series*, Washington DC.
Nolan, M.P., C. Oppenheim and K.A. Winters, 1980, "Patenting, Profitability and Marketing Characteristics of the Pharmaceutical Industry", *World Patent Information* 2, 169-176.
Norris, K. and J. Vaizey, 1973, *The Economics of Research and Technology*, Allen and Unwin, London.
Odagiri, H., 1985, "Research Activity, Output Growth and Productivity Increase in Japanese Manufacturing Industries", *Research Policy* 14, 117-130.
Odagiri, H. and H. Iwata, "The Impact of R & D on Productivity Increase in Japanese Manufacturing Companies", *Research Policy* 15, 13-19.
Olin, J., 1973, "R and D Management Practices", *R and D Management* 3, 125-135.
Ong, C.H. and A.W. Pearson, "The Impact of Technical Characteristics on Export Activity: A Study of Small and Medium-Sized UK Electronic Firms", *R and D Management* 12, 189-196.
OECD, 1981, *The Measurement of Scientific and Technical Activities - 'Frascati Manual' 1980*, OECD, Paris.
OECD 1982a, *Innovation in Small and Medium Firms: Background Reports*, OECD, Paris.
OECD, 1982b, *Innovation Policy: Trends and Perspectives*, OECD, Paris.
OECD, 1984, *Reviews of National Science Policy: Greece*, OECD, Paris.
OECD,1986a, *OECD Science and Technology Indicators: No 2 - R & D, Invention and Competitiveness,* OECD, Paris.
OECD, 1986b, *STI Indicators Newsletter* 9, Scientific Technological and Industrial Indicators Division (DSTI), OECD, Paris.
Parker, J.E., 1978, *The Economics of Innovation*, Second Edition, Longman, London.
Pavitt, K., 1982, R & D, "Patenting and Innovative Activities", *Research Policy* 11, 33-51.
Pavitt, K., M. Robson and J. Townsend, 1987, "The Size Distribution of Innovating Firms in the UK: 1945-1983", *Journal of Industrial Economics* 35, 207-316.
Pearson, R. and D. Parsons, 1983, *The Biotechnology Brain Drain.* SERC/IMS, Swindon/Brighton.
Rothwell, R., 1976, "The Characteristics of Successful Innovators and Technically Progressive Firms", *R and D Management* 7, 191-206.
Rothwell, R., 1982, Evaluating the Effectiveness of Government Innovation Policies, *Report to the Six Countries Programme on Innovation*, Sic Countries Secretariate, TNO, Delft, The Netherlands.
Rothwell, R., 1986a, "Evaluating Innovation Policies: Some Examples, Methods and Guidelines", *Paper for Technology Division, United Nations Conference on Trade and Development*, January.
Rothwell, R., 1986b, "Public Innovation Policies: Some International Trends and Comparisons", *Paper for Thirteenth Imperial College/SPRU/TCC-Science,* Technology and Public Policy Lecture, Imperial College, February, London.

Rothwell, R. and W. Zegveld, 1982, *Innovation and the Small and Medium Sized Firm*, Frances Pinter, London.

Rubenstein, H.H., 1964, "Organizational Factors Affecting Research and Development Decision-Making in Large Decentralized Companies", *Management Science* 10, 618-633.

Rugman, A., 1981, "Research and Development by Multinational and Domestic Firms in Canada", *Canadian Public Policy* 7, 604-616.

Scherer, F.M., 1970, *Industrial Market Structure and Economic Performance*, Rand McNally, Chicago.

Schock, G, 1974, *Innovation Processes in Dutch Industry*, T.N.O. Industrial Research Organisation, Apeldoorn.

Schwartzman, D., 1976, *Innovation in the Pharmaceutical Industry*, John Hopkins University Press, Baltimore.

Shen, T.Y., 1968, "Competition, Technology and Market Shares", *Review of Economics and Statistics* 50, 96-106.

Simon, H.A., 1965, *The Shape of Automation for Men and Management*, Harper and Row, New York.

Taylor, M.J., 1977, "Spatial Dimensions of Inventiveness in New Zealand: The Role of Individuals and Institutions", *Tijdschrift voor Economische en Sociale Geografie* 68, 330-340.

Thorngren, B., 1970, "How do Contact Systems Affect Regional Development?", *Environment and Planning A* 2, 409-427.

Thwaites, A.T., A. Edwards and D. Gibbs, 1982, "Inter-Regional Diffusion of Product Innovation in Great Britain", *Final Report to the Department of Industry and the EEC*, Centre for Urban and Regional Development Studies, University of Newcastle upon Tyne.

Thwaites, A.T., R. Oakey and P. Nash, 1981, "Innovation and Regional Development", *Final Report to the Department of the Environment*, Centre for Urban and Regional Development Studies, University of Newcastle upon Tyne.

Tomlin, B., 1981, "Inter-location Technical Communications in a Geographically Dispersed Research Organisation", *R and D Management* 11, 19-23.

Törnqvist, G., 1970, "Contact Systems and Regional Development", *Studies in Geography, Series B*, 35, Lund.

Twiss, B.C., 1974, *Managing Technological Innovation*, Longman, London.

Wever, E., 1986, "New Firm Formation in the Netherlands" in Keeble, D. E. Wever, *New Firms and Regional Development in Europe*, 54-74, Croom Helm, London.

Williams, B.R., 1967, *Technology, Investment and Growth*, Chapman and Hall, London.

Wood, P.A., 1978, "Industrial Organisation, Location and Planning", *Regional Studies* 12, 143-152.

CHAPTER 8

Diffusion of Technological Change and Economic Growth

Yannis M. Ioannides

8.1 INTRODUCTION

One of the most significant developments in neoclassical economic theory is the theory of economic growth. While economic growth as a subject has received a lot of attention from the classical economists, the pioneering work by Solow (1956) gave it new impetus and elevated the subject to great prominence. The literature that has been produced is vast, and it has been eloquently reviewed in many survey articles and books (e.g., Burmeister and Dobell 1970; Dixit 1976; Wan 1971).

A crucial development within growth theory is models of technological change. One of several concepts of technological change (or progress) which have been developed is the notion that innovations may be embodied in new capital goods. Johansen (1959) is the seminal paper that launched the topic. The key idea of that theory is differences in the substitutability between factors of production before and after investment takes place. One of the most interesting features of Johansen's model is that, under certain conditions, per capita production may grow, stagnate, or even decrease over time. Even though the growth rate of production and output per capita do not depend on the savings rate, the latter affects the time-independent factor in these growth rates. This, in particular, implies for countries with different savings rates, which however start at the same initial position, the country with a higher propensity to save will start out with the higher relative growth rate (before "the asymptote" is reached).

We are currently witnessing a renewed interest from economists in the fundamental determinants of economic growth. Lucas (1985), in a provocative reexamination of the performance of neoclassical growth theory in explaining economic development, has concluded that allowing for increasing returns is the only way for that theory to explain observed economic growth across countries and time. An essential element of Lucas' reexamination is that the steady state growth rate per capita magnitudes in standard versions of the neoclassical growth model does not depend on those behavioral parameters which may be interpreted as determinants of the propensity to save. E.g., in the model in Cass (1965), per capita consumption and per capita capital grow at a rate equal to the (exogenous) rate of neutral technological change divided by the share of labor in aggregate production.

The special notion of increasing returns that is invoked by Lucas, namely that they are external to the industry but internal to the economy as a whole, [1] has received support from Romer (1986), who argues that (technological) knowledge is the key factor responsible for such form for increasing returns.[2] Romer (1987) offers a critical review of the traditional growth accounting literature. He interprets the empirical evidence as implying an aggregate production function which contains no autonomous technological change factor, and exhibits capital elasticity close to unity and labour elasticity close to zero.[3] [4]

Several researchers, such as Baumol, et.al. (1985) and Baumol (1986), have emphasized that there are inherent factors in the economy which constrain the economy-wide propagation of cost improvements accomplished in some sectors - such as those associated with information technology. Baumol, et.al. (1985) emphasize the significance of what they call asymptotically stagnant sectors. There inputs from "very progressive sectors" are used jointly with inputs from "stagnant" sectors. They show that the share in total costs of progressive components diminishes continually while that of stagnant components increases both in real terms and as a share of total cost. Computers have brought about an extraordinary increase in labor productivity, but they are intermediate products and thus not consumed directly. They are used with complementary inputs to produce computational services. The argument is the cost of these complementary inputs comes to dominate the total cost - "the progressive component is innovating itself out of its cost-dominating positions" (ibid) - and ultimately the activity assumes all the characteristics of stagnant services.

This paper addresses the impact of the diffusion of technological change on the dynamics of economic growth. This topic has been recognized as worthy of attention by the growth literature of the 1960's, but has received surprisingly little attention. An exception is a recent paper by Shleifer (1986), in which the timing of innovations is endogenous. The innovating firm captures the whole market but prices its product at the marginal cost of inefficient firms. However, the advantages of innovation are rather short-lived. An interesting feature of Shleifer's model is that it leads to cyclical activity. Firms choose to innovate at time of high aggregate demand. Therefore, the resulting synchronization of innovations gives rise to a multiplicity of perfect-foresight equilibria.

---

[1]Actually, this concept has been utilized by trade theorists [Chipman (1980)] and is central to the modern urban economics literature [see Henderson and Ioannides (1981) for an application in modeling growth of a system of cities]. Lucas, *op.cit.* also emphasizes the urban aspects of the operation of such externalities - motivated both by the significance of urbanization for economic growth and the presumed importance of externalities for understanding the development of urban economies. Production of knowledge (inventive activity), though not by increasing returns, was investigated by Shell (1967).

[2]The essence of the methodology applied by Romer (1986) is that knowledge is treated as an externality. Given the aggregate level of knowledge, firms' production functions are assumed to be concave in inputs - say firms' own technology and labor input. Then equilibrium is defined by equating the sum of individual firms' demand for knowledge to aggregate technology, which enters into individual production functions. A *critical* assumption is that from a planner's point of view, by treating all firms as identical, an individual firm's production function is convex in knowledge. No accumulation of physical capital, but only accumulation of knowledge takes place. Accumulated knowledge has increasing external effects, but is produced with diminishing returns.

Capital investment can have positive external effects through the creation of new knowledge, which spills over throughout the economy and may be propagated by the introduction of new intermediate goods. The latter is related to the total amount of invested capital, so that investment has an additional effect resembling that of an externality.

[3]Romer, ibid. does not explain why labor's coefficient should be so small, except that it could be explained by a putty-clay capital type of theory. Romer also corroborates the evidence in Baumol et.al. (1985) that computerization has not brought forth the magnitudes of benefits one would expect on the basis of the sheer reduction in the cost of computing machinery.

[4]A small labor elasticity is consistent with putty-clay theories of economic growth [Johansen (1959)].

One of them is always the steady-growth acyclical equilibrium, in which inventions are implemented immediately.[5]

Chari and Hopenhayn (1986) also endogenize the adoption of new technologies. A key, and innovative, characteristic of their model is that new technologies are embodied in the form of *human* capital. Young individuals choose the vintage of capital with which to work and thus acquire capital vintage-specific skills. The paper emphasizes the determination of the distribution of workers over vintage-specific skills, which they show to be single-peaked at the steady state. Thus, in general, it takes time for newer technologies to become popular, and technologies associated with different vintages are used at the same time.

From a modern perspective, the technological change of particular interest is the one associated with computerization. Certain stylized facts about computerization may be constructed as embodied, as in the case of digitally controlled machine tools, or disembodied, as in the case of software improvements. It takes a long time and substantial start-up costs to fully computerize a particular establishment, time during which the new technology cannot be utilized to full advantage. A related issue is that of flexibility.[6] Technological change associated with computerization may improve flexibility, but such flexibility may be utilized in a economy-wide sense only if all firms are computerized. Another crucial stylized fact about computerization is the significance of the network effect (Chow 1969; Katz and Shapiro 1986). The more computers have been installed, the easier it is to utilize an additional computer. However, what is significant here is the adoption of the same standard, or else there is no network effect. The advantages of computerization do not appear only within firms. E.g. firms may pass to one another masses of data in electronic form, but such communication is otherwise very costly.

This paper uses an aggregative neoclassical growth model, which includes explicit assumptions about the diffusion of technological change in order to study the ensuing dynamics. We show that the consideration of non-instantaneous diffusion of technological change in conjunction with embodied capital-augmenting technological change leads to a second-order differential equation in aggregate capital per capita. Even though this equation is difficult to solve, it allows us to conclude that the long-run growth rates of all magnitudes of interest do not depend upon the propensity to save. That conclusion is based on two alternative specific assumptions about the process of diffusion of technological change. Our result depends critically on the assumption that technological change is capital - augmenting, for which it is well-known that it is possible to aggregate across capital vintages. That along with the equalization of effective factor intensities across vintages ensures the validity of our results. The second-order differential equation that we obtain implies interesting dynamics, whose precise properties depend on the characteristics of the process of diffusion. The study of these dynamics is left for future work.

---

[5]Cost reductions associated with economy-wide computerization may be assumed to diminish at an increasing rate with the percentage of the economy which is computerized. To see this, consider the cost of having bills paid directly from bank accounts. The cost is miniscule in a fully computerized economy, and may be substantial in a partly computerized one.

[6]The issue of flexibility is of particular interest in its own terms [Kulatilaka (1986)]. Very much like a clay-putty model, embodied innovations may allow greater substitution ex post than the substitution possibilities embodied in the technology available *ex ante*. Flexible manufacturing systems are a case in point here. As far as modelling flexibility is concerned, dual models like that of Fuss (1977) [see Ansar, et.al. (1986) for a recent application] are superior for the purposes of empirical investigations to those utilized by Mizon (1974) and Malcomson and Prior (1979).

8.2. NEOCLASSICAL ECONOMIC GROWTH WITH DIFFUSION OF TECHNOLOGICAL CHANGE

8.2.1 Diffusion of Technological Change

Consider a neoclassical economy consisting of an infinity of firms with an aggregate measure of unity. At the beginning of time all firms have identical technologies. Their production function exhibits constant returns to scale. As time evolves some firms innovate. The proportion of firms which have innovated by time t is given by $y(t)$. The function $y(\cdot)$ satisfies the following conditions:

$$\dot{y} > 0; \lim_{t \to \infty} y(t) = 1.$$

We have now yet made any specific assumption about the nature of the process which described the diffusion of technological change through the economy. For the purposes of comparison we could consider a number of alternative processes. An assumption made frequently is that the spread of innovations across the economy is described by the so-called logistic growth equation:[7]

$$\frac{\dot{y}(t)}{y(t)} = c(1-y(t)),$$

where c is a positive constant. That is, the growth rate of $y(t)$, the proportion of all firms which have adopted the innovation by time t, is proportional to the number of firms which have not yet innovated. The higher that number the higher is the growth rate. It is an interesting property of logistic growth that innovation spread initially at increasing rate but ultimately at a decreasing one. Integrating the logistic growth equation yields:

$$y(t) = \frac{y(0)e^{ct}}{1-y(0) + y(0)e^{ct}}.$$

If innovations were to spread at a constant growth rate, then they would spread through the entire economy in finite time. This case is not particularly interesting in our context.

However, the case of a continuously decreasing growth rate is particularly interesting. In this negative exponential case we have:

$$y(t) = 1 - e^{-ct}.$$

Innovation takes the form of knowledge that is newly acquired and allows better utilization of productive factors.

8.2.2 Embodied Technological Change With Diffusion

As we described in Section A above, innovation takes the form of knowledge about how to implement capital-augmenting technological progress. For a firm which innovates at

---

[7]See Allen (1982) for a justification of the logistic growth equation by means of a behavioral model of information transmission (involving Gibbs states with nearest neighbor potential), and Chow (1969) for an application on the growth of demand for computers).

time θ, one unit of capital yields $e^{\gamma\theta}$ efficiency units of capital, where γ is an exogenous rate of technological change. We further assume that technological progress is embodied in capital goods and that the technology is putty-putty. That is, a firm that has innovated at time θ and invests $I(\theta)$ units of capital, has an effective capital equal to $e^{\gamma\theta}I(\theta)e^{-\delta(t-\theta)}$ at any time $t \geq \theta$, where δ denotes a constant exponential rate of deterioration. Thus, the rate of output at t by a firm that innovated at time θ is given by:

$$F(\theta,t) = F(e^{-\delta t}e^{\beta\theta}I(\theta),L_i(\theta,t)), \qquad t \geq \theta. \tag{1}$$

where $\beta = \gamma + \delta$, and $(F\cdot,\cdot)$ is a variable-proportions production function which exhibits constant returns to scale. The subscript i is a mnemonic for *innovation*. After it has innovated, a firm may vary its output only by varying its labor input at time t, $L_i(\theta,t)$, $t \geq \theta$. We assume throughout that all existing vintages of capital are used at any point in time. This in effect, is an assumption about the curvature of the production function.[8]

Equilibrium in the labor market implies that all firms face the same wage rate. Thi requires that the marginal product of labor be equalized across all firms, that is, firms which have already innovated and thus operate with capital stocks of different vintages, as well as firms which have yet to innovate. This, in turn, along with the constant returns to scale assumption implies that the capital-labor ratio in firms which have not yet innovated, $k_s(t) = K_s(t)/L_s(t)$, is equal to the effective capital-labor ratios across all currently existing vintages:

$$k_s(t) = \frac{K_s(t)}{L_s(t)} = \frac{e^{-\delta t}e^{\beta\theta}I(\theta)}{L_i(\theta,t)}, \qquad \theta \leq t. \tag{2}$$

The subscript s is a mnemonic for *stagnant*.

Equilibrium in the labor market requires that the labor supply, $N(t)$, be equal to the demand for labor by all firms, that is by those which have yet to innovate and by those which have already innovated. A proportion $1\text{-}y(t)$ belong to the former category and $y(t)$ to the latter. To aggregate the labor demands by the firms that belong to the latter group we must take into consideration when they innovated. Therefore, we have:

$$N(t) = (1-y(t))L_s(t) + \int_{\theta\leq t} L_i\,(\theta,t)\dot{y}(\theta)d\theta. \tag{3}$$

Let $N(t)$, the exogenous labor force (which is equal to total labor supply), grow at a constant exponential rate n. Equations (2) and (3) yield the total demand for capital:

$$k_s(t)N(t) = (1-y(t)K_s(t) + e^{-\delta t} \int_{\theta\leq t} e^{\beta\theta}I(\theta)\dot{y}(\theta)d\theta. \tag{4}$$

The total rate of output is obtained as follows:

---

[8]That capital becomes obsolete in finite time is an assumption which has crucial consequences for the dynamics of the model in Johansen (1959).

Kurz (1963) criticises Johansen for failing to account for the endogeneity of obsolescence. As labor productivity might rise over time, certain vintages of capital might have to be retired before they have physically depreciated. Kurz defines as a terminal path the situation where each variable develops through time at a constant relative rate of change. He shows that on a terminal path, capital becomes obsolete after a constant length of time has elapsed from when first installed, $\theta = -\frac{1-\omega}{\lambda}\log(1-\omega)$, where $1-\omega$ is labor's relative share of output and λ is equal to the rate of technical progress in the vintage production function.

$$(1-y(t))F(K_s(t),L_s(t)) + \int_{\theta\leq t} F(e^{-\delta t}e^{\beta\theta}I(\theta),L_i(\theta,t))\dot{y}(\theta)d\theta$$

$$= f(1,k_s^{-1}(t))[(1-y(t))L_s(t) + e^{-\delta t} \int_{\theta\leq t} e^{\beta\theta}I(\theta)\dot{y}(\theta)d\theta]$$

$$= k_s(t)N(t)F(1,k_s^{-1}(t)). \tag{5}$$

Thus, output per capita is equal to $F(k_s(t),1)$ and depends only on the capital-labor ratio in the stagnant sector, $k_s(t)$. (5) exploits a well-known result of aggregation from growth theory, namely that constant returns to scale together with capital-augmenting embodied technological progress allow aggregation of capital of all vintages (Solow, 1959; Dixit 1976). A new feature in our analysis is that this aggregation still holds when a certain portion of the economy operates under different production conditions.

In order to complete the analysis, we must develop the equation of motion of the system and the determination of the rate of investment when innovation occurs. Even though it pays to adopt the innovation, the process of competition bids away all rents, reducing profits of innovating firms to zero, in the current period as well as in all future ones. Were it not for this, we should characterize the optimal amount of investment in terms of the trade off between the future stream of extra profits and the present cost of investment. It is thus natural to assume that the innovating firms simply convert their current capital into one that embodies the new technology. We assume that this conversion process is costless. Therefore, for the innovating firms we have for all t:

$$K_s(t) = I(t). \tag{6}$$

We retain the neoclassical growth theory assumption of a constant gross savings rate. We may obtain the equation of motion of the system by equating savings per unit of time to additional investment needed by the stagnating sector, $(1-y(t))\dot{K}_s(t)$, plus what is needed to make up for depreciation of capital in the stagnating sector, $\delta K_s(t)$. Capital in the stagnating sector that is freed by the diffusion of technological change is transformed into capital that embodies technological change. We thus have:

$$(1-y(t))\dot{K}_s + \delta K_s(t) = sN(t)f(k_s(t)), \tag{7}$$

where $f(k_s) \equiv F(k_s,1)$ denotes output per capita. Note that because technological change is embodied, no additional investment is made by firms which have already innovated, nor is depreciated capital replaced.

We shall transform (7) in order to make it tractable. First we divide both sides of (7) by $N(t)$ and then use the auxiliary transformation $\psi = K_s/N$ and (6) to get:

$$(1-y(t))\dot{\psi}(t) + (\delta + (1-y(t))\psi$$

$$= sf((1-y(t))\psi(t) + e^{-(n+\delta)t} \int_{\theta\leq t} e^{\beta\theta}\psi(\theta)\dot{y}(\theta)d\theta). \tag{8}$$

Next we perform an integral transformation in order to convert the integro-differential equation (8) into a differential one. We define:

$$J(t) = e^{-(n+\delta)t} \int_{\theta \le t} e^{(\beta+n)\theta} \psi(\theta)\dot{y}(\theta)d\theta. \tag{9}$$

From (9) by differentiating we obtain an expression for $\psi(\cdot)$ in the terms of $J(\cdot)$:

$$\psi(t) = Y(t)(\dot{J}(t) + (n+\delta)J(t)),$$

where $Y(t) \equiv (1/\dot{y}(t)e^{(\beta-\delta)t})$, a known function.

Furthermore,

$$\dot{\psi}(t) = \dot{Y}(t)(\dot{J}(t) + (n+\delta)J(t)) + Y(t)(\ddot{J}(t) + (n+\delta)\dot{J}(t)).$$

$J(\cdot)$ plays here the same role as aggregate capital per capita in vintage capital models (*c.f.*, Dixit *op.cit.*). It is now clear that (8), the equation of motion of the system, may be transformed into a second-order differential equation in $J(t)$ with non-linear coefficients:

$$\begin{aligned}(1-y(t))Y(t)(\ddot{J}(t) \; &+ [(1-y(t))(n+\delta)Y(t) + (1-y(t))\dot{Y}(t) \\ &+ (\delta + (1-y(t))n]\dot{J}(t) + [(n+\delta)\dot{Y}(t)(1-y(t)) + (\delta + (1-y(t))n]J(t) \\ = sf[(1-y(t))Y(t)\dot{J}(t) &+ ((1-y(t))Y(t)(n+\delta) + 1)J(t)].\end{aligned} \tag{10}$$

We have thus obtained a generalization of the law of motion in the neoclassical model of economic growth when the diffusion of technological change is not instantaneous (*c.f.*, Dixit, *op.cit..*, 91-94). This equation is expressed in terms of an aggregator for the capitals of different vintages, where the aggregation is weighted by the efficiency factors, multiplied by the speed of innovation diffusion. The fact that the law of motion is a second-order rather than a first-order differential equation is, of course, entirely due the non-instantaneous diffusion of technological change. Therefore, it could, in principle, give rise to much more complicated dynamics than the standard neoclassical growth model. This differential equation, however, is rather difficult to solve because it is not linear in terms of the derivatives of $J(t)$.

It would be interesting to compare the above solution with the case where technological change is disembodied. Unfortunately, as it is clarified in the Appendix, this case is intractable. Therefore in the following section we restrict ourselves to a comparison with the textbook case of embodied technological change, that is when diffusion is instantaneous.

The differential equation (10) is linear only if $f(\cdot)$ is linear in its argument. This may occur only if $F(K,N) \equiv AK + BN$, where A and B are constants. In that case, (10) becomes:

$$\begin{aligned}&(1-y(t))Y(t)(\ddot{J}(t) + \{[(1-y(t))(n+\delta-sA) + (\delta + (1-y(t))n)]Y(t) \\ &+ \; (1-y(t))\dot{Y}(t)\}\dot{J}(t) + [(n+\delta)(1-y(t))\dot{Y}(t) + \delta(n+\delta)Y(t) \\ &+ \; (n-sA)(1-y(t))(n+\delta)Y(t)]J(t) = sB.\end{aligned} \tag{10'}$$

For particular specifications of $y(t)$, this second-order differential equation which describes the law of motion of the system may be solved in closed form. (10') is a linear second-order differential equation with non-linear coefficients.

It is easy to see that the long-run rate of growth of aggregate capital depends upon the specification of the process which describes the diffusion of technological change.

However, it can be shown, by working with (10), that both for the logistic as well as the negative exponential class the asymptotic growth rate of aggregate capital per capita is independent of the propensity to save. This demonstration is rather tedious and uninteresting and is thus deleted here. It involves assuming a long-run rate of growth for capital per capita and substituting in (10) so as to compute the unknown asymptotic growth rate.

8.2.3 Embodied Technological Change and Instantaneous Diffusion

In order to assess the impact of the slow diffusion and adoption of technological change, we compare with the vintage capital model when the diffusion occurs uniformly throughout the economy. The equation of motion for the putty-putty model with embodied technological change is:

$$\psi(t) = sf(J(t)), \tag{11}$$

where $\psi(t) \equiv I(t)/N(t)$ denotes investment per capita, and $J(t)$ is defined as:

$$J(t) = e^{-(\delta+n)t} \int_{\theta\leq t} e^{(n+\beta)\theta}\psi(\theta)d\theta.$$

$J(t)$ is, again, an aggregate measure of capital per capita across vintages. As we mentioned earlier, such an aggregation is possible because technological change is capital-augmenting. Working as before, we obtain the equation of motion in terms of $J(t)$:

$$\dot{J}(t) + (n+\delta)J(t) = e^{\gamma t}sf(J(t)). \tag{12}$$

Equation (12) is the counterpart of Equation (8. It is again easy to solve (12) for a variety of specifications of the aggregate production function. If $F(\cdot,\cdot)$ is Cobb-Douglas, then (12) may be solved as a Bernoulli equation. Specifically, if $f(J) \equiv J^{\alpha}$, then by multiplying both sides of (12) by $(1-\alpha)J^{-\alpha}$ we obtain a linear differential equation in terms of $u \equiv J^{1-\alpha}$ which may then be solved in closed form.

By working in this fashion we have:

$$\dot{u}(t) + (1-\alpha)(n+\delta)u(t) = (1-\alpha)se^{\gamma t}. \tag{13}$$

The general solution of (13) has the form:

$$u(t) = (1-\alpha)\left[\bar{k} + s\int_0^t e^{\gamma t}e^{(1-\alpha)(n+\delta)t}dt\right]e^{-(1-\alpha)(n+\delta)t},$$

which yields the aggregate capital per capita as

$$J(t) = (1-\alpha)^{1/1-\alpha}\left[\bar{k} - \frac{s}{\gamma+(1-\alpha)(n+\delta)} + se^{(\gamma+(1-\alpha)(n+\delta))t}\right]^{1/1-\alpha} e^{-(n+\delta)t}. \tag{14}$$

It readily follows from (14) that the asymptotic rate of growth of aggregate capital per capita does not depend on the savings rate.

Therefore we conclude that the complicated dynamics characterizing the growth model with slow diffusion of embodied technological change are entirely due to our assumption about how innovations spread in the economy.

APPENDIX

Disembodied Technological Change

It is somewhat surprising that the case where technological change is disembodied and diffusion of technological changes is not instantaneous is intractable. To show this, consider that technological progress is implemented as soon as news of it reaches firms. Any additional investment made after the innovation has been adopted also benefits from it. The assumption means that the adoption of technological change augments capital by a factor $e^{\gamma(t-\theta)}$, where θ is the point in time when the innovation was adopted. Let $L_s(t)$ and $L_i(\theta,t)$ denote the quantities of labor employed by a stagnant and an innovating firm, respectively. For equilibrium in the labor market, it is required that the marginal product of labor be equalized across both kinds of firms. Thus we have:

$$k(t) = \frac{K_s(t)}{L_s(t)} = \frac{e^{\gamma(t-\theta)}K_i(\theta,t)}{L_i(\theta,t)}, \ t \geq \theta. \tag{A.1}$$

and

$$N(t) = (1-y(t))L_s(t) + \int_{\theta \leq t} \dot{y}(\theta)L_i(\theta,t)d\theta. \tag{A.2}$$

As before, we assume that firms convert their capital into a form suitable for production after the innovation has been adopted:

$$K_i(t,t) = K_s(t). \tag{A.3}$$

For the total amount of capital in efficiency units we have:

$$K^*(t) = (1-y(t))K_s(t) + \int_{\theta \leq t} \dot{y}(\theta)e^{\gamma(t-\theta)}K_i(\theta,t)d\theta = k(t)N(t). \tag{A.4}$$

As for the amount of physical capital we have

$$K(t) = (1-y(t))K_s(t) + \int_{\theta \leq t} K_i(\theta,t)\ \dot{y}(\theta)d\theta. \tag{A.5}$$

The equation of motion for the system is obtained by equating savings per unit of time to additional investment needed by the stagnating sector, $(1-y(t))\dot{K}_s(t) - \dot{y}(t)K_s(t)$, plus new investment in the innovating sector, $\dot{y}(t)K_s(t)$, plus what is need to make up for depreciation. That is:

$$(1-y(t))\dot{K}_s(t) + \int_{\theta\leq t} \frac{\partial K_i(\theta,t)}{\partial t}\dot{y}(\theta)d\theta + \delta K(t) = sN(t)f(k(t)). \quad \text{(A.6)}$$

If the dynamics of this model could be analyzed in a simple manner, then we could analyze the economic value of flexibility which is implied by disembodied technological change relative to embodied. To the extent that the value of flexibility is positive, it would be interesting to require that firms could costlessly implement the embodied technology but would have to incur an extra cost in order to adopt the disembodied technology. Unfortunately, the only tractable way in which disembodied technological change may be handled is to assume that the technological change capital-augmenting factor for firms which have adopted the innovation takes the form $e^{\gamma t}$. Consequently, factor use by such firms is independent of the point in time in the past when they adopted the innovation.

REFERENCES

Allen, B., 1982, "A Stochastic Interactive Model for the Diffusion of Information", *Journal of Mathematical Sociology* 8, 265-281.
Ansar, J., A Ingham, G. Leon, M. Toker and A. Ulph, 1986, "A Vintage Model of Demand for Energy in U.K. Manufacturing", University of Southampton, (mimeo, March.
Baumol, W.J., et.al., 1985, "Unbalanced Growth Revisited: Asymptotic Stagnancy and New Evidence", *American Economic Review* 75, 906-817.
Baumol, W.J., 1986, "Information, Computers and the Structure of Industry", R.R. ≠86-29, C.V. Starr Center for Applied Economics, October.
Burmeister, E. and A.R. Dobell, 1970, *Mathematical theories of Economic Growth*, MacMillan, New York.
Cass, D., 1965, "Optimum Growth in an Aggregative Model of Capital Accumulation", *Review of Economic Studies* 32, 233-240.
Chari, V.V. and H.A. Hopenhayn, 1986, "Vintage Human Capital and the Diffusion of New Technology", Federal Reserve Bank of Minneapolis, Working Paper No. 237, December.
Chow, G., 2969, "Technological Change and the Demand for Computers", *American Economic Review*, 1117-1130.
Dixit, A., 1976, *The Theory of Equilibrium Growth*, Oxford University Press, Oxford.
Fuss, M.A., 1977, "The Structure of Technology Over Time: A Model for Testing the 'Putty-Clay' Hypothesis", *Econometrica* 45, 8, November, 1797-1821.
Henderson, J.V. and Y.M. Ioannides, 1981, "Aspects of Growth in a System of Cities", *Journal of Urban Economics* 10, 117-139.
Johansen, L., 1959, "Substitution vs. Fixed Production Coefficients in the Theory of Economic Growth: A Synthesis", *Econometrica* 27, 2 April, 364-383.
Katz, M.L. and C. Shapiro, 1986, "Technology Adoption and the Presence of Network Externalities", *Journal of Political Economy* 94, 4, 822-841.
Kulatilaka, N., 1986, "The Economic Value of Flexibility", School of Management, Boston University (mimeo).
Kurz, M., 1963, "Substitution vs. Fixed Production Coefficients: A Comment", *Econometrica* 31, 1-2, January, 209-217.
Lucas, Jr., R.E., 1986, "On the Mechanics of Economic Development", Marshall Lectures, reprinted by The Institute of Economics, Academia Sinica, Taipei, April.
Malcomson, J.M. and M.J. Prior, 1979, "The Estimation of a Vintage Model of Production for U.K. Manufacturing", *Review of Economic Studies*, October, 719-736.

Mizon, G.E., 1974, "The Estimation of Non-Linear Econometric Equations: An Application to the Specification and Estimation of an Aggregate Putty-Clay Relation for the United Kingdom", *Review of Economic Studies*, 41, 3, July, 353-369.

Romer, P., 1986, "Increasing Returns and Long-Run Growth", *Journal of Political Economy* 94, October, 1002-1037.

Romer, P., 1987, "Crazy Explanations for the Productivity Slowdown", presented at the *Macroeconomics Annual*, NBER, Cambridge, March.

Shell, K., 1967, "A Model of Inventive Activity and Capital Accumulation", in K. Shell, ed., *Essays on the Theory of Optimal Economic Growth*, MIT Press, Cambridge, Massachusetts, 67-85.

Shleifer, A., 1986, "Implementation Cycles", *Journal of Political Economy* 94, 6, December, 1163-1190.

Solow, R.M., 1956, "A Contribution to the Theory of Economics Growth", *Quarterly Journal of Economics* 32, 65-84.

Solow, R.M., 1959, "Investment and Technical Progress", 89-104., K.J. Arrow, et al., eds, *Mathematical Methods in the Social Sciences*, Stanford University Press.

Wan, Jr., H.J., 1971, *Economic Growth*, Harcourt, Brace and Jovanovitch, New York.

PART II

TECHNOLOGICAL AND ECONOMIC INTERACTIONS: SOME EMPIRIAL STUDIES

CHAPTER 9

The Barrier-Breakthrough Model of Innovation and the Life Cycle Model of Industrial Evolution as Applied to the U.S. Electrical Industry

Robert U. Ayres

9.1 INTRODUCTION

In a previous paper I argued for a somewhat new perspective on the mechanisms responsible for technological innovation in economic theory (Ayres, 1988). This perspective is essentially Schumpeterian, but puts special emphasis on the role of perceived technological barriers and the opportunities created by a breakthrough, when a major barrier is overcome. An important implication of this model is that major inventions do not occur at random. The model also predicts a clustering of innovations following and related to any major breakthrough. Thus the rapid progress in particular fields at particular times is not accidental. The identities of inventors are, of course, unpredictable, along with the technical details of their inventions. On the other hand the technological functions (i.e. The problems solved) are often well understood in advance.

This paper reviews the history of the electrical industry from the perspective of the barrier-breakthrough model. A question one might well choose to ask is: why did this industry develop predominantly in the U.S., though virtually all of the significant research and invention until the mid 1870's was carried out in Europe?[1]

9.2 THE BARRIERS

The major barrier to the widespread application of electricity was its unavailability in useful quantities or voltages at a reasonable cost. Until the mid-1840's the only practical source of electricity, even for experimental purposes, was a voltaic 'pile', a crude battery consisting of a stack of 'cells'. Each cell was a cathodic disk of silver or copper, a disk of

[1]The major exception was the important work of the physicist Joseph Henry (1797-1878) who devised the first crude motor (1829), the first electromagnet (1829), and the first practical electromagnetic telegraph (1830-31) at Albany, NY. In 1832, he discovered the principle of self-induction, (which is credited to him). He also independently discovered the principle of electromagnetic induction's simultaneously with Faraday, though the latter published first and is given the credit. In 1835, Henry invented the first electrical relay and also first used the earth as a return conductor to complete a circuit. he discovered in 1842 the oscillatory character of electrical discharges. But he made no effort to develop or exploit any of his discoveries apart from advising Samuel F.B. Morse during the latter's successful efforts to commercialize the telegraph in the late 1830's and 1840's. (See Sharlin, 1961, 1967).

paper soaked in electrolyte (acid or alkali) and an anodic disk of zinc. In 1808 the Royal Institute of London obtained a battery of 2000 cells for experimental use, and in following years quite large batteries were built. Efficiency and lifetime were gradually improved over time, but the fundamental limitation of all such primary cells[2] was - and still is - that the electrical energy is obtained by chemical dissolution of the anodic material (usually zinc). The only way to recharge a primary battery is by physically replacing the anodes when they are consumed.

Thus, the major technological challenge was to develop a means of obtaining electricity without chemical reactions. They key to doing this was electromagnetic induction, discovered almost simultaneously in 1832 by Faraday in the U.K. and Henry in the U.S. The induction principle states that an electric current can be produced in a conductor by a *changing* magnetic field. The effect can be produced by passing a conductor (wire) through a magnetic filed (e.g. between the poles of a permanent magnet) or, conversely, by moving a magnet near a stationary conductor. The energy embodied in the magnet is not depleted: it is the energy of motion - kinetic energy - that is converted into electricity. Thus, was born the concept of an electric 'dynamo', which could convert mechanical energy (from a steam engine or a waterwheel) into electric power.

Dynamos were conceived almost as soon as Faraday's discoveries were announced, but the first ones were very inefficient, because the fundamental principles were not clearly understood in quantitative and geometrical terms. Thus process occurred largely by trial and error for many years. Improvements were made from time to time but dynamos did not even compete effectively with voltaic piles until the early 1840's and then only for a very limited application: 'electrotyping' a means of electroplating copper on moldings of engravings for printing purposes. But this one application spurred the activities of inventors such as Wheatstone, Varley and Wilde (U.K.) and Siemens (Germany) and Pacinotti (Italy).

The availability of crude but practical dynamos also led people to start considering practical application of the carbon-arc light - till then a laboratory curiosity. The first lighthouse using carbon-arc lights was built in 1857 by Frederick Hale Holmes (U.K.). Arc lights spread much more widely in the late 1860's and 1870's with the invention of the Jablochkoff "candle" in Paris, and Zenobe Theophile Gramme's improved ring-wound dynamos (ibid).

While arc lights were appropriate for exterior purposes and public places, they were inherently too bright and required too much power (at least, in the 1860's and 1870's) to be applicable for most illumination purposes. By mid-19th century candles (mainly from whale oil) had been largely superceded by oil lamps. (In fact, whales were becoming scarce by 1850 and the price of whale oil had risen sharply.[3]) Gas lighting was the other alternative. Experiments with coal gas as an illuminant can be traced back to the 1760's. The first public demonstration seems to have been in Paris (Lebon, 1801). Murdock illuminated the engine house at Boulton & Watt's factory in 1806. Of course, gas lighting is only feasible where gas is available. The creation of a complete system including gas plants, pipelines, meters and safe but efficient burners required several decades. The first gas company was chartered in London (1812). Baltimore was the first U.S. city to install a gas lighting system, starting in 1816. But the gas distribution and lighting system was largely in place by 1880. By that time it was the established competitor against which the upstart electric illumination industry had to compete. (In fact, gas-lighting experienced its greatest period of growth between 1880 and 1900). In his efforts to develop a practical incandescent lamp (1877-79) Edison specifically recognized the identity of his competitor. In a newspaper interview he commented "I have an idea that I can make the electric light

---

[2]Rechareable (secondary) cells such as the lead-acid automotive battery were developed much later, for applications where electric current from some other source is available.

[3]Whale-oil scarcity provided the motive for increased interest in petroleum, resulting in the initiation of petroleum drilling in Pennsylvania (1859).

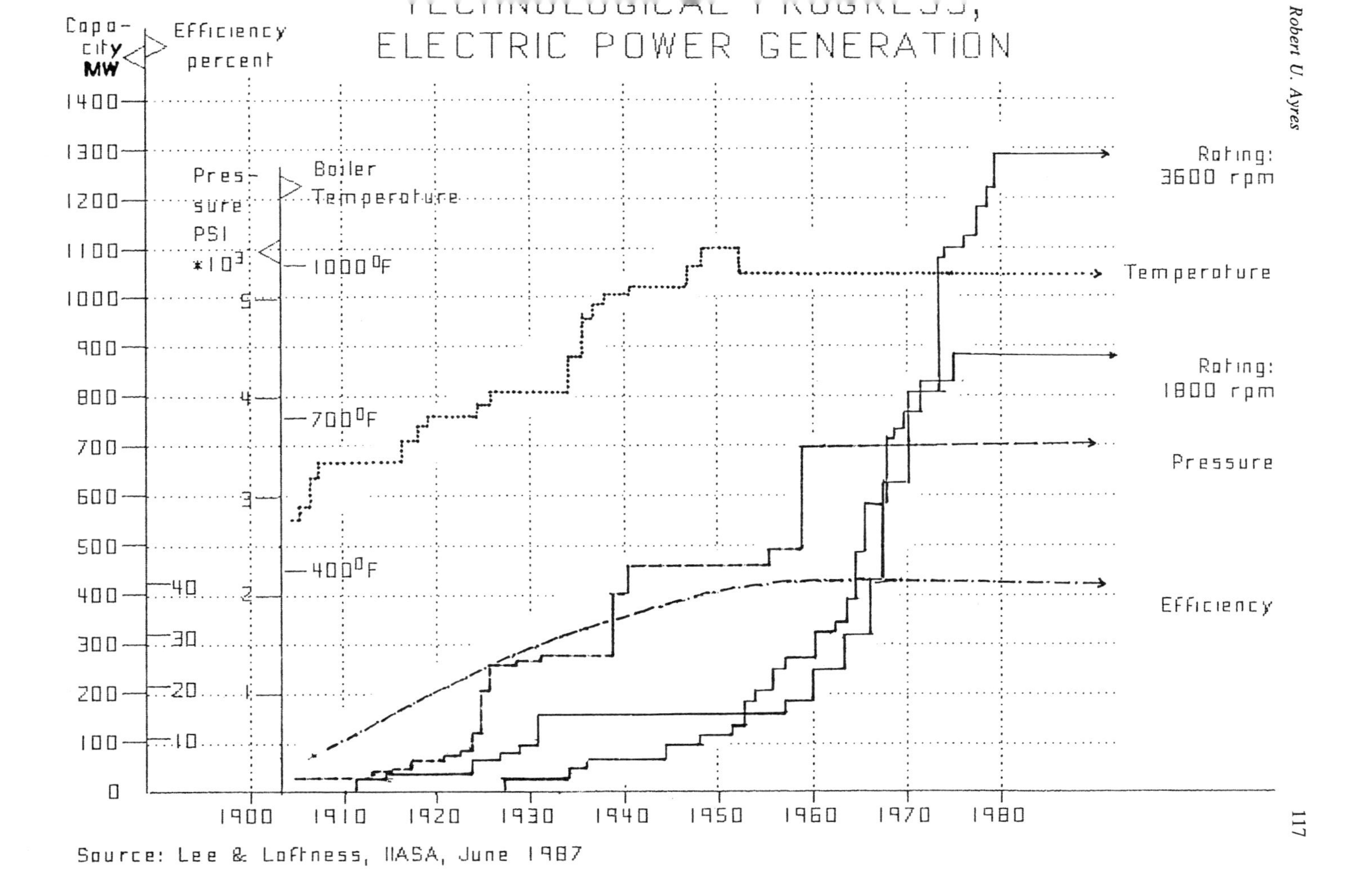

ELECTRIC POWER GENERATION
Capa-city MW
Efficiency percent
Pres-sure PSI *10³
Boiler Temperature
1000°F
700°F
400°F
Rating: 3600 rpm
Temperature
Rating: 1800 rpm
Pressure
Efficiency
1400
1300
1200
1100
1000
900
800
700
600
500
400
300
200
100
0
40
30
20
10
5
4
3
2
1
1900
1910
1920
1930
1940
1950
1960
1970
1980
Source: Lee & Loftness, IIASA, June 1987

available for all common uses, and supply it at a trifling cost *compared with that of gas*[4] (Sharlin, op cit).

From 1800 to 1872 efforts to develop practical means of producing electricity and practical applications of electromagnetism - especially to lighting and traction - proceeded slowly and almost exclusively in Europe, as shown in the timetable (Appendix I).

9.3 THE BREAKTHROUGHS

Technically speaking, the biggest breakthrough in generator design was the self-excitation concept which was apparently discovered independently several times between 1857 and 1866. The formulation and demonstration of the dynamo-electric principle by Werner V. Siemens in 1866 was crucial, since further development was accelerated within the Siemens & Halske company. This led to the improved drum winding design by Hefner-Altaneck, the chief engineer of Siemens & Halske. This design may be said to have established the 'technology trajectory' for all subsequent D.C. generators. The first commercial success was Gramme's dynamo (1872). Nevertheless, it was Thomas Alva Edison's bi-polar generator (1878) that 'put it all together'. Before Edison, the best dynamo's achieved around 40% efficiency (Josephson, 1959). By scaling up in size, carefully taking into account all known resistive loss mechanisms and greatly increasing the iron to copper ration (for superior containment of the magnetic field) Edison achieved a 90% conversion efficiency (ibid). This was the economic breakthrough that made the electrification of industry, urban transportation, and, later, of households, economically feasible.

A second key breakthrough was Gramme's discovery (1873) that a dynamo could be operated in reverse as a motor, by linking it to another (larger) dynamo at a distance. This immediately created the possibility of dynamo-motor combinations. This subsequently proved to be the basis of the electrical traction industry and, later on, the substitution of electric motors for steam engines in factories. Another important invention in this field was the Tesla's AC motor, which appeared rather late (1887), *after* the industry was already rapidly expanding and well into its "childhood" phase.

As regards electric illumination, the key prerequisite was Sprengel's mercury vapor vacuum pump (1865). Of course, it took some time for this technology to spread beyond the physics laboratory, which only ocurred after 'Crookes' radiometer development (1875). By 1877 the availability of good vacuum pumps made electric lights commercially feasible. Joseph Swan and Thomas Edison were among a number of inventors who saw this opportunity, along with William Sawyer and his partner Albion Man (Brooklyn) and others (Sharlin, op cit). Edison's great achievement was to produce a commercially manufacturable carbon filament incandescent lamp, which he did in 1879. The production of lamps began almost immediately (1880) at Edison Lamp Co., Harrison, NJ. Edison moved quickly to consolidate his advantages by combining dynamo technology and lighting technology in a complete system, with the Pearl Street generating plant (NYC 1882). From a technical point of view, however, Edison was scarcely ahead of his contemporaries and even in the field of electric lighting his monopoly was not absolute. By 1885 , 250,000 incandescent lamps were in use of which Edison companies had produced 80% (Passer, 1953). But scores of competitors sprang up to exploit the burgeoning demand for electric lighting systems.

What Edison achieved for the incandescent lamp was nearly matched by Charles Brush (1878) and Elihu Thomson with his partner Edwin Houston (1879) for the arc lamp. Both men (independently) developed commercially successful arc lighting systems, one using DC and the other AC. Both systems were rapidly commercialized and, by 1886, three

[4] Italics added.

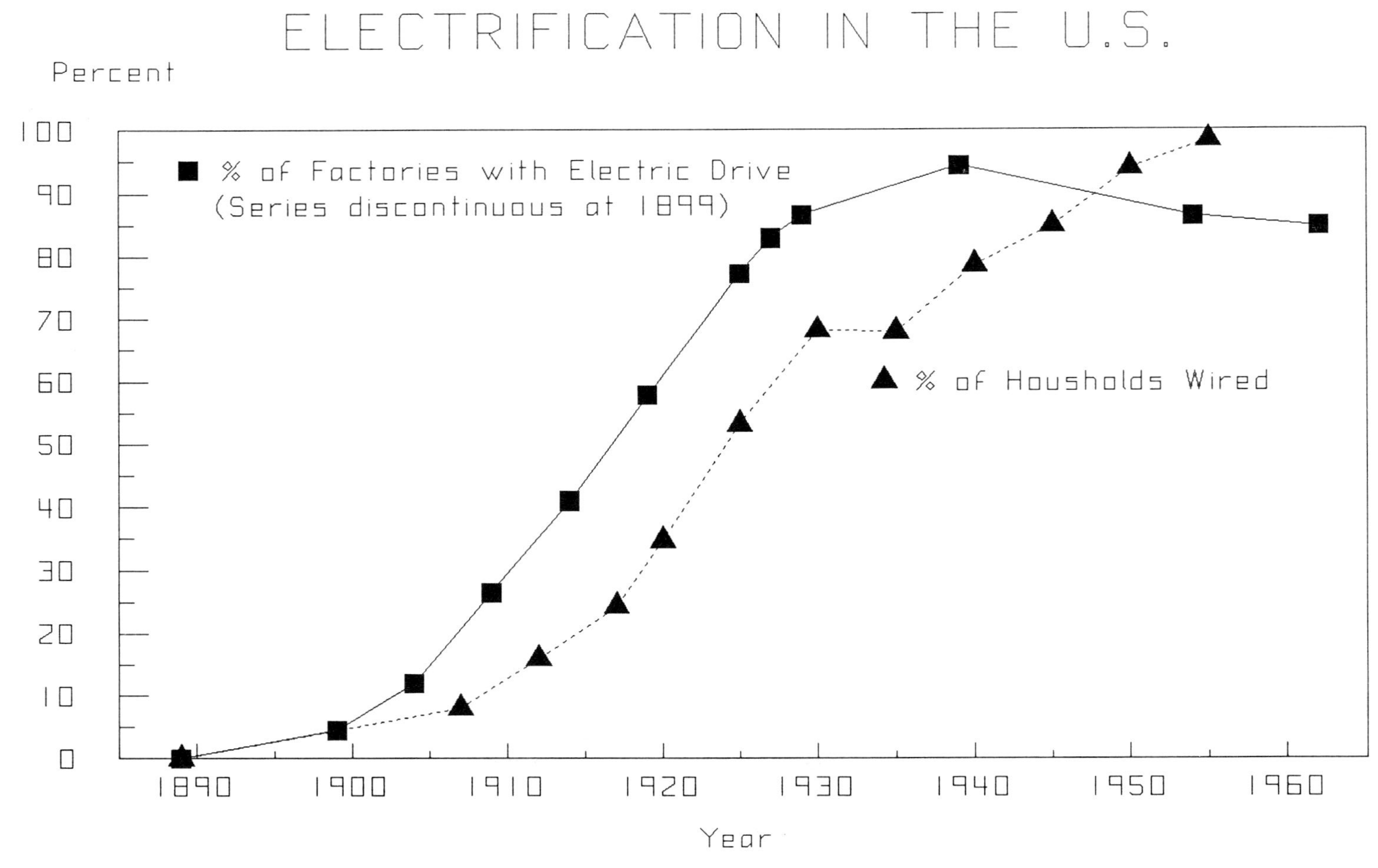

Source: Historical Statistics of the United States

quarters of the world's arc lighting was supplies by Brush Electric Company (Cleveland, Ohio) or the Thomson-Houston Company (Lynn, Mass.). One of the other significant producers of dynamos was Brady Manufacturing Company (Brooklyn), which manufactured a "spark free" dynamo patented by James J. Wood (1879). Brady Manufacturing was purchased by Ft. Wayne Electric Company (1890) which was, in turn, acquired by General Electric Company in 1898, becoming GE's Fort Wayne Works (Gorowitz et al., 1981). It is worth noting that, while arc lighting has largely been supplanted by incandescent lights over the past century, it was a major growth sector until well after 1900. Moreover, arc lighting technology was a direct precursor of arc welding (pioneered by Thomson) and electric arc furnaces for high temperature metallurgical and other industrial purposes. Moreover, the phenomena associated with electrical discharges in a vacuum, first investigated by Crookes (1875) and Edison (1883) led on to the enormously important applications of 'electron tubes' after 1900, which made possible radio, television, radar and finally computers.

9.4 CHILDHOOD PHASE (1880-1892)

The first phase of the industrial life cycle, following a breakthrough, is characterized by a tremendous proliferation of new entries, both products and firms, offering alternative configurations, variations and improvements. Technological progress is rapid, standardization of products is difficult, markets are wide open and customers are fickle. The childhood phase ends when the subsequent technological "trajectories" are well-defined.

In the case of the electrical industry, this phase began with Edison's two big breakthroughs (1878, 1879) and included the lesser one of Brush (1877), Thomson (1878), Sprague (1884), Stanley-Westinghouse (1884) and the big breakthroughs of Tesla-Westinghouse (1888) which opened the door to AC. A great many companies were formed, but a wave of mergers and consolidations began in 1888 and culminated in 1892 with the creation of General Electric Company as one of the two surviving giants. This year also marked the final resolution of the AC-DC controversy which was effectively settled by the decision by the Board of Directors of the Niagara Cataract Construction Company in December 1891 (following an extensive international competition and public debate) to select AC for its generators at the falls and its transmission system to Buffalo, New York (Sharlin, op cit). The significance of this decision will be discussed later.

Edison's commercial success was by no means a random event. He was an investor-backed professional inventor, who picked his targets with great care and with every intention of commercial application. His first company, Pope, Edison & Company, was formed in 1875. It changes its name to Edison Electric Light Company in 1878 backed by G.P. Lowrey, Chief Councel for Western Union, and several other senior executives of that company, along with E.P. Fabbri, a partner of J.P Morgan, the banker, among others. Initial capitalization was $300.000. Over the next decade he also formed or helped form Edison Lamp Company (1880), Sigmund Bergmann & Co. (1880), Edison Machine Works (1881), Edison Electric Tube Company (1881), Edison Company for Isolated Lighting (1882), and Edison Shafting Company (1884), mostly in New York City or New Jersey. Edison Machine, Edison Tube and Edison Shafting Company combined as United Edison Company and moved to Schenectady in 1886. All the Edison companies merged in 1890 to form Edison General Electric Company. In toto, these companies represented initial venture capital investments of well over a million dollars.

The second major new-application of electric power opened up by Edison's economic breakthrough of 1878-1882 was electric traction motors. Their first practical application was to streetcars. Here the Europeans kept pace. In fact, the first public demonstration was by Siemens and Halske at the Berlin exhibition (1879). Edison developed an electric railway and tested it at Menlo Park, NJ (1880) but it was never built. Others, notably E.

Julien (Belgium), E. Bentley, W.H. Knight, Charles van Depoele, Frank Sprague and Elihu Thomson (U.S.) pioneered the field. Van Depoele began engineering electric street railways in Chicago in 1874. By 1887 his company had built 12 street railways aggregating 160 miles of track and 100 cars (ibid). The first complete commercial electric streetcar system in the U.S. (12 miles, 40 cars) was built in 1887-88 by Frank Sprague (Richmond, VA), using an efficient DC motor of his design. Numerous engineering problem arose and were solved in the period 1887-1892. By 1892, the electric streetcar was established, substantially in its modern form. From 1888 to 1902, 98% of the street railways in the U.S. were electrified (Sharlin, op cit; Passer, op cit).

Thomson and Houston formed American Electric Works in 1880 (New Britain, Conn.). The inventors lost (or gave up) control of their company as expansion required ever more capital. In 1882, the company moved from Connecticut to Lynn, Massachusetts and was taken over by a group headed by Charles Coffin, a former shoe machinery industry executive with Wall Street connections, even as it changed its name to Thomson-Houston Company. Brush Electric Company was acquired by Thomson-Houston in 1889 creating an effective monopoly of the arc lighting business [Gorowitz, op cit,]. (T-H subsequently acquired a number of other arc-light manufacturers).

The Thomson-Houston Company acquired several of the important U.S. electric traction companies, including the Van Depoele Electric Railway Company of Chicago (1888) the Bentley-Knight Electric Railway Company of Cleveland (1889). The arc light and electric traction monopolies were completed by the 1892 merger between Edison General Electric Company and Thomson-Houston Company, master-minded by J.P. Morgan (ibid). The primary motivation for this major merger, however, was probably to combine the AC power generating technology of T-H with the incandescent lighting technology of Edison.

The electrification of factories was somewhat slower than the electrification of streetcars, because electric motors were competing not with horses, but with large steam engines of advanced design. A few factories were electrified before 1900, but for the most part, factory electrification occurred later, during the adolescent phase of the electrical industry.

Key American inventors and entrepreneurs of the 1875-1895 period were Brush, Edison, Sprague, Tesla Thomson, van De Poele and Westinghouse. Biographical details are given in Appendix II.[5]

Other important inventors of the period could also be cited , including James J. Wood (spark-free dynamo, electric fan), Edward Weston (hydrocarbon "flashing" process for filaments, magnetic speedmeter). Charles Bradley (AC-DC convertor), William Stanley (self-regulating AC transformer), J.W. Howell (automatic filament treating machine, glassworking machine for lamp manufacturing, improved phosphorus vapor exhaust process) and so on. Stanley and Weston were also successful entrepreneurs. Wood[6], Bradley and Howell were employed as inventors.

The identification of 1892 as the "end of childhood" is clearly indicated by the end of the wave of mergers and the creation of the duopoly which effectively characterized the U.S. electrical industry for many decades thereafter. It also corresponds to the effective end of the AC-DC controversy and the irreversible choice of AC as the preferred mode of transmission and distribution. What is perhaps surprising is that this choice occurred so soon after the key Tesla inventions.

---

[5]Biographical data have been taken from [Gorowitz et al. 1981], the encyclopedia Brittanica (1955), and [Asimov 82].

[6]Wood had 240 patents in his career.

9.5 ADOLESCENT PHASE 1892-1930[7]

This phase is sometimes termed the 'growth' phase. In the case of the U.S. electrical industry it corresponds roughly to the period of universal electrification as shown in Figures 9.1 and 9.2. By 1930, almost all U.S. factories and most households (except remote farms) were 'electrified'. Stationary reciprocating steam engines and gas lights had been essentially completely displaced by 1930. Steam turbines replaced reciprocating steam engines for power generation and arc lights were largely displaced by incandescent mercury vapor lights. During the 1920's there was also a wave of new domestic applications of electricity, including radios, irons, toasters, kitchen ranges, washing machines and refrigerators. This trend continued in the 1930's despite the depression. It accelerated after WWII, with the advent of TV and air-conditioning propelling a new period of demand growth, although the basic electric power industry was fully mature at that stage.

A key characteristic of the adolescent phase of the life cycle is product standardization and increased automation of production. However, in the electrical industry such technologies were frequently interrupted by major technological changes. For example, the case of incandescent lights, major improvement occurred as a result of the shift from carbon filaments to tantalum (C. 1906) and then to tungsten (1910-11). The sintered tungsten filament was first developed in Germany and Austria (1904), but major improvements were made by Coolidge et al. at G.E. (1908) largely based on their prior research with X-ray equipment. Of course, the improved tungsten lamp with its higher output and longer life permitted sharp price reductions that greatly increased the market. Coolidge's invention of a new method for treating tungsten to increase its ductility has a great impact on quality and cost. The new tungsten filament lamps of 1911 produced three times more visible light per watt than the carbon filament lamps they replaced. Coiling the filament and introducing an inert gas (rather than a vacuum) increased efficiency by a further 100% (C. 1913). There were no major improvements thereafter in incandescent lamp technology until 1937, although mass production technology continued to evolve.

The efficiency of generation and transmission continued to improve marginally, mainly by exploiting higher voltages, and economies of scale. For example, typical transmission voltages increased rapidly from 2400 V. (Mill Creek Cal., 1983) to 10,000 V. (1984, Niagara Falls) to 15,000 V. (Ogden-Salt Lake, 1896) to 33,000 V. (Santa Ana to Los Angeles, 1897). By 1907, improved insulation permitted 100,000 V. (Niagara Falls) and this had further doubled to 220,000 V. by 1921. Further increases did not occur till after WWII. William Stanley was the pioneer of high voltage transmission.

From 1882 (Pearl St. New York) to 1893, steam-powered DC dynamos of the "Jumbo" type increased six-fold in capacity from roughly 330 hp to 2000 hp. Hydroelectric generating units of 5000 KW were first installed at Niagara Falls (1895). This increased to 25,000 KW in the second phase (1904). By 1929, the world's largest hydroelectric powered generator was a 77,500 KW unit (Dneiper R., USSR).

The first large action-reaction steam turbo generators in the U.S.[8] were 5000 KW vertical-shaft units designed by Curtis, installed by General Electric in Chicago (1903). By 1918, a 45,000 KW steam turbo-generator had been supplied to Philadelphia Electric Company. The substitution of steam turbines for reciprocating steam engines for electric power generation began in 1903 in the U.S. and was substantially complete only a decade

---

[7]Technical data in this section is mainly from Gorowitz (op cit).

[8]Charles Parsons, who successfully innovated both the axial-flow steam turbine (1884) and the radial flow turbine (1891). The first radial unit was supplied to the Cambridge Electric Lighting Company (United Kingdom). By 1898, Parsons had built a 1000 KW unit for a power station in Eberfeld, Germany. It was then the largest in the world. Most of the early steam turbine innovations were European, including the first high pressure impulse turbine (de Laval, 1888-Stockholm), the first multi-stage axial turbine, (Rateau, Paris, 1896) and the compound radial-flow turbine (Ljungstrom, Sweden, 1906) (Strandh, 1979).

later. Both General Electric Company and Westinghouse began to manufacture large steam turbines as well as generators, transformers and control systems.

DC and AC electric motor technology, by contrast, was largely stabilized by 1893. The DC motor had a fairly long history of gradual development, but the AC induction motor as designed by Tesla in 1888-89 and refined by Westinghouse engineers has scarcely changed in subsequent decades. Similarly, Stanley's original transformer design is still standard. In these cases, improvements since that time are measurable primarily in terms of increased power outputs and reduced weight-to-power ratio. These improvements, while incremental, have been cumulatively dramatic. For instance, in 1893, 65 hp AC motors were installed in a textile plant (Columbia S.C.). By 1908, 6000 hp AC motors were supplied to a steel rolling mill (Gary, Indiana). In 1933, the French oceanliner "Normandie" was provided with electric motors delivering 45,000 hp to the propeller shafts.

As regards power output per unit weight an AC motor rated at a given output in 1900, weighed about 1/3 as much in 1930. The iron frame for a 7.5 hp induction motor (C. 1900) was assembled from many individual parts. In 1903, the equivalent frame was cast as a single unit, with substantial space saving permitting 10 hp output. The invention of centrifugal casting (C. 1920) permitted a doubling in motor output from the same size frame. By 1945, a variety of other improvements chiefly in insulation, permitted a further doubling of output from the same size motor frame, to 40 hp.

9.6 HOW WELL DO THE MODELS FIT?

In general, it seems that the barrier-breakthrough model can be fitted to the historical reality, in the sense that the major breakthroughs seem to be quite well-defined and can be identified without difficulty, nonwithstanding possible disputes with regard to authorship. The acceleration in the rate of invention and the proliferation of variations and new applications are easily documented.

The 'life-cycle' model fits the post-breakthrough period somewhat less well. The model depicts a 'childhood' phase of rapid technological change and easy market entry. This phase is followed by an 'adolescent' phase of slower technological change, product standardization, rapid market growth, increasing difficulty of entry, and rapid consolidation.

In the case of the U.S. electrical industry, the major consolidation appears to have occurred "too soon", i.e. during the first phase, long before the pace of technological change had significantly slowed down. It is known, of course, that the consolidations were artificially induced, largely by the influence of J.P. Morgan and his associates. In the absence of an individual "orchestrator" of Morgan's unique power and influence, it is possible that the process or consolidation would have occurred somewhat later than it did. However, it would almost certainly have come about eventually.

It is also possible, incidentally, that in the absence of J.P. Morgan and his chief lieutenants (e.g. Charles Coffin and Henry Villard) the burden of patent litigation would have greatly slowed down the rate of technological progress by swallowing up the profits that financed R&D. Had the Edison companies remained independent of each other, and had Thomson-Houston not absorbed most of the competing arc light dynamo manufacturers, the likelihood is that there would have been many incompatible power generating systems for a much longer time. It was a lucky accident that permitted Westinghouse to 'standardize' AC systems frequency of 60 cycles at a time (1891) when most of the potential competition was in the process of being consolidated into a single giant company still strongly influenced by Edison and his personal commitment to DC power. Had J.P. Morgan not existed, the battle would have been much more protracted, and many other AC systems would doubtless have emerged, though Westinghouse would probably have been the ultimate victor due to the robustness of the Tesla patents.

In any case, the standardization to 60-cycle AC did occur rather earlier than it might have, and this was probably very fortunate for the industry. After 1896, both GE and Westinghouse became highly profitable, yet still strongly influenced by their inventive founders. Consequently, they were able and willing to invest heavily in R&D which enabled them to quickly adopt and improve on key inventions from the outside, notably in the field of radio and electronics.

Returning finally to the question raised at the outset, the rapid development of the electrical industry in the U.S. from 1880 on cannot be attributed to the individual genius of even so great an inventor as Edison. Europe produced as many talented inventors as the U.S., including Tesla who conceived his greatest invention (the induction motor) while still living in his native Austria-Hungary (Cheney, 1981). The difference cannot be attributed to market demand, either, since demand for effective interior illumination in European cities was as great (or greater) than in the U.S. due to the longer hours of winter darkness. Nor was the U.S. in 1880 significantly wealthier than Great Britain, at least. The early customers in both Europe and the U.S. were shops, factories, hotels, steamships, and public places.

The major factor favoring the U.S. seems to have been the greater availability of 'venture capital', mainly from entrepreneurs who had accumulated some wealth from successful businesses in the previous generation or two. Edison's career was launched by the profitable sale of his stock ticker ($90,000), carbon telephone transmitter and non-metallic telegraph sounder ($200,000) and non-magnetic telephone receiver (£ 30,000) (Passer, op cit, 1953). It is probably not an accident that much of Edison's early backing came from executives of the Western Union Telegraph Company which had purchased Edison's telephone inventions. However, many of the backers were local business men on a much smaller scale. The banker J.P. Morgan was also a major investor, though he is better remembered as an orchestrator of acquisitions and mergers. The greater availability of venture capital in the U.S. is most probably attributable to the relative ease with which newly formed companies could be formed and 'go public', with their stock traded: Edison, Tesla and others were able to raise hundreds of thousands of dollars for R&D, with relative ease. Executives and stockholders of previously successful firms (such as Western Union) provided much of this money.

By contrast to the U.S. case, Siemens & Halske in Germany was probably the world's first electrical engineering firm (founded 1847) but it was privately financed by a loan of 6842 Thalers from a cousin, and it remained an ordinary partnership (OHG) until 1890, when it became a limited partnership (KG). Siemens only became a limited company (Aktiengesellschaft or AG) in 1897 when it already had nearly 11,000 employees and a turnover of 50,000,000 DM. [von Weiher 80].

APPENDIX I

1800 Alessandro Volta (Italy) perfected and described the 'voltic pile'; the first practical battery.

1809 Humphrey Davy (U.K.) discovered the carbon-arc light, but did not exploit the device.

1820 Hans Christian Oersted (Denmark) discovered that a current flowing through a wire produces a magnetic field.

1831 Michael Faraday (U.K.) discovered electromagnetic induction, described its basic principles (electric current induced by a *changing* magnetic field) and carried out several experiments to demonstrate it. He also built a continuous homopolar (DC) generator, consisting of a copper disk rotating between the poles of a permanent magnet. Current was induced radically in the disk (from center too rim). Faraday made no attempt to improve his invention or seek practical applications of it.

1832 Hyppolyte Pixii (France) exhibited a simple bipolar AC generator,using spool (bobbin) wound coils and a rotating permanent magnet. At Ampere's suggestion, he then devised a rotating reversing switch (commutator) to obtain DC current..

1833 Joseph Saxton (U.S.) built a dynamo similar to Pixii's but with a stationary magnet and rotating bobbins.

1839 The invention of electrotyping (electroplating of copper on the mold of an engraving) created a demand for more and higher voltage DC current than batteries could conveniently supply. Unlike the telegraph, this provided a significant incentive to inventors. Dynamo's gradually came into use in the 1840's.

1842 John Stephen Woolrich (U.K.) improved Pixii's simple dynamo for electrotyping applications by stacking several flat horseshoe magnets. He also improved the commutator.

1845 Charles Wheatstone (U.K.) was the first to build a dynamo using battery-energized electromagnets in place of permanent magnets. However, his scheme was forgotten and reinvented later by Hjorth (Denmark) (1855) and patented by Wilde (U.K.) 1863.

1849-53 F. Nollet (Belgium) tried to develop a form of lighthouse, using heat from an oxyhydrogen flame to heat a block of lime to incandescence ("limelight"). He tried to utilize a dynamo to obtain the oxygen and hydrogen by electrolysis of water. The scheme failed.

1855 Soren Hjort (Denmark) reinvents a self-excited dynamo.

1856 Ernst Werner von Siemens (Germany) invented the so-called shuttle winding, a precursor of the drum winding.

1857 Frederick Hale Holmes (U.K.) built a large "magneto-electric" generator designed to power an arc light. His generator weighed 4000 lbs. and produced 1500 watts. It was demonstrated at the London Great Exhibition (1862) and finally installed in Souter Point Lighthouse (U.K.), 1871.

1857 Charles Wheatstone (U.K.) was apparently the first to suggest the principle of self-excitation, where the rotating electromagnets are energized by the output current of the dynamo itself.

1860 Antonio Pacinotti (Italy) devised a new type of "ring" winding for a self-excited dynamo, including the use of slots for containment and an iron core which greatly increased the efficiency of the dynamo. This scheme was described in a scientific journal but was ignored until rediscovered by Gramme ten years later.

1860 Joseph Swan (U.K.) demonstrated an electric glow lamp with a carbon filament in an evacuated glass globe, heated by current from a battery.

Swan's filament was a strip of pyrolized paper. His lamp would glow at low voltages but would burn out quickly at higher voltages because the vacuum pumps of the time were inadequate and the filament was too fragile. Swan and other experimenters dropped the idea of an incandescent lamp.

1863 Henry Wilde (U.K.) successfully patented the self-excited generator (incorporating ideas suggested by Wheatstone, Hjorth and others).

1864 James Clerk Maxwell (U.K.) published his mathematical theory of electromagnetism.This contributed substantially to the general understanding of the magnetic field and the subsequent development of practical rules for designing efficient generators and minimizing losses. This way done by John and Edward Hopkinson at the U.K. (1886) who devised practical rules for predicting the performance of a magnetic circuit.

1865 Hermann Sprengel (Germany) invented a greatly improved mercury vacuum pump.

1866 Werner von Siemens again formulated the "Dynamo-electric" principle for a self-excited generator in a published paper and demonstrated a working model.

1866 Moses G. Farmer (U.S.) also "discovered" the principle of self-excitation in a letter.

1866 Henry Wilde (U.K.) described in a paper and demonstrated a self-excited dynamo.

1866 S. Alfred Varley (U.K.) applied for a patent on the principle of self-excitation.

1868-69? Paul Jablochkoff (France) invented a simple unenclosed AC arclight consisting of two carbon rods separated by an insulating layer which could be volatized by the arc. (It resembled an arc-welder.) The so-called Jablochkoff "candle" consumed a pair of 10 inch carbon rods in 1 1/2 hours.

1871 Zenobie Theophile Gramme (France) combined the self-excitation feature with Pacinotti's ring-windling scheme. It was the first to produce continuous currents and was successfully commercialized. It was widely used in Europe together with Jablochkoff's "candles".

1872 F. von Hefner-Altaneck (Germany) working for Siemens & Halske invented the so-called drum-winding, which superceded the ring-winding and is close to the modern form.

1873 Z.T. Gramme (Fr.) discovered that a dynamo could act as an electric motor, if driven by a larger dynamo. This was the key to electric street railway systems and factory electrification systems.

1875 William Crookes (U.K.) successfully used Sprengel's mercury vacuum pump to evaluate glass globes to make a radiometer.

APPENDIX II

Thomas Alva Edison (1847-1931)

1876 Carbon microphone (for telephone) patents purchased by American Speaking Telegraph Company, subsidiary of Western Union. Western Union installs Edison Quadruplex telegraph system, (which doubled the capacity of existing lines).

1879 Demonstration and patent of carbon filament incandescent lamp, using carbonized thread in an evacuated glass bulb (40 hour life). U.S. patent 223.898 received in 1880.

1880 Carbonized bamboo increases the life of the incandescent lamp to 600 hours. Manufacturing begins at Menlo Park. First installation on steamship "Columbia". Patents filed for large-scale distribution system (granted in 1887). Begins trials of an electric railway at Menlo Park.

1881 Patents issued on current regulator for dynamo; apparatus for producing high vacuum, a method of treating carbon for lamp filaments. "Jumbo" dynamo exhibited in Paris at exposition. First incandescent lamp installation (N.Y.C).

1882 First hydroelectric installation using Edison bipolar generator (Appleton, Wisc.). Patent applied for on 3-wire distribution system (reducing copper requirements by 2/3). 2 "Jumbo" dynamos installed at Holborn Viaduct London. 6 "Jumbo" dynamos installed at Pearl Street to supply power for 7200 (16 candlepower) incandescent lamps over 1/6 mi^2.

1883 Discovery and patent of "Edison effect", induced current flowing between carbon points in a vacuum. Precursor of vacuum tube. First central station using 3-wire systems (Sunbury; PA); first underground 3-wire system (Brockton, MA).

1889 Paris exhibition uses 10,000 incandescent lamps to stay open in the evening.

1891 Courts sustain Edison patents on incandescent lamps. Manufacturing begins in Europe.

1892 Merger of Edison companies with Thomson-Houston Company to create General Electric. Edison's financial interest was bought out when GE was formed. After 1892, Edison worked in other fields mainly sound recording (he formed the Victor talking Machine company, later merged with RCA), moving pictures and iron ore concentration. He invested the proceeds of the GE buyout in an ore concentration plant which failed in 1900.

Charles F. Brush (1849-1929)

1875 Cleveland Telegraph & Supply Company formed.

1876 Constructs and demonstrates a practical, economical dynamo for Cleveland Telegraph & Supply Company, with exclusive rights to his invention.

1877 patents copper coating for carbon arc electrodes.

1878 patents first series arc lamp.

1879 patents compound electric wiring.
patents secondary distribution system for arc lighting.
demonstrates his arc lighting system in Cleveland, (Monumental Park), and Niagara Falls.

1879 promotes formation of California Electric Light Supply Company (San Francisco).

1880 patents automatic cutout for arc lamps.
Cleveland Telegraph & Supply Company changes name to Brush Electric Company. (By this time over 5,000 Brush dynamos and arc-lighting systems had been sold).

1881 Invests first compound DC generator by adding a second field coil, resulting in constant voltage output independent of load.

1884 First electric submarine in America uses Brush storage batteries and DC motors.

1885 First industrial electric arc furnace built (by Alfred and Eugene Cowles) using Brush dynamo; later used to manufacture first synthetic rubies & sapphires.

In 1889, Brush lost patent protection on key parts of his arc lighting system (due to a lawsuit), which led to the absorption of Brush Electric by Thomson-Houston Company. After 1889 he retired from business and worked independently from his private home laboratory. Altogether Brush received over 100 patents in his lifetime, mostly in the electrical field.

Elihu Thomson (1853-1937)

1877 demonstrates use of "German silver" for use in a resistive heating element, while at Central High School, Philadelphia.

1878 demonstrates AC dynamo (alternator) and transformer at Franklin Institute, Philadelphia - with spherical 3-coil armature (with Edwin J. Houston) invents vibrating arc lamp powered by Thomson dynamo.

1879 (with E.J. Houston) demonstrates T-H arc lighting system at Fuller's Bakery, Philadelphia, using 3-phase AC.

1880 (with E.J. Houston) forms American Electric Works in New Britain, Connecticut to manufacture T-H dynamos and arc-lights. Patent on AC dynamo is received.

1881 develops constant-current regulator for arc lighting.

1882 develops air-blast arrestor for protection of high current switches.

American Electric changes ownership, moves to Lynn, Massachusetts and changes name to Thomson-Houston Company.

1883 patents magnetic surge protection systems.

1885 devises grounded secondary for high voltage transformers to cope with problems of insulation breakdown.
develops DC dynamo for incandescent lamps.
adopts magnetic blowout surge protection system for arc lighting.

1886 patents electrical resistance welding machine and transformer.

1887 devises an AC motor based on "repulsion-induction" phenomenon.
patents use of oil as a transformer coolant.

1888 forms Thomson Welding Company to commercialize his arc-welding system.

1889 invents integrating wattmeter, precursor of meter used to measure power consumption by utility customers. (Over 30,000,000 had been built by the time Thomson died in 1937.)

1890 invents a high frequency dynamo.

Elihu Thomson remained as director of the Lynn Research Laboratories (renamed Thomson Laboratories) after the formation of General Electric Company. He continued to invent profilically after 1892. He built the first practical X-ray machine (1896), and invented the constant current transformer (1897), among other achievements. Altogether, Thomson received more than 700 patents in his lifetime, mostly electrical devices.

Charles Van Depoele (1846-1892)

1877 forms Van Depoele Electric Light Company in Hamtramck, Michigan.

1879 moves Van Depoele Electric Light Company to Chicago to specialize in Street railways.

1881 changes to Van Depoele Electric Railway Company, Chicago.

1883 Electrical Railway Company build first elevated electric railway (Chicago) for exhibition. (By 1887, the firm had installed 12 systems with 60 miles of track,)

1885 invents an electric drill. Thomson-Van Depoele Electric Mining Company (Chicago), a subsidary of T-H, begins production.

1888 invents carbon brush commutator for DC motors sharply reducing sparking and wear. Van Depoele Electric Railways company is merged with Thomson-Houston.

He continued to work for the T-H company until his death. Van Depoele was also a very prolific inventor with over 250 patents in his lifetime, mostly related to electric tradition systems.

Frank J. Sprague (1859-1934)

1882 invents basic upward-pressing contacting system for a trolley and overhead cable.

1884 develops the first practical DC motor.

1884 organizes Sprague Electric Railway and Motor Company.

1885 Sprague motor is recommended by Edison Electric Company to all its licenses.

1886 First 220 volt DC motor for an electric freight elevator.

1887 awarded contract to build streetcar system for Richmond, Virginia. By 1890, 110 streetcar systems had been built in Sprague.

1890 Sprague Electric Railways by Edison General Electric Company.

In 1892, Sprague formed the Sprague Electric Elevator Company. He subsequently developed many elements of remote control systems both for elevators and electric trains, including the "multiple unit system" of electric train control, which was first installed on the South Side Elevated Railway in Chicago (1897). In 1903, Sprague Electric Elevator Company was also acquired by General Electric, but Sprague continued as an independent entrepreneur-inventor and created several more companies. He received more than 100 patents, mainly in the electric traction and control field.

George Westinghouse (1846-1914)

Prior to 1882, Westinghouse was primarily interested in railways, having invented a system for replacing derailed cars, (1865) a reversible "frog" for switching, the air-brake (1869) and the automatic air brake (1872). He also organized the Westinghouse Air brake Company (WABCO) in 1869 and the Union Switch and Signal Company (1881). It was the latter which got Westinghouse interested in electricity.

1882 forms Westinghouse Electric Company (WECO) which acquired U.S. rights from Lucien Gaulard (France) and John Dixon Gibbs (U.K.) for an AC transformer system. Also works with William Stanley on further development of AC systems. Stanley designed the first practicial transformer and developed a technique for manufacturing it.

1885 WECO installed first commercial AC lighting plan (Buffalo, NY) using Stanley technology.

1888-89 WECO rights to Nikola Tesla's patents on AC introduction ("squirrel cage") motor transformers and polyphase distribution system. Tesla becomes a consultant to Westinghouse.

1886-90 WECO builds 300 Central Stations supplying an estimated 500,000 16 candlepower incandescent lamps.

1891 WECO installs electric equipment for Edgar Thompson Works of Carnegie Steel in Pittsburgh.
WECO introduces 60 cycle frequency standard for AC.
installs first industrial AC power system in U.S., Telluride, Colorado.

The years 1889-93 were the peak years of the great AC-DC controversy, which became a public vendetta between Edison and Westinghouse. Edison used unscrupulous tactics in attempting to discredit AC on the grounds that it was "too dangerous", but lost out to Tesla and Westinghouse because of the benefits of high voltages for power transmission, which required AC. General Electric began manufacturing its own AC systems (based on Elihu Thomson's inventions) and by 1896 General Electric and Westinghouse had over 300 patent litigations pending.

It appears likely that J.P. Morgan intended to include Westinghouse in the great electrical merger of 1982, and Westinghouse was under heavy pressure from his financial backers until 1897, at least. Westinghouse seems to have resisted primarily because of the AC-DC feud with Edison. He was able to remain independent because of the spectacular success of the Westinghouse AC system demonstrated at the Columbian Exhibition (Chicago, 1893) and Westinghouse's selection to construct the generators for Niagara Falls (to a new design by George Forbes), which were successfully completed in 1895.

But even these victories were not sufficient to rescue Westinghouse Electric from financial difficulties. Only Tesla's agreement to sell his patent rights outright for an absurdly modes sum ($216,000) forgoing millions of dollars in already-earned royalties, saved the company from a forced merger at the time of the 1894-5 recession. The problem of patent litigation with GE was resolved by the creation of a joint patent pool in 1896, which gave GE 62.5% of the combined royalties and Westinghouse 37.5%. The rivalry between the two firms continued, however, as long as George Westinghouse remained in control. This ended and Westinghouse lost his personal fortune in the panic of 1907, when Standard Oil Company interests (which then controlled GE) exerted "a financial pressure that soon drove the Westinghouse Company into a position from which it escaped only by becoming an auxiliary to those interests". [Myers 36, p625].

Nikola Tesla (1856-1943)

1882 conceives the AC induction motor (in Budapest) and the principle of the rotary magnetic field.

1883 moves to Paris and worked for Edison's European Telephone subsidiary. Demonstrated and experimental induction motor in Paris.

1884 arrives in U.S. and worked for Edison but left after a quarrel.

1885 forms Tesla Electric Light Company, developed and patented Tesla arc lamp. (Company failed)

1887 creates Tesla Electronic Company with help of A.K. Brown of Western Union (capitalized at $500,000) to develop AC systems: 1-phase, 2-phase, and 3-phase systems together with dynamos, motors, transformers, and automatic controls. In all 40 patents were applied for and received in this field between 1887 and 1891.

1888 (May 16) Tesla gives an invited lecture on his AC system at the AIEE in New York.
Signs a royalty agreement with Westinghouse electric Company under which Tesla was to receive $2.50 per electric horsepower sold using his technology. (A further agreement was signed in 1889). Patent rights were later sold outright to save Westinghouse from bankruptcy.

1889 moves to Pittsburgh (temporarily) as a consultant to Westinghouse.

Tesla moved back to New York City in 1890 after his year with Westinghouse, but he actively led the battle against Edison to prove the value of AC. He continued for the rest of his life as an independent inventor, best known for his work in high-frequency, high voltage phenomena. Tesla later feuded with Marconi over the invention of radio. In 1900, Westinghouse finally won a major legal victory when all Tesla's patents were upheld by the U.S. Circuit Court, although by that time the Westinghouse contest with G.E. was moot.

REFERENCES

Asimov, I., 1982, *Biographical Encyclopedia of Science and Technology*, Doubleday, New York.

Ayres, R.U., 1987, "Barriers & Breakthroughs: An Expanding Frontiers Model of the Technology Industry Life Cycle" *Technovation*, 7:87-115.

Cheney, M., 1981, *Tesla, Man Out of Time*, Laurel, New York.

Gorowitz, B., ed., 1981, *A Century of Progress: The General Electric Story*, Hall of History Foundation, Schenectady NY.

Josephson, M., 1959, "The Invention of the Electric Light", *Scientific American*, November.

Passer, H.C., *Electrical Manufacturers 1875-1900*, Harvard University Press, Cambridge, Ma,

Sharlin, H.I., 1961, "From Faraday to the Dynamo", *Scientic American*, May.

Sharlin, H.I., 1967, "Applications of Electricity", in Kranzberg, M. and C. Pursell, eds., *Technology in Western Civilization* 1, Oxford University Press, New York.

Sharlin, H.I., 1967, "Electrical Generation and Transmission", in Kranzberg, M. and C. Pursell, eds., *Technology in Western Civilization* 1, Oxford University Press, New York.

Strandh, S., 1979, *History of the Machine*, A&W Publisher.

von Weiher, S., 1980, "The Rise & Development of Electrical Engineering & Industry in German in the 19th Century: A Case Study - Siemens & Halske", in Okochi, Akio & Hoshimi Uchida, eds., *Development and Diffusion of Technology*, Chapter 2, Series: International Conference on Business History, 6, University of Tokyo Press, Tokyo.

CHAPTER 10

The Evolution of High Technology in the Boston Region 1920-1980

Sam Bass Warner, Jr.

10.1 INTRODUCTION

The theme of my Boston history is unpredictability. Although at the present moment the Boston metropolis seems to be prospering on the high technology branches of the electronics industry, no one in 1920 and probably no one in 1950 could have predicted this prosperity. Moreover, the Boston region had once before ridden to prosperity on the success of a new industry only to have it crest, break, and ebb away. I see no reason to believe that such flooding and ebbing will not recur. Indeed I think such cycles of fresh advantage and later obsolescence are the rhythm of modern industrial society and will become a world-wide experience (Warner, 1984). Therefore, the question my historical narrative poses for us all is: what are the appropriate public policy responses to the inevitable cycle of growth and obsolescence when it is known that the future is unpredictable and the economy inherently unstable?

I came to this question while studying the writings of two outstandingly successful science administrators: Vannevar Bush and James Bryant Conant (Bush, 1970; Conant, 1970). The former was an electrical engineer, the latter a chemist; the one from Massachusetts Institute of Technology, the other from Harvard University. Together they ran the United States weapon development programs of World War II, and subsequently laid the foundations for the scientific policies of the federal government.

From these two I learned of the long distance which separates science from useful products, and of the many uncertainties and hazards which intervene between good ideas and successful new market items. One of their associates, Edward Linley Bowles, summarized the situation with the observation that in his experience "No one could waste money faster than a research scientist".

For the past half century, or, at least since the work of Joseph A. Schumpeter (Schumpeter, 1934), economists, historians, and geographers have been trying to find the regularities amidst the surprise of innovation. They have quite properly turned their attention to centers of activity, places which seemed to be hives of progress. Here they have detected some common elements: plentiful capital, skilled labor, access to scientific information, and an active demand which would support prices sufficient to allow experimentation (Rosenberg, 1976). Such social ingredients can be found in the inventive years of Manchester, England, Pittsburgh, Pennsylvania, and Los Angeles, California. A Swedish geographer has even been at work trying to map private information exchanges in order to understand how such hiving might work (Hägerstrand, 1967).

Yet the regularities don't seem to work by themselves. Rather, the uncertainties and risks which my science administrators spoke of, seem to dominate our experience with technical progress. For example, seventeenth and eighteenth century England was rich in capital, at the forefront of technological improvement, and somnolent in science after Newton. nineteenth century French science led the world, the nation was rich, yet its technical progresses lagged. In both the United States and Japan technological progress preceded excellence in science, and nineteenth century America was chronically short of capital.

The best estimates of scholarship today combine the major background conditions with attention to men and institutions. There seem to be times and places when and where creative groups and clusters of supportive institutions are strong enough to overcome the inevitable failures, and flexible enough to find successful pathways. Alfred Chandler's corporate histories are full of adaptations to mistakes, losses, and failures (Chandler, 1962, 1977), and economists seem to be turning to institutions as important variables (Rosenberg & , 1986). My hope is that there lie buried in these institutional histories key factors which have been neglected, and relationships which, if better attended to might reduce the costs, even the suffering, of economic development.

10.2 THE SETTING

In 1920 Boston, and its New England hinterland, stood on the brink of disaster. New England was American's oldest industrialized region, the companion of Belgium, the two places first to imitate England's new industrial ways. The region's economy rested on textiles and shoes and the related machine-building, transport, finance, and marketing institutions appropriate to such a rich provincial economy.

Boston's advantages had slipped away, one by one. First the metropolis lost its market centrality as settlers moved ever farther from the Atlantic cost. New York, with its connections with London, rose to be the nation's financial capital, and as such, its financiers seized control of many Boston institutions: its railroads, its wharves and shipping, its textile marketing, and even successful Boston innovations. The telephone had been invented in Boston and the utility company had been put together by Boston capitalists. Soon it was lost to New York management. So too the start made in electrical machinery by a group of leather and shoe manufacturers in Lynn. Their Thompson, Houston Company was quickly absorbed by the General Electric consolidation fashioned by the New York investment banker J.P. Morgan.

Of course, by 1920 the techniques of textile and shoe manufacture had become known across the world so that Boston manufacturers controlled few secrets of the trade. Scholars now think the long sustained prosperity of these early Boston industries was in fact maintained not by entrepreneurial skill but by a plentiful supply of cheap immigrant labor from Europe and Canada. These low-paid machine tenders had kept the firms profitable for a century, but now the First World War and the succeeding anti-immigration legislation cut off that supply. Boston firms were also on the move. Ever since the 1890's New England textile firms had been building plants in the southern states, and shoe firms had been relocating in the midwest in order to tap new pools of cheap labor and cheap raw materials. Indeed, one big Boston firm had built a plant in India.

The world collapse of wartime prices in 1921 inaugurated a regional depression in New England, which continued unabated except for the wars of 1941-45, and 1951-53. Only in 1960 did the Boston metropolis find its new economic base and return to a level of prosperity somewhat like the one it had enjoyed during the nineteenth century (Eisenmenger, 1967),

During this forty-year depression there was much that worked against recovery. Boston had been a high-wage region. Now it had to go though a long process in which all wages declined relatively and some absolutely. Wage cuts, speed-ups and short-time provoked strikes and endless labor troubles. Manufacturers abandoned factories, and as they withdrew, the values of houses and stores in mill towns collapsed. The decline of real estate values in turn reduced the tax revenues of cities and towns, and thereby hobbled local efforts for public aid. The whole region, its human capital, its real property, and its public capital, all underwent a forty year write-down. By 1950, in the Boston region, you could buy a physicist, a lawyer, a machinist, a house, a mill, or an office building at below the average national prices.

In the setting of declining expectations people scratched to hold on to what they had. Employees struggled for tenure and seniority, banding together with friends and struggling to keep strangers out. An extremely nasty attitude invaded both private and public institutions. The ethnic and religious antipathies of native against immigrant, Protestant against Catholic, Christian against Jew, white against black dominated politics and became a fruitful source of corruption and incompetence in government and business firms. Meanwhile liberal and generous leaders could do little more than preach patience. One group who assembled to review the situation thought some hope might lie in imitating Henry Ford's mass production methods, but such ways only worsened textile and shoe operations, which were more and more tied to short-run fashions (Filene, 1933). No formal institutions of research and development existed to bring together academic science with practical innovators, or to join the innovators with businessmen and investors. Research and development depended on the causal social networks of the city. This is the way the telephone had been invented and financed and such were the expected pathways.

The phenomenal success of the Gillette Safety Razor offered a good example of the old Boston's practices. The ideal for a razor with a thin throw-away blade and a rigid frame was the product of a bottle cap salesman, King Gillette. His invention had been made practical by a mechanical engineer, William E. Nickerson (M.I.T. Class of 1876). Nickerson was an independent consultant who had previously invented elevator equipment and automatic weighing machines for food packagers. He carried out the development process for Gillette's razor by working with two assistants in a downtown loft. During the years from 1900 to 1903 he perfected a machine to hone the thin razor blades (the blade steel was imported from Sweden!). Because Boston's established bankers had refused credit, the costs of this development and production start-up had been financed by an Irish immigrant brewer (Baldwin, 1951; Bezanson, 1961).

The textile and shoe industries carried on without basic research (Bush, 1932). The only researchers were the machinery builders, companies like the Draper Loom Company and United Shoe Machinery Corporation, who kept altering their machines so they could refresh their patents and thereby continue their business of leasing patented equipment to manufacturers. The only university-industry links came from individual consultants. For example, the chairman of the M.I.T. electrical engineering department, and expert on power and street railways, ran a very successful consulting firm, but his was not a research organization (Bush, 1951).

The development of nylon thread showed the failure of Boston's old research linkages. After World War I chemists at Harvard University had made considerable progress with polymer chemistry, but it remained for the wealthy DuPont Corporation of Wilmington, Delaware, to sustain the development research and thereby turn academic investigations into the commercial product, nylon (Conant, 1952).

10.3 THE STORY

It was in such a fragmented urban environment that Boston's first giant electronics company was born. Today, the Raytheon Corporation is the forty-eighth largest industrial corporation in the United States, a firm with sales of over seven billion dollars. It employs 30,000 men and women in the Boston region and altogether 70,000 across the nation and the world. (Its newer, post World War II neighbor, Digital Corporation, is slightly larger, ≠44 on the *Fortune* list.) It began as Spencer Thermostat Company, an organization founded by three men in 1920 to manufacture a small safety switch that prevented clothes irons from overheating.

The founders embodied the varieties of human capital which Boston then afforded, and the accidents of urban acquaintance connected one of them with the accumulated wealth of the region's earlier textile prosperity.

Vannevar Bush (1890-1974) was the academic. A brilliant mathematician and engineer of electric power transmission, he followed his mentor, Prof. Dugald Jackson, by combining college teaching with commercial consulting. Bush, a city minister's son, had graduated from Tufts College in suburban Boston in 1913. After a year of short-term jobs with General Electric and the U.S. Navy, he returned to Tufts to teach mathematics and to earn a PhD in electrical engineering from Harvard and M.I.T. During World War I he did a lot of consulting to an nearby manufacturer who made radios for amateurs. The firm was trying to develop a device for the magnetic detection of submarines. As chance would have it, the manager of the company was the former radio operator on the financier J.P. Morgan's yacht. Morgan hoped that in peacetime such a device would detect ores in the ground. In any case, the connection with this powerful banker proved a key linkage in the subsequent development of the little thermostat firm (Bush, 1970).

The second partner was the entrepreneur, a man with the old-fashioned desire to build his own business. Laurence Marshall (1889-1980) was the son of a wholesale butcher. He was the enthusiast, the gambler, the encourager of men, and the promoter of new products. Bush and Marshall had been Tufts College classmates, but since Marshall was very short of money, he had to rush through college in three years, and thereafter he worked as a civil engineer for a contractor who drilled railway tunnels and erected factories. He was on a factory job in Rhode Island when the United States entered World War I in 1917. He immediately enlisted and subsequently served as a second lieutenant in the artillery in France. His commanding officer, Major William Gammell, Jr., was the son of a Providence banker and cotton manufacturer and in peacetime served, himself, as treasurer of a big textile company.

After his discharge Marshall wanted desperately to get out of tunnel digging and factory construction. He wanted to find a place in the new field of radio, which he had touched upon in college. He returned to Boston and talked with Bush, who had by now become an associate professor of electrical engineering at M.I.T. Bush told Marshall about the promising invention of a machinist who worked at the radio company Bush consulted to.

The third partner, John A. Spencer, the inventor, was an old-style Yankee tinkerer. He had been born in Maine and was a self-trained man. After his father had been killed in a lumber mill accident when the boy was only three years old, the family had dispersed. Spencer was raised by an uncle and he supported himself by working in mills which turned wooden spools for thread manufacturers. Somehow he got to New York City as a young man, and without formal apprenticeship or much previous schooling, he mastered the machinist's trade. Spencer's invention took the form of a metal disc which snapped in and out under heat, much the way the bottom of an oil can snaps back and forth when you press upon it.

Marshall went to work on the invention, improving it so that it could be used on alternating as well as direct current. He then applied to his old commanding officer who put together a group of Providence, R.I. investors who soon included, in addition to himself, Russell Grinnel, a successful manufacturer of fire alarm equipment, and Richard S. Aldrich, a man whose sister had married John D. Rockefeller, Jr., the son of the famous oil millionaire.

Although the new product had a big market awaiting it, a patent pool stood in the way of its immediate acceptance. In 1920 the American manufacturers of clothes irons were collectively linked in a patent agreement whereby each member was obliged to share any patented innovation with the other pool members. Therefore, no one member wanted to incur the costs of changing to a new design which would give no long-term competitive advantage. Marshall, however, succeeded in bluffing Westinghouse Electric into purchasing the thermostat by mentioning his well-known and wealthy backers. He said that they intended to enter into clothes iron manufacturing themselves if the thermostat could not break through the patent pool barrier. The Westinghouse agreement made the little company an instant success and its financiers eagerly awaited the next venture.

The young men's next idea proved a total failure. Thcy imagined an electric refrigerator with no moving parts. A new man, Charles C. Smith, had the idea. He was also a former employee of the radio company Bush consulted to, and Bush regarded Smith as a genius. Smith was a Texas physicist with only a college degree. He had come to Boston to study at Harvard but never took many courses. Instead he worked with Bush on experiments for radio tubes which would transform alternating current to direct current - what came to be called "power tubes" because of their role in the function of radios. These tubes, once perfected, transformed the alternating current of the house electricity into direct current to power the radio itself.

In 1921 J.P. Morgan realized that his radio company was ill-managed, and he closed it up, only retaining its patents. In the retrenchment Smith was thrown out of work so he moved over to the little thermostat company bringing his unworkable refrigerator idea with him. Smith, Marshall, and Bush worked away on a prototype in a spare bedroom of Smith's house, but it came to nothing except exciting the investors' interest. They put more money in the firm and enabled it to rent space in the factory district behind M.I.T. in Cambridge. There it continued to manufacture thermostats, but could not get its refrigerator to function properly.

Then, in 1924 Marshall visited J.P. Morgan in New York City and there arranged to purchase the patents to Smith's power tubes. Morgan accepted $10,000 in cash and $40,000 in stock. Thereafter the Morgan interest continued to be an important element in the business. For many years Thomas Nelson Perkins, a distinguished Boston corporate lawyer, represented Morgan in the company. Subsequently Charles Francis Adams, an investment banker with the large national firm of Paine, Weber, Jackson & Curtis, assumed this role.

In 1924 there were two and a half million radios in the United States, all battery powered. Smith's rectifier tubes could transform alternating current into direct current, thereby making it possible to dispense with the batteries and to plug the radio directly into an electric outlet. The tubes proved an instant success, the very thing Marshall had been searching for, a major new product in the new radio industry. The Champion Lamp Co. of nearby Salem, an electric light company, blew the glass and assembled the tubes, while the Cambridge shop tested and packed them. In 1925 the young firm took the distinctive name Raytheon (Scott, 1974).

Then in its moment of triumph Raytheon faced its first life-threatening disaster. This time another patent pool of giant corporations. The pool originated in the confusion of patents which followed Marconi's invention of radio. Many inventors had rushed into the field and one of the most successful was Lee DeForest, a Yale-trained engineer, who invented the three-element vacuum tube which proved to be the basis for all subsequent tube design. In the midst of this rush of innovation, the American Telephone and Telegraph Company grew fearful that it would lose its control over voice communication. In 1911 it established its famous Bell Laboratories in New Jersey, and soon had 900 men at work on fundamental and applied research. It also bought the DeForest patents. Yet so rapid and diffused was the pace of innovation that by 1917 there were eight major companies holding significant radio patents in the United States. During World War I conflict was held at bay by the U.S. government which encouraged manufacturers to make radio equipment without regard to patent ownership.

Thereafter a contest developed between General Electric and the telephone company. General Electric bought the patents to British Marconi and, in 1919, established its own radio subsidiary, Radio Corporation of America. There followed years of negotiations between the two giants, and among the major companies. Slowly they came to terms among themselves. First they divided the field into telephonic and non-telephonic communications: A T & T being given the rights to all the telephonic-related patents, it agreeing to share its non-telephonic ones with the others. Then, in the final agreement of 1926, the pool members agreed to make their sharing exclusive. That is, no outside manufacturer would be allowed to use any of their patents unless all the patented elements of the product were based either on the pool patents or patents of the outsider himself. In a word, no manufacturer could use Raytheon's power tube if he wanted to use any of the pool's many patents. Raytheon was frozen out (Danielian, 1939).

Marshall counterattacked by starting a lawsuit and by going into the radio tube business in a big way. Raytheon leased an old textile mill in suburban Newton, Massachusetts and began making tubes of all kinds. The lawsuit dragged on for twelve years and in the end brought money only to the lawyers. When the Great Depression of the thirties struck, the weakness of Raytheon's position revealed itself. It had a full line of good tubes, but it did not have the marketing capabilities to compete with manufacturers like RCA. The firm limped along, saved only by Marshall's seemingly limitless enthusiasm and a brief marketing contract with American Carbon (a flashlight battery company) which brought in a lot of cash.

Marshall desperately sought new products for this company to manufacture. He bought up a little firm of M.I.T. engineers who had invented the "excito tube", the tube which translates the visual track on the edge of movie film into sound. The group was then supporting itself building police radios for local police forces. Raytheon also tried to break into electric welding, but when it marketed its new product the established firms cut the price to destroy all the profit. It seems likely that Raytheon would have disappeared had it not be for an accident of wartime procurement.

It was not the opening up of the radio tube market which saved Raytheon. Rather war brought on a new generation of scientists and a new generation of interests: research attention shifted from radio communications to the use of radio for the location of distant objects.

In 1938 Vannevar Bush had moved to Washington to be the head of Carnegie Foundation. From that office Bush pushed forward aggressively to seize the role of director of weapons development for the United States government. He had, however, severed his ties with his old Raytheon associates. Success came without his aid. In fact, Raytheon even won a competition against its old adversary, American Telegraph and Telephone Company. The win had all the unpredictability of the firm's earlier successes.

Just the prior to the German invasion of Poland, in September 1939, British physicists had developed a high powered, high frequency signal system for detecting aircraft - what we now call radar. The British were well in advance of American experimenters, but they faced very serious problems in the manufacture of their equipment. The magnetron, the core element that generated the energy for the radar device, was cumbersome, complicated, and exceedingly difficult to make. In September 1940 the British brought the device to the United States and requested that the Bell Laboratories take on the problem. They did.

Marshall knew nothing of these secret inventions. He was, however, supplying an M.I.T. telephone engineer with high-frequency tubes in an endeavor to keep up with new developments. The professor, Edward Linley Bowles, and his associates were trying to find ways do direct aircraft in the fog. Bowles, also, was a former student of Vannevar Bush, and Bush soon appointed him to his military research committee. There Bowles learned of the British radar problem, and he urged the British to let Raytheon work on the task as well as the telephone company. He argued that sometimes a small group can make faster progress than a large one. The British agreed, and showed the magnetron to Marshall and his engineers.

At this moment another old-fashioned Yankee tinkerer entered the story, Percy Lebaron Spencer (1894-1970), the older brother of the thermostat inventor. After the death of the father, Percy had been raised by an aunt, an itinerant weaver who moved from mill to mill in Maine, seeking the factories that were hiring. Percy, therefore, received little formal schooling. He went to work in a machine shop and later was introduced to electricity by a contractor who had been hired to electrify a paper pulp mill. Thereafter, in 1912, Percy, seeking the poor boy's education, enlisted in the Navy and from 1912 to 1915 served as a radio technician. During the war he served as the superintendent of a Boston radio manufacturing firm that was later absorbed by RCA. Then he went to work for a firm that had been created by Boston yachtsmen and bankers to explore sonic detection. When the firm wanted to move Percy to Norfolk, Va., he quit and joined his brother at Raytheon. During the twenties and thirties he rose in the firm as an expert in tube manufacture. It was he who figured out how to make the magnetron.

At the outset he visited the Western Electric plant of the telephone company to observe its methods. They were using very elaborate milling machines and highly skilled machinists to fashion the copper cores of the magnetron, but they were having difficulty turning out even eight per day. Spencer realized that Raytheon could not command the necessary priorities to obtain such machinery, nor could it find the machinists to run the tools even if it could purchase them. Instead of reworking a solid block of copper, Spencer proposed to stamp out thin sheets of metal to conform to the changing contours of the magnetron's pattern. That could be done by unskilled operators running available punch presses. Next, the sheets would be coated with solder, piled one atop the other, and baked in an oven until they fused into a solid unit. The scheme proved a breakthrough, and Raytheon's production zoomed, reaching a peak of 2,600 magnetrons in one day.

The magnetron saved Raytheon. As a military contractor specializing in radar, anti-aircraft, and related naval equipment, the company took off. At its 1945 peak the number of employees reached 16,000, its sales $173 million, and its profits $3.4 million.

Two years later the company faced bankruptcy. Immediately after the war Laurence Marshall plowed all his surplus cash into seeking new products. He purchased a company in Chicago which had been successful in making radios and phonographs for Montgomery Ward and other large retail chains. Captured by the potential of high-frequency wireless transmission, he bought a line of hilltops from Boston to New York - City upon which he hoped to put his relay towers. His designs were faulty, however, and he had not secured any agreement to link his network to existing telephone exchanges. In another attempt he personally baked hundreds of batches of gingerbread, trying to perfect what in time became Raytheon's successful microwave oven. Everything he attempted required much more time and money than Raytheon could afford, or it faced stiff competition from large, entrenched companies. The automobile radio field was then dominated by Bendix, while RCA and others defeated Marshall's attempt at breaking into the manufacture of phonographs, radios, and television. Finally, much of the cash Marshall spent from his wartime profits was subsequently withdrawn by a post-war U.S. government review of contract costs and profits.

In 1948 the Morgans of New York moved to protect their interests. Henry S. Morgan, son of J.P. Morgan, persuaded his brother-in-law, the Boston investment banker and yachtsman, Charles Francis Adams, Jr., to take over the management of Raytheon. Adams, in turn, arranged for the First National Bank of Boston to advance a large loan. Laurence Marshall was forced to resign. It is doubtful, if Adam's management, and his stress on cost containment and strict procedures, would have saved the firm had it not been for the discovery of yet another new product line.

The final success came out of one of Marshall's wild hunches. During World War II, when he could at last afford to hire all the talent he could find, Marshall hired three disgruntled young scientists from RCA: Royden C. Sanders, James Ludwig, and William Mercer. The trio had invented a fine airplane altimeter at RCA, but the research director there would not allow them to follow this lead. Marshall told them to go to work on their ideas. Later he recalled that he didn't fully understand what they were about, but that they seemed full of fire and promise.

After the war the perfection of radar raced on, and the trio continued in the field. They first took up the problem of detecting low-flying aircraft, and then the direction of rockets. By 1950, after a tangled series of military development contracts which they shared with other firms, the Raytheon team succeeded in perfecting a test rocket that intercepted and destroyed a low-flying aircraft.

The Korean War and the ceaseless arms race with the Soviet Union put Raytheon back on its feet once more as a military contractor. The Bell Laboratories went on to long-range missiles, Raytheon perfected the short-range missile systems, like the ones on the frigate *Stark* which were not turned on in the Persian Gulf incident last June. However, to avoid becoming the prisoner of the Defense Department, Adams used the money made from arms contracts to purchase civilian firms. Today half Raytheon's business is military, half civilian. It publishes books, makes phonograph records, manufactures stoves and refrigerators, erects industrial plants, and still make tubes, transistors, switches, cables, and all manner of electronic gear. It is also served as the nursery of at least two generations of engineers who have worked there and gone on to other companies in the region (Kidder, 1981).

10.4 CONCLUSIONS

What are the conclusions to be drawn from such a narrative? I find several:

First, there was no way for the Boston region to know in the midst of its long textile and shoe depression that the electronics industry would develop as the base of a new prosperity.

Second, the sequence of transitions from electric power, to radio communications, to high frequency electronics was also unpredictable, therefore all small firms in these industries faced sudden obsolescence and competitive wipe-out.

Third, from a regional point of view the abundance of capital proved extremely important. In this case Raytheon was able to draw on the accumulated fortunes of the textile era, New York risk capital (the Morgans), capital from military contracts, and Boston bank capital.

Fourth, patents were always a problem. Patents provided Raytheon with its successful entry into radio with its power tube, but patents also magnified the advantage of large national firms and thereby tended to limit competition and perhaps slow the pace of innovation.

Fifth, the openness of class lines seems especially important. In both the invention and the development of products the tinkerer, the machinist and the enthusiast without much formal education proved to be very important. Marshall's management style, and the primitive nature of the radio and electronics industry, seem to have kept the class lines open. Perhaps in a large old corporation, like General Electric, bureaucratic rigidities close off these important lines of communication and innovation.

Sixth, the long process from science to product development seems to have been subsidized by the region's universities. Men like Bush and Bowles made their living as academics and followed their specialities in the academy even without research grants. This secure home base then enabled them to work part-time for regional companies seeking new products, whether the company succeeded or not. Today the large research grants to academic laboratories and the big private corporate research and development programs, have substantially altered these old-fashioned part-time relationships. It is, perhaps, a possibility that these highly structured academic and corporate arrangements will not prove as open and creative as the informal urban structure that preceded them.

REFERENCES

Baldwin,. G.B., 1951, "The Invention of the Modern Safety Razor: A Case Study of Industrial Innovation", *Explorations in Entrepreneurial History* 4, 74-77.
Bezanson, A., 1952, The Invention of the Modern Safety Razor: Further Comments. *Explorations in Entrepreneurial History* 4, 193.
Bush, V., 1932, "The Key to Accomplishment", *Textile Research* 3, 7-13.
Bush, V., 1951, "A Tribute to Dugald C. Jackson", *Electrical Engineering* 70, 1063-1064.
Bush, V., 1970, *Pieces of the Action*, Morrow, New York.
Chandler, A.D., Jr., 1961, *Strategy and Structure: Chapters in the History of American Industrial Enterprise*. M.I.T. Press, Cambridge, Mass.
Chandler, A.D., Jr., 1977, *Visible Hand: The Managerial Revolution in American Business*, Harvard University Press, Cambridge, Mass.
Conant, J.B., 1952, *Modern Science and Modern Man*, Columbia University Press, New York.
Conant, J.B., 1979, *My Several Lives, Memoirs of a Social Inventor*, Harper & Row, New York.
Danielian, N.R., 1939, *The Story of Industrial Conquest,* The Vanguard Press, New York.
Eisenmenger, R.W., 1967, *The Dynamics of Growth in the New England Economy*, Wesleyan University Press, Middletown, Connecticut.
Filene, E.A., 1933, "Unemployment in New England, Some Fundamental Factors", in American Geographical Society: *New England's Prospect*, 65-95, American Geographical Society, Special Publications ≠ 10.
"The Fortune 500 Industrial Corporations", 1987, *Fortune* 115, 364-365.
Hägerstrand, T., 1967, *Innovation Diffusion as a Spatial Process*, University of Chicago Press, Chicago.
Kidder, T., 1981, *The Soul of a New Machine*, Little Brown, Boston.
Rosenberg, N., 1976, *Perspectives on Technology*, Cambridge University Press, Cambridge.
Rosenberg, N. and L.E. Bridzell, Jr., 1986, *How the West Grew Rich: The Economic Transformation of the Industrial World*, Basic Books, New York.
Schumpeter, J.A., 1934, *The Theory of Economic Development*, Harvard University Press, Cambridge, Mass.
Scott, O.J., 1974, *Creative Ordeal. The Story of the Raytheon Corporation*, Athanaeum, New York.
My narrative of the history of the Raytheon company rests upon this excellent history, which was commissioned by Charles Francis Adams.
Warner, S.B., 1984, *Provice of Reason*, Harvard University Press, Cambridge, Mass.
The argument of this essay is a condensation of a much lengthier treatment set forth in this book of regional biographies.

CHAPTER 11

Innovation, R and D, and Firm Growth in Robotics: Japan and the United States*

Edwin Mansfield

11.1 INTRODUCTION

Robotics is viewed by economists, technologists, managers and government officials as a key technology of the late twentieth century, one that will potentially affect large segments of manufacturing (and other economic activities) throughout the world.[1] At present, two of the leading nations in robotics are Japan and the United States. The government of Japan has targeted robots as one of a small number of industries that it will help in various ways. In the United States, there has been considerable concern that, while many of the basic inventions occurred here, the Japanese robot industry has grown more rapidly and in many respects is ahead of our own.[2]

To understand the evolution and dynamic behavior of both the Japanese and American robot industries, it is necessary to compare the quickness and efficiency of Japanese and American robot producers as innovators, and the amounts spent by Japanese and American robot producers on various kinds of R and D and other innovative activities. Also, it is important to study carefully the processes of firm growth in the robot industry in both countries, and to see how a firm's growth rate is related to its R and D and innovative activities. My purpose in this paper is to present the results of what seems to be the first such study of this sort.

---

* The research on which this paper is based was supported by a grant from the Division of Information Science and Technology of the National Science Foundation, which, of course, is not responsible for the views expressed here. My thanks go to Laurence Rosenberg and Charles Brownstein of the Foundation for helpful comments, as well as to the many Japanese and American firms that provided data.

[1]For example, see Office of Technology Assessment (1984).

[2]See Baranson (1983) and International Trade Commission (1983). At the outset, it should be noted that the definition of a robot put forth by the Robot Institute of America (now the Robotics Industry Association) is used in this paper. According to this definition, a robot is a reprogrammable multi-functional manipulator designed to move material, parts, tools, and specialized devices through various programmed motions for the performance of a variety of tasks. In their statistics, the Japanese often include manual manipulators and fixed sequence "robots" as well. The only place where the Japanese definition is used is in Section 3 and 4, where the sales data for the individual Japanese firms could only be obtained on this basis.

11.2 THE ROBOT INDUSTRY: U.S. AND JAPAN

In 1961, Unimation, a subsidiary of Condec Corporation, installed the first commercial robot in the American auto industry. Founded to exploit George C. Devol's patent, Unimation had sold about 200 robots by 1970. In 1968, it assigned license rights to Kawasaki Heavy Industries, which became the leading Japanese producer of robots. By the mid-1970s, robot sales in the United States exceeded $10 million per year, and although Unimation remained the leader, firms like Cincinnati Milacron, DeVilbiss, ASEA, and Prab entered the industry. During the late 1970s, robot sales began to grow more rapidly, as the auto industry increased its demand for robots, particularly for spot welding. By 1980, sales were over $60 million (Table 11.1).[3]

Table 11.1: Sales of robots, United States and Japan, 1976-85

Year	United States (millions of dollars)	Japan (millions of dollar)[a]	Japan (billions of yen)
1976	12	N.A.	N.A.
1977	23	30	7.9
1978	28	59	12.4
1979	28	89	19.5
1980	64	218	49.3
1981	113	339	74.6
1982	143	455	113.0
1983	169	620	147.1
1984	332	759	180.2
1985	443	N.A.	N.A.

Source: The U.S. figures come from the Robotics Industry Association (1984 and 1985), the International Trade Commission (1979-83), and *Business Week*, June 9, 1980 (1976-78). Because of differences in concepts, the figures are not entirely comparable, but sufficiently accurate for present purposes. The Japanese figures come from Paul Aron (1983, 1985), who has based them on the U.S. definition of a robot.

[a]To convert from yen to dollars, each year's average exchange rate, as given in the 1985 Annual Report of the Council of Economic Advisors, is used.

During the 1970s, the robot population grew more rapidly in Japan than in the United States, due in part to differences between American and Japanese firms in the minimum rates of return required to justify investing in robots, to Japanese government programs to encourage the use of robots, to concerns about a prospective labor shortage in Japan, and to a variety of other factors. In both countries, welding and materials handling have been the most important applications of robots; assembly seems to have been more important in Japan than in the United States. In both countries, the automobile and electrical equipment industries have been among the most prominent users of robots, although many also are used in the machinery, metals, and aerospace industries.[4]

From 1980 to 1985, while the U.S. robot industry's growth did not keep pace with many overly optimistic forecasts, it nonetheless was impressive. By 1985, total sales exceeded $400 million. At the same time, there was considerable entry, both by large firms like IBM and General Electric and by small start-up firms like Automatix and

---

[3]See Ayres, Lynn, and Miller (1981) and Office of Technology Assessment (1984).

[4]In 1986, robot sales in the United States dropped, due partly to scaled-back investment plans of General Motors. See *New York Times*, November 25, 1986.

Intelledex. Moreover, General Motors and Fujitsu Janic established a joint venture in 1982, which by 1984 had become the new industry leader. Nonetheless, in the 1980s as in the 1970s, both robot sales and the robot population remained substantially lower in the United States than in Japan.[5]

Whereas the American robot industry contains about 50 firms, there are about 150 members of the Japanese robot industry, some very large (like Hitachi), some very small (like Dainichi Kiko). The four-firm concentration ratio in Japan has been much lower than in the United States (in the early 1980s, about 25 percent versus 70 percent).[6] While major American producers have tended to focus attention on general purpose robots, many Japanese firms have developed specialized robots, frequently for their own use. (Japanese producers have often begun by making robots for themselves.) For many Japanese firms (as well as American firms), robots constitute a small percentage of their total sales.[7]

11.3 GIBRAT'S LAW

Economists have long been interested in the processes of firm growth, but few studies have focused on the nature of these processes in young, science-based industries. To see how the processes of firm growth in the Japanese and American robot industries compare with those in mature industries, we begin by testing whether Gibrat's law prevails. According to Gibrat's law, which plays a central role in many stochastic models of firm growth,[8] the probability of a given proportionate change in a firm's size during a specified period is the same for all firms in a particular industry, regardless of their size at the beginning of the period. For example, a firm with sales of $50 million is as likely to double in size during a given period as a firm with sales of $5 million. Put differently, Gibrat's law states that:

$$s_{ij}^{t+\Delta} = U_{ij}(t,\Delta)s_{ij}^{t} , \tag{1}$$

where s_{ij}^{t} is the size of the jth firm in the ith industry at time t, $s_{ij}^{t+\Delta}$ is its size at time $t+\Delta$, and $U_{ij}(t,\Delta)$ is a random variable distributed independently of s_{ij}^{t} .

In mature industries, Gibrat's law is often violated in two ways: (1) Small firms tend to grow more rapidly than large ones. (2) The variance of s_{ij}^{t+1}/s_{ij}^{t} tends to be inversely related to s_{ij}^{t}.[9] To test whether Gibrat's law is a good approximation in the Japanese and American robot industries, note that equation (1) implies that

$$\ln s_{ij}^{t+\Delta} = V_i(t,\Delta) + \ln s_{ij}^{t} + z_{ij}(t,\Delta), \tag{2}$$

---

[5]See Mansfield (1986a).

[6]The data on which these concentration ratios are based come from Paul Aron (1983) and Prudential-Bache (1984). It is interesting to note that, although some economists have hypothesized that concentration rises rapidly in industries with rapid rates of technological change, this has not been the case in robots. In the United States, the four-firm concentration ratio has fallen from about 100 percent in 1976 to about 55 percent in 1985. In Japan, concentration seems to have remained quite low. This appears to buttress the argument in Mansfield (1983) that technological change frequently reduces, not increases, concentration.

[7]See Aron (1983).

[8]For example, see Ijiri and Simon (1977).

[9]See Mansfield (1962).

where $V_i(t,\Delta)$ is the mean of $\ln U_{ij}(t,\Delta)$ and $z_{ij}(t,\Delta)$ is a homoscedastic random variable with zero mean. Thus, if $\ln s_{ij}^{t+\Delta}$ is plotted against $\ln s_{ij}^{t}$, the data should be scattered with constant variance about a line with slope of one.

For each successive pair of years during 1980-85 in the United States and 1979-83 in Japan, $\ln s_{ij}^{t+1}$ was regressed on $\ln s_{ij}^{t}$. In all nine cases, the estimated slope is less than one, the difference (from one) being statistically significant in six cases (Table 11.2). Also, in all but one case, the variance of s_{ij}^{t+1}/s_{ij}^{t} among small producers exceeds that among large producers, and in over half of the cases, the difference is statistically significant (Table 11.3). Thus, Gibrat's law tends to be violated in the same two ways in the robot industries of both Japan and the United States as in many mature industries.

Table 11.2: Estimated slope of regression of $\ln s_{ij}^{t+1}$ on $\ln s_{ij}^{t}$, Japanese and American robot producers, 1979-85

Nation and Time Period[a]	Estimated Slope	Standard Error of Slope	n	Standard Error of Estimate	R^2
Japan					
1979-80	0.919	0.102	37	0.522	0.70
1980-81	0.928	0.063	57	0.478	0.80
1981-82	0.844***	0.053	58	0.502	0.82
1982-83	0.715***	0.062	43	0.537	0.77
United States					
1980-81	0.849	0.191	10	0.828	0.71
1981-82	0.724*	0.152	11	0.682	0.72
1982-83	0.397***	0.182	16	0.989	0.25
1983-84	0.722**	0.132	16	0.497	0.68
1984-85	0.644***	0.113	18	0.486	0.67

***Significantly different from 1.00 at .01 level.
**Significantly different from 1.00 at .05 level.
*Significantly different from 1.00 at .10 level.

[a]For each time period shown in this table, the natural logarithm of each firm's sales in the later year (time t+1) is regressed on the natural logarithm of its sales in the earlier year (time t). For example, the top line of the table pertains to the regression of the natural logarithm of each Japanese firm's 1980 sales on the natural logarithm of its 1979 sales. The basic data, which come from Aron (1983) and Prudential-Bache, pertain to a large proportion of the robot producers in both countries, but the coverage is not the same from year to year. Obviously, only firms for which data are available for consecutive years can be included in the present analysis. The resulting sample size (n) for each time period and nation is shown in the table. In some cases, the basic data, which were provided by the robot producers, are only rough, but they should be adequate for present purposes.

Table 11.3: Standard deviation of s_{ij}^{t+1}/s_{ij}^{t}, large and small robot producers, Japan and the United States, 1979-85

Nation and Time Period[a]	Standard Deviation		Variance of Small Producers ÷ Variance of Large Producers
	Large Producers[b]	Small Producers[c]	
Japan			
1979-80	.569	.708	1.55
1980-81	.730	.917	1.58
1981-82	.578	1.706	8.71*
1982-83	.515	23.703	2 118.76*
United States			
1980-81	.014	2.858	41 656.81*
1981-82	.205	2.542	153.76*
1982-83	.522	2.735	27.46*
1983-84	1.726	1.187	0.47
1984-85	.399	0.416	1.09

[a]For each time period shown in this table, the earlier year is time t and the later year is time t+1. Thus, the top line of the table contains the standard deviation of the ratios of 1980 to 1979 sales for large Japanese robot producers and the standard deviation of the ratios of 1980 to 1979 sales for small Japanese producers. The basic data come from Aron (1983, 1985) and Prudential-Bache.

[b]In Japan, large producers in 1979-80 are firms with over 1 billion yen of robot sales; in 1980-81, 1981-82, and 1982-83, they are firms with over 2 billion yen of robot sales. In the United States, they are firms with over $10 million of robot sales in 1980-81 and 1981-82; in 1982-83 and 1983-84, they are firms with over $20 million in robot sales; and in 1984-85, they are firms with over $40 million in robot sales. Of course, these changes in the definition of a larger producer reflect the overall growth of the firms in the industry.

[c]In Japan, small producers are firms with under 200 million yen in robot sales. In the United States, they are firms with under $3 million in robot sales ($10 million in 1984-85).

*Significant at .05 level.

11.3 MOBILITY AND SERIAL CORRELATION OF GROWTH RATE

Based on hypotheses put forth by Caves and Porter (1978) and Mansfield (1962), one would expect that the amount of mobility in an industry - that is, the extent to which firms change their relative positions in the size distribution - will be higher in young industries where there is a rapid pace of design change due to technological advance than in mature industries. To measure the amount of mobility in the robot industry in a given time interval, suppose that we have a list of firms in existence at the beginning and end of the interval. Suppose that we choose at random a pair of firms where one firm was 60-70 percent as large as the second firm at the beginning of the interval. The probability that the first (initially smaller) firm will be bigger than the second (initially larger) firm at the end of the interval is a rough measure of the amount of mobility.[10]

This measure was calculated for time intervals of various lengths in Japan (during 1979-83) and the United States (during 1980-85). The results indicate that the amount of mobility in the robot industry was much the same in the two countries and that, in accord with the above hypotheses, it was relatively high (Table 11.4). In both countries, there was about a 0.45 probability that a robot producer initially 60-70 percent as large as another producer would be larger than the other producer at the end of a 3-year interval.

[10]This sort of measure was used in Mansfield (1962).

This probability is greater than in mature industries like steel, oil, and tires, even when the time interval considered for these other industries is much longer than 3 years.[11]

Table 11.4: Probability that a firm initially 60-70 percent of the size of another firm will be larger than the other firm at the end of the time interval, robot and other industries, Japan and U.S.

Industry. Nation, and Length of Time Interval	Probability
Japanese Robot Producers	
1 year	0.27
2 years	0.33
3 years	0.44
U.S. Robot Producers	
1 year	0.25
2 years	0.31
3 years	0.50
Other U.S. Industries:	
Steel (10 year intervals)	0.20
Oil (9 years intervals)	0.35
Tires (7 years intervals)	0.28

It has also been hypothesized by Mansfield (1962) and Nelson and Winter (1982) that there will be some positive serial correlation between successive residuals of the regression of $\ln s_{ij}^{t+1}$ on $\ln s_{ij}^{t}$. For example, a successful technological innovation by a firm is likely to raise its growth rate for a number of consecutive years. In the robot industry, the data for both countries are in accord with this hypothesis, although the correlation is generally quite low. When each firm's residual from the regression of $\ln s_{ij}^{t+2}$ on $\ln s_{ij}^{t+1}$ is correlated with its residual from the regression of $\ln s_{ij}^{t+1}$ on $\ln s_{ij}^{t}$, the resulting correlation coefficients average 0.24 in both the United States and Japan (Table 11.5).

11.5 INNOVATION TIME AND COST: JAPAN AND THE UNITED STATES

Turning from the history and processes of firm growth in the robot industry, we consider the rivalry between Japanese and American robot producers. Because product innovation is so important in this industry, one of the key questions is: Do American robot producers develop and introduce new products as quickly and economically as the Japanese? To obtain evidence on this score, a random sample of 20 American and 15 Japanese robot producers was chosen. This sample includes over 80 percent of American firms with 1984 sales of $10 million or more, as well as a substantial number of the smaller firms, and about 20 percent of all Japanese production.[12] Each firm in the sample was asked to

[11]The data for Japan in Table 11.4 are based on 1979-80 (1 year), 1979-81 (2 years), and 1979-82 (3 years). The data for the U.S. are based on all 1-year, 2-year, and 3-year periods during 1980-85.

[12]In 1984, almost 90 percent of all robots produced in the United States were accounted for by our sample. The Japanese sample also contains a wide range of firm sizes. The data were collected through

estimate the percentage difference, if any, between the average time and cost of developing and commercially introducing a new robot (of comparable novelty, importance, and complexity) by Japanese and American firms.

Table 11.5: Correlation coefficient between successive residuals of regression of $\ln s_{ij}^{t+1}$ on $\ln s_{ij}^{t}$.

Nation and Years of Successive Residuals	Correlation Coefficient
United States	
1981 and 1982	.14
1982 and 1983	.10
1983 and 1984	.41
1984 and 1985	.60***
Japan	
1980 and 1981	.13
1981 and 1982	.26*
1981 and 1983	.32**

***Significant at .01 level.
**Significant at .05 level.
*Significant at .10 level.

[a]The years given in this table are the successive values of time t+1. Thus, the top line of the table shows the correlation coefficient between (1) an American firm's residual from the regression of the natural logarithm of 1982 sales on the natural logarithm of 1981 sales, and (2) this firm's residual from the regression of the natural logarithm of 1981 sales on the natural logarithm of 1980 sales.

Both the Japanese and American firms report that the Japanese tend to be faster in this regard. The Japanese put the difference at almost 30 percent; the Americans put it at almost 20 percent. In comparing costs, it is essential to recognize that the official exchange rate does not adequately reflect the relative price of R and D and other inputs to the innovation process in the two countries. Thus, as the Organization for Economic Cooperation and Development (1979) and others have pointed out, one must construct an exchange rate reflecting purchasing power parities for resources used in the innovation process.[13] To do so, we obtained data from a sample of firms concerning the relative price of R and D and other relevant inputs in the two countries.[14] Using this exchange rate, together with the cost data described in the previous paragraph, it appears that the Japanese firms tend to be more economical innovators than the Americans. Based on both the Japanese and the American estimates, the difference between the two countries in the amount of resources used to develop and commercialize a new robot is about 10 percent.[15]

---

correspondence and interviews. The firms were chosen at random from the lists of robot producers in Aron (1983. 1985) and Prudential-Bache (supplemented by the Thomas Register). Problems of non-response were minor.

[13]See Organization for Economic Cooperation and Development (1979).

[14]These data were obtained from leading firms that carry out R and D and related innovative activities in relevant fields in both countries. The results for these firms were averaged. Because the interfirm differences turned out to be small, the sampling errors in the resulting R and D exchange rate seem to be negligible.

[15]For estimates of this sort for other industries, see Mansfield (1988a). Note that the results are really an average over a variety of types of innovations. There is evidence that the Japanese time and cost

11.5 THE INNOVATION PROCESS: RESOURCE ALLOCATION IN THE TWO COUNTRIES

To understand why Japanese robot producers seem to have this advantage over American producers, we investigated how robot firms in each country allocate resources in the innovation process, which we divide into the following six phases:[16]

1. *Applied research*, which, according to the National Science Foundation, includes research projects that "represent investigation directed to discovery of new scientific knowledge and which have specific commercial objectives with respect to either products or processes."[17]
2. *Preparation of project requirements and basic specifications*, which includes the specification of the product's characteristics, often in coordination with marketing and other non-R and D personnel, as well as planning and scheduling of the project.
3. *Prototype*, which includes the design, production, and testing of the prototype.
4. *Tooling and manufacturing equipment and facilities*, which includes preparation for manufacturing, and the design, construction, and acquisition of manufacturing facilities for the new product, as well as tooling and equipment.
5. *Manufacturing startup*, which includes the training of production workers, the "debugging" of the production facilities, and the resources required to attain an acceptable quality level.
6. *Marketing startup*, which includes marketing studies, advertising, sales promotion, and other marketing activities before the sale of any appreciable amount of the new product.

Detailed data were obtained from the firms in our sample concerning the proportion of the total cost of developing and introducing a new product that was incurred in each of these six stages. The results indicate that, in carrying out an innovation, Japanese robot firms allocate their resources quite differently than do American firms. In particular, the American producers devote a much larger percentage of total innovation cost to marketing startup than do Japanese firms. Whereas this percentage is about 10 percent in Japan, it is almost 40 percent in the United States.[18] The surprising fact seems to be that, if the American firms were to reduce their marketing startup costs by one-third, which would mean that the percentage of total innovation costs devoted to marketing startup would still be more than double the level in Japan, the Japanese edge with regard to the total resource cost of innovation in this industry would essentially disappear.[19]

No single factor seems to account for the much smaller percentage of total innovation cost devoted to marketing startup in Japan than in the United States. Because (as noted above) many Japanese robot producers have developed robots for their own use, these firms have not devoted much marketing effort to them in their early stages. Also, American robot producers have devoted unusually large percentages of total innovation cost to marketing startup, since the diffusion of robots has been slower than many firms forecasted, and there has been considerable pressure to educate and woo potential

---

advantages tend to be less for more technologically ambitious innovations and for innovations based largely on internally developed technology than for other types.

[16]This division of the innovation process was used in Mansfield et al. (1971), where each phase is discussed in more detail.

[17]Sanow (1959), p. 124.

[18]Based on a t test, this difference is statistically significant at the .025 level.

[19]If American firms could reduce marketing startup costs by about one-third, these costs would be about 25 percent of total innovation costs (which is more than double the Japanese percentage), and total innovation costs in the U.S. would be cut by about 12 percent, which would make them approximately equal to those in Japan. Of course, this calculation assumes that the costs of phases of the innovation process other than marketing startup remain constant.

customers.[20] In addition, there seem to be other factors that operate generally, not in robotics alone. In other industries, not just robotics, the Japanese seem to devote relatively fewer resources to marketing startup than American firms, due either to greater efficiency in this regard or to the feeling that fewer resources are required.[21]

Another striking difference between Japanese and American robot producers is that the former devote a much larger percentage of total innovation cost to tooling and manufacturing equipment and facilities than do the latter. The emphasis of Japanese firms on process engineering and efficient manufacturing facilities is well known. But nonetheless it is surprising that the Japanese percentage is about five times that in the United States. This difference, like that regarding marketing startup, is found in many other industries, not just robots (although it is larger in robots than in other industries for which we have data). It helps to explain why Japanese production costs tend to be relatively low.[22]

11.7 R & D EXPENDITURES IN JAPAN AND THE UNITED STATES

Particularly since the robot industry is still very young, the relative position of each country in this industry will be influenced heavily by the size and quality of its R and D programs. Given the fact that American robot producers have devoted more of their efforts to the development and introduction of relatively sophisticated robots, while the Japanese have focused their attention on simpler types of robots, one might expect that the percentage of sales devoted to research and development would be higher in the United States than in Japan. Based on our sample of firms, this is true. In 1985, American robots producers devoted about 17 percent of sales to R and D, an extremely high figure. But it is worth noting that Japanese robot producers devoted about 10 percent of sales to R and D, and since their sales have tended to be more than double those of American firms (Table 11.1), their total 1985 R and D investment in absolute (real) terms may have been greater than that of their American rivals.[23]

The composition of a firm's or industry's R and D expenditure can be as important as their total size.[24] As pointed out by the U.S. Department of Commerce (1983), "In the past, Japan's R and D efforts were geared primarily toward modifying, adapting, and improving imported technologies. Now, however, it is targeted at the high-risk stage or areas with the potential for major technological advances".[25] Based on detailed data obtained from our sample of firms, it appears that in robotics the Japanese firms devote about as large a percentage of their R and D expenditures to basic and applied research (rather than development) as do American firms. Moreover, there is no significant difference between the American and Japanese firms in the percentage of R and D

---

[20]See International Trade Commission (1983).

[21]However, the difference between American and Japanese firms with regard to the percentage of innovation cost devoted to marketing startup seems greater in robots than in other industries for which we have data. See Mansfield (1988a). The importance of the interface between R and D and marketing is well known. Whether higher marketing startup costs are worthwhile depends, of course, on the extent to which they promote better decisions concerning the design and introduction of the innovations. According to some U.S. executives in the robot industry, Japanese producers are more likely than American producers to put forth innovations that are not well geared to the market. Others do not share this view.

[22]Based on a t test, that difference is statistically significant at the .05 level.

[23]It is reassuring to note that our figure of 17 percent for R and D expenditure as a percent of sales for all U.S. firms is very close to the International Trade Commission's figure of 18 percent. See International Trade Commission (1983). Also note that R and D inputs tended to be cheaper (at official exchange rates) in Japan than in the United States, which makes it even more likely that R and D spending (in real terms) by the Japanese robot industry exceeded that by the American robot industry.

[24]See Mansfield (1981).

[25]U.S. Department of Commerce (1983), p. 30.

expenditures reportedly aimed at the development of entirely new products and processes (rather than the improvement or adaptation of existing products and processes). Apparently, the Japanese robot firms are betting as heavily as the Americans on making big advances, rather than incremental improvements.[26]

Surprisingly, the share of R and D expenditures devoted to new products and product improvements (rather than new processes and process improvements) is larger for Japanese robot firms than for American robot firms.[27] In most industries, this is not the case; instead the Japanese emphasize process development more than the Americans do.[28] Also, given the oft-stated assertion that the Japanese are more patient in this regard than American firms, it is interesting that the percentage of R and D expenditure devoted to relatively long-term projects (those expected to last more than five years) seems to be lower in Japan than in the United States. However, the difference is not statistically significant.[29]

11.8 RESOURCE ALLOCATION IN HIGH-GROWTH AND LOW-GROWTH FIRMS

Returning to the processes of firm growth, economists have shown considerable interest in the ways in which high-growth firms tend to differ from low-growth firms. However, relatively little information has been obtained concerning the differences between them in the nature of their R and D and innovative activities. Based on our results, there are marked differences of this sort, and they are similar in both Japan and the United States.[30]

In both countries, high-growth robot producers tend to devote more resources to tooling and manufacturing equipment and facilities and less to marketing startup than low-growth robot producers. The differences are striking. Both in Japan and the United States, the percentage of innovation cost devoted to tooling and manufacturing equipment and facilities was about four times as large among high-growth as among low-growth producers, and the percentage devoted to marketing startup was about one-half as large among high-growth as among low-growth producers. Apparently, in the innovation

---

[26]Of course, it is difficult to distinguish between an entirely new product or process and an improvement of an existing product or process, so this finding should be viewed with caution. However, it is worth noting that in other industries the Japanese generally report a lower percentage of R and D expenditures aimed at entirely new products and processes than the Americans do.

[27]However, this difference is significant only at the .10 level.

[28]See Mansfield (1988b).

[29]In both countries, it appears that small producers devote less of their R and D expenditures to applied and basic research, rather than development, than large producers. This is not surprising, since small producers frequently do not find such work worthwhile. More surprising is the fact that in both countries the percentage of R and D expenditures devoted to relatively long-term projects seems to be much higher among the small producers than the large ones. But it is important to note that some small robot producers are subsidiaries of larger firms, and that not all of the observed differences between large and small producers are statistically significant. In particular, the difference between large and small firms in the percent of R and D expenditures devoted to applied and basic research is not significant, and the difference in the percentage devoted to relatively long-term projects is significant only at the .10 level.

[30]In the United States, high-growth producers are defined as those that had more than a 50 percent average annual increase in robot sales in 1982-85; low-growth producers are those that had a 50 percent increase or less. (Of course, this is a short period, but the robot industry is very young. In one case where data were unavailable for 1982-85, the growth rate had to be based on only part of the period.) In Japan, high-growth producers are those that had an average annual growth rate of sales of more than 50 percent during 1979-84; low-growth producers are those that had an average annual growth rate of 50 percent or less. (In cases where data were unavailable for 1979-84, the growth rates had to be based on only part of the period.)

process in this industry, the more successful firms tend to emphasize investments in efficient production of a new product, rather than in ways to market it.

To prevent confusion, note the distinction between (1) the percentage of a firm's innovation cost devoted to tooling and manufacturing equipment and facilities, and (2) the firm's total investment in tooling and manufacturing equipment and facilities. Obviously, one would expect that, all other things equal, the latter would be higher among high-growth than among low-growth firms. Our finding in the previous paragraph pertains to the former, and is entirely different (and far more interesting). What our finding highlights is the emphasis that high-growth firms put on tooling and manufacturing equipment and facilities *before the initial commercialization of the product* (when its sales volume is still very uncertain).

High-growth robot producers in both countries also tend to be more research-intensive and technologically ambitious in their R and D programs than low-growth robot producers. Specifically, the percentage of sales devoted to R and D was about two or three times as great among high-growth as among low-growth producers. Also, the percentage of R and D expenditures devoted to research (rather than development), and the percentage aimed at entirely new products and processes, was at least twice as large among high-growth as among low-growth producers.[31] Clearly, the more successful firms seem to devote a larger share of their R and D to more fundamental and technologically ambitious projects - and this is likely to have been one reason for their success.[32]

11.9 SUMMARY AND CONCLUSIONS

In science-based industries, whether a nation can compete effectively may depend on how quickly and economically it can develop and commercially introduce new products. In the robot industry, it appears that the United States is currently at a disadvantage in this regard vis-a-vis Japan, whose firms seem to develop and introduce new products about 20-30 percent faster and with about 10 percent less resources than do American firms. In a a considerable extent, this disadvantage with regard to the resource cost of innovation reflects the much greater allocation of resources to marketing startup in the United States than in Japan, which is due in part to factors peculiar to the robot industry and in part to general factors that operate in a variety of other manufacturing industries. With regard to the R and D required to develop and introduce a new robot, there is no evidence that American firms have higher resource costs than do the Japanese.[33] This finding is

[31]We carried out t tests to determine the statistical significance of these differences. The difference between high-growth and low-growth producers in the percentage of R and D expenditures aimed at entirely new products and processes is significant at the .025 level. The difference in the percentage devoted to applied and basic research is significant at the .05 level. There also seems to be a tendency for the percentage devoted to product technology to be higher among the high-growth producers than among the low-growth ones, but it is not significant.

[32]It is likely, of course, that the lines of causation run in both directions, since high-growth firms often are more profitable and better able to support relatively high-risk R and D than low-growth firms. Also, the data used to classify firms as high-growth or low-growth pertain to the period before or up to 1985, whereas the data on R and D as a percent of sales (and those on the composition of R and D) pertain to 1985. However, because interfirm differences in the extent and composition of R and D tend to change slowly, it seems likely that the results would be roughly similar if the latter data pertained to some year prior to the growth data. Nonetheless, for lack of adequate data, we cannot make a precise estimate of how much of the observed relationship between the extent and composition of a firm's R and D and its growth rate is due to each of these lines of causation.

[33]R and D consists of applied research, preparation of product specifications, and prototype in our data. According to our results, R and D accounts for 56 percent of the cost of developing and commercially introducing a new robot in Japan and 45 percent of this cost in the United States. Based on our data, the resource cost of developing and commercially introducing a new robot in Japan is about 89 percent of that in the United States. Consequently, the resource cost of the R and D for a new robot in the United States

important, given the recent tendency in policy discussions to assume that R and D efficiency is higher in Japan than in the United States.[34]

In both countries, our findings are consistent with hypotheses put forth by Caves and Porter (1978) and Mansfield (1962) that imply that young, science-based industries like robotics exhibit substantially greater mobility within the size distribution of firms than do more traditional, mature industries. However, this does not mean that many basic patterns of firm growth observed in more traditional, mature industries do not prevail in young, science-based industries as well. In robots, as in steel, oil, and tires, Gibrat's law often is violated because small firms tend to grow more rapidly than large ones, and because the variance of firm growth rates tends to be inversely related to firm size. These patterns of firm growth, because they are not affected materially by the rate of technological change, are likely to be much the same in young, science-based industries as in more traditional, mature ones.

Firm growth and technological change are very closely intertwined in science-based industries like robots. Successful innovators tend to grow more rapidly than other firms,[35] and high-growth firms tend to carry out different sorts of R and D and related innovative activities than low-growth firms.

This paper has presented for the first time detailed data concerning the differences in this regard between high-growth and low-growth firms. In robots, the high-growth firms spend a much larger percentage of sales on R and D, and devote a much larger percentage of their R and D expenditure to research (rather than development) and to technologically ambitious projects than do low-growth firms.[36] (Also, they emphasize tooling and manufacturing equipment and de-emphasize marketing startup in the innovation process). These differences may exist in many young, science-based industries, and economists should incorporate them into models designed to explain the processes of growth and technological change in such industries.

REFERENCES

Aron, P., 1983, "The Robot Scene in Japan: An Update", *Daiwa Securities*, September 7.

Aron, P., 1985, "The Robot Scene in Japan: The Second Update", *Daiwa Securities*, April 15.

Ayres, R., L. Lynn and S. Miller, 1981, "Technology Transfer in Robotics Between the U.S. and Japan", in Uyehara, C., ed., *Technological Exchange: The U.S.-Japanese Experience*, 77-15, Japan-American Society, Washington, D.C.

Baranson, J., 1983, *Robots in Manufacturing*, Lomond, Mt. Airy, MD.

Caves, R. and M. Porter, 1978, "Market Structure, Oligopoly, and Stability of Market Shares", *Journal of Industrial Economics* 26, 289-313.

Ijiri, Y., and H. Simon, 1977, *Skew Distributions and the Sizes of Business Firms*, North-Holland, Amsterdam.

International Trade Commission, 1983, *Competitive Position of U.S. Producers of Robotics in Domestic and World Markets,* December.

Japan Industrial Robot Association, 1982, *The Robotics Industry of Japan*, Fuji Corporation, Tokyo.

---

must average about $\frac{.45}{.89 \times .56}$, or about 90 percent, of that in Japan for a comparable new robot. Of course, these figures are rough.

[34]For example, see L. Sumney and R. Burger (1987).

[35]See Mansfield (1962).

[36]Because the robot industry is still young, these results pertain necessarily only to relatively recent years. (See note 30.) It will be interesting to see how these relationships change as the industry matures.

Mansfield, E., 1962, "Entry, Gibrat's Law, Innovation, and the Growth of Firms", *American Economic Review* 52, 1023-1051.
Mansfield, E., 1981, "Composition of R and D Expenditures: Relationship to Size of Firm, Concentration, and Innovative Output", *Review of Economics and Statistics* 63, 610-615.
Mansfield, E., 1983, "Technological Change and Market Structure: An Empirical Study", *American Economic Review* 73, 141-145.
Mansfield, E., forthcoming, "The Diffusion of Industrial Robots in Japan and the United States", *Research Policy.*
Mansfield, E., 1988a, "The Speed and Cost of Industrial Innovation: Japan and the United States", *Management Science* 34, 1159-1168.
Mansfield, E., 1988b, "Industrial R and D in Japan and the United States: A Comparative Study", *American Economic Review* 78, 223-228.
Mansfield, E. et al., 1971, *Research and Innovation in the Modern Corporation*, W.W. Norton, New York.
Nelson, R., and S. Winter, 1982, *An Evolutionary Theory of Economic Change*, Harvard, Cambridge.
Office of Technology Assessment, 1984, *Computerized Manufacturing Automation: Employment, Automation, and the Workplace*, Government Printing Office, Washington, D.C.
Okimoto, D., T. Sugano and F. Weinstein, eds., 1984, *Competitive Edge*, Stanford University, Stanford.
Organization for Economic Cooperation and Development, 1979, *Trends in Industrial R and D, 1967-75*, OECD, Paris.
Peck, M. and S. Tamura, 1976, "Technology", in Patrick, H. and H. Rosovsky, eds., *Asia's New Giant,* Brookings, Washington, D.C.
Prudential-Bache, Computer Integrated Manufacturing, (November 29, 1985, April 27, 1984 and other dates).
Riggs, H., 1985, *Innovation: A U.S.-Japan Perspective. High Technology Research Project.* Stanford University, Stanford, September.
Sakamoto, K., ed., 1981, *Robots in the Japanese Eonomy, Survey Japan*, Tokyo.
Sanow, K, 1959, *Development of Statistics Relating to Research and Development Activities in Private Industry. Methodology of Statistics on Research and Development*, National Science Foundation, Washington, D.C.
Sumney, L. and R. Burger, 1987, "Revitalizing the U.S. Semiconductor Industry", *Issues in Science and Technology* 3, 32-41.
U.S. Department of Commerce, 1983, *An Assessment of U.S. Competitiveness in High Technology Industries*, Government Printing Office, Washington, D.C.

CHAPTER 12

Spatial Diffusion of Information Technology in Sweden

Christer Anderstig and Charlie Karlsson

12.1 INTRODUCTION

It is common knowledge that technical progress is an important element in economic growth and that its contribution to the national growth rate tends to be at least as important as the growth of factor inputs. In Sweden, technical progress is mainly the result of adopting, and adapting innovations generated and also first introduced elsewhere. Analyzing the spatial diffusion of innovations and new techniques thus seems essential for the study of regional growth rate disparities.

Against the background of the information technology revolution, the spatial diffusion of specific innovations is a matter of great significance, and this problem is the origin of the present paper. Does the diffusion of product and process innovations, based on information technology, follow a predictable spatial pattern? Is there any difference between the spatial diffusion of product and process innovations? Do peripheral regions lag behind metropolitan regions in adopting the new techniques?

The purpose of this paper is to analyze the spread of selected new techniques, based on information technology, across different types of regions in Sweden. The empirical analysis is based on a survey of 263 manufacturing plants in parts of the engineering industry, located in 45 municipalities of varying size, from Stockholm to small peripheral municipalities. The paper is organized as follows. Section 2 gives a short survey of the theory of spatial innovation diffusion and briefly reports on some related studies. The model framework is presented in Section 3, and in this context various analytical problems are discussed. Section 4 describes the empirical basis for the study and the results are presented in Section 5.

12.2 A THEORETICAL AND EMPIRICAL BACKGROUND.

To analyze the spatial diffusion of innovations, it is necessary to distinguish the following sub-processes: (i) Generation, (ii) Adoption, and (iii) Diffusion of innovations. We limit our discussion to entrepreneurial innovations, i.e. innovations with application to enterprises.

On innovation generation, Thompson (1965) has advanced the hypothesis that inventions tend to generate and become innovations in large metropolitan areas of 'high-technology' regions. Thompson's hypothesis is supported by McNeill (1963), who reports that inventions and innovation generation all through history have taken place in the larger metropolises open to the interchange of newer products and ideas (cf., also, Jane Jacobs, 1969). It is also consistent with the findings of Pred (1966). There are several reasons why metropolitan areas are the best "seed-beds" for new products. For one thing we may notice that in the early stages of product development, there is a heavy reliance on highly skilled designers, engineers, etc, resources which are concentrated to larger metropolitan areas due to agglomeration economies.

Thompson's hypothesis, however, can be made more precise. Present and historical data show that innovations are generated not just in any large open city but particularly in those in the most developed countries of the time. Today, most innovations in the industrialized market economies originate in five large megapolises: the eastern and western United States' seabords, the area of the Ruhr-Rhine valleys, the London-Midland plain and the Shinkansen region in Japan (cf., Lasuen, 1973).

Once inventions have become innovations, knowledge about them is diffused to other countries and regions. The diffusion process follows different communication mechanisms depending on the kind of innovation. Entrepreneurial innovations are of interest mainly to enterprises directly or indirectly affected by the innovations. Consequently, the diffusion of entrepreneurial innovations is carried out through specialized communication channels mostly directed to the enterprises in the trade.

Thus, with respect to Sweden, we may assume that the vast majority of innovations are generated exogenously and in a few relatively constant areas. Consequently, in our analysis, the critical subprocesses are those of innovation adoption and diffusion. For a small country like Sweden, not being a major innovation generating region, we may conclude that the import channels, i.e. the distribution networks of the innovation generating enterprises, are vital in the innovation diffusion process. Information about the innovation may, of course, reach the potential adopter through a number of information channels, as for example, journals, fairs, exhibitions, and customers but the actual transfer of embodied innovations takes place within an import channel. The distribution network within Sweden then consists of selling companies owned by the innovating enterprises but also of Swedish enterprises specialized in importing innovations, often working as agents or representatives for the innovating enterprises.

A study of the spatial pattern of innovation diffusion, within a country like Sweden, should start by emphasizing that the processes of innovation diffusion and industrial location are intertwined. The distinguishing feature of industries using new advanced technology is that some significant part of the design or manufacture of their product is still in the introduction stage.

Viewed from the perspective of traditional location theory these industries tend to be "foot-loose". However, their most important locational factor is the supply of high skills and scientific workers, and the location of these specialized resources is a determinant in the location process. These industries are normally dominated by large multi-functional/multi-product/multi-plant/multi-location enterprises (*M-enterprises*) (Singleplant/-single-location enterprises we name *S-enterprises*). The separate facets of the activities (R&D, design, assembly, testing, component manufacturing, etc) of these M-enterprises can be performed in different locations as the products have passed the introduction stages. The functional breakdown of production along cost-minimizing lines means that the different parts of the production activities will be carried out in different locations. Thus, the plants in one industry located in different areas may be producing quite different products.

As to spatial innovation diffusion, a distinction has been made between two kinds of spatial diffusion - *neighborhood* and *hierarchical*. It was the former, with its emphasis on the over-riding importance of physical distance in the diffusion process, that was emphazised by Hägerstrand (1952, 1967) in his pioneering analysis of the diffusion of 'household' innovations. The distinction has been expressed by Cohen (1972, pp.14-5) as follows:

"The neighbourhood effect means that.... the closer a potential adoption unit to the source of innovation or to another unit that has already adopted .... the greater the probability that it will adopt .... The hierarchical effect implies that .... the higher the ranking of a potential adoption unit in a hierarchy, the greater the chance that it will adopt".

The hierarchical model is here of greater interest, because it illustrates the 'short-circuiting' of the neighborhood diffusion process and appears more applicable to entrepreneurial innovations (cf., Moseley, 1974). As early as 1955 Perroux stated that growth manifests itself at points or poles of growth and diffuses through the economy in definite channels. Growth cannot in this view be separated from the diffusion of innovations, a process involving the acceptance over time of some specific idea or practice by individuals, groups or other adopting units such as enterprises linked to specific channels of communication, to a social structure, and to a given system of values or culture. Berry (1972) claimed that the role played by growth centers in regional development was a particular case of the general process of innovation diffusion. He also argued that the development role of growth centers involves the simultaneous *filtering* of the innovations that brings growth down the urban hierarchy and the *spreading* of the benefits accruing from the resulting growth, both nationally from core to hinterland regions and within these regions from their metropolitan centers outward to the inter-metropolitan periphery. This means that what is important here is the role of urban size in the innovation diffusion process.

Berry based his work on the contribution of Thompson (1965), who had hypothesized the way innovations spread out spatially from the large urban areas in the industrial heartlands to the peripheries. Thompson's reasoning was the following: the adoption of new innovations requires a flexible and diversified force of skilled labour, imaginative management, highly research-oriented technical manpower and significant external complementarities in the fields of finance, commerce, etc. This is why the site of innovation-based production is likely to be the large metropolitan area where its main plant, R&D facilities, product design facilities and principal management functions are normally located (cf., Thompson, 1969; Oakey, 1979). This means that the use of new advanced input innovations, for example, new advanced microelectronic circuits, will first be adopted by enterprises in large, advanced urban areas. However, as the most advanced and qualified functions of the enterprises are located in the same areas we may also hypothesize that new "process" innovations for those functions, as for example the use of computer support in R&D, computer-aided design, computer-aided production planning, computer- based material and production systems, and computer-based management resource planning systems, will be first adopted in the same large, advanced urban areas.

During its early life the new product is introduced into new market areas by means of export. However, as the product ages it moves along the product cycle curve and its product and process technology matures, skill requirements fall, and the dependency on agglomeration ceases. The high skill levels and advanced infrastructure of the more industrially sophisticated metropolitan areas actually become more and more superfluous. The ageing of the product means that it becomes more standardized. At this stage products start to filter through the hierarchical system of urban areas, to cities with less industrial sophistication. Decentralization of production might in this context be seen as the spatial manifestation of changes which occur in production and industrial organization in accordance with the product life cycle (cf. Thomas, 1974). The lower an area lies in the skill and wage hierarchy, the older a product tends to be when it arrives. Even if not all authors fully agree on hierarchical filtering (for example, Pred, 1974), several reasons for such a process can be posited (cf., Mansfield, 1968):

(i) *Information flows*. In much of the pioneer work on the spatial diffusion of innovations (notably Hägerstrand, 1952 and 1967), the role of information availability in neighborhood diffusion was afforded paramount importance, i.e. the diffusion of innovations was viewed as a learning process. But this approach is also valid for the problem of hierarchical diffusion, considering that there exist networks of social communications which connect certain places (central places in this case) to the exclusion of others. Innovations are then transmitted through a hierarchy of information networks (international, interregional, intraregional, etc). Enterprises with spatially abundant information networks may learn of, and be convinced earlier of, the relevance of a particular technique, compared to enterprises which obtain technical information from a narrower base and at a later date. 'The diffusion of information therefore becomes a key variable in the diffusion of new techniques' (Thwaites, 1978, p.452). However, the strength of information flows may vary over space and information networks may be stronger in certain regions (Pred, 1966; Feller, 1971; Goddard and Morris, 1976). Given this background, we may, for example, think that a hierarchical diffusion pattern is generated by a simple probability mechanism in which the probability of adoption depends upon the chance that an enterprise residing in a given center will learn of the new innovation - a probability which declines with the size of the center (Hudson, 1969). Of course, the mass media, the mushrooming of technical journals, etc, have brought more rapid and more widely spread of information about the existence of entrepreneurial innovations than ever before. Still, however, information about the characteristics of innovations, must perhaps be transferred interpersonally and thus flow hierarchically, generating a tendency for innovations to diffuse down the urban hierarchy.

(ii) *Markets.* A second reason for a hierarchical diffusion of innovations is concerned with the need to minimize risk. The introduction of an innovation on a new market is an inherently risky business for its propagator and as noted by Pred (1966, p.99), "uncertainty has a spatial dimension". Problems of a technical nature, of factor supply or of marketing may arise, and each is likely to be more easily overcome the larger the city. The argument is not simply that technical expertise, skilled labour and a ready market are most abundant in cities, but also that the information in these respects is greater. Thus, in a sense, the argument here too is one of abundance of information. Hierarchical diffusion may thus be the result of a "market-searching" process in which the enterprises propagating the new innovations exploit market opportunities in a larger-to-smaller sequence (Harris, 1954). The idea is that the most profitable (or least risky) locations (=markets) are exploited first, and successively less profitable or more risky locations are exploited later (Brown and Cox, 1971).

(iii) *Factors of production.* The argument here is that highly educated labour, capital and entrepreneurial skills are in short supply and most readily available in the largest urban places. This means, for example, that the absolute number of people, ready, willing and able to perform a certain job is likely to vary closely with size of urban center (cf., Anderstig and Hårsman, 1986). When a product-innovation becomes standardized, it pays the M-enterprises to shift the manufacturing of the product (or at least the assembly and manufacturing of standardized parts) to branchplants in the more peripheral areas of low-cost unskilled labour. Thus, the innovation diffusion process in this case takes the form of a relocation process (cf., Vernon, 1966 and 1971). This process is, of course, stimulated by price competition from other producers. Typically, manufacturing in a M-enterprise is relocated to a branch plant in a smaller community, where labour and other production cost savings are possible. Oakey (1979) suggests that branch plants will receive more or less exclusively mature products. Thus, the manufacturing of products containing the product innovations of yesterday is today relocated to peripheral non-metropolitan areas. This means that, when products reach the later life cycle stages the producers in the metropolitan areas seek for the industrial backwaters of non-metropolitan areas where cheaper unskilled labour is apt to the now simplified manufacturing processes. This relocation process is known as the "trickle-down" process (Thompson, 1968).

(iv) As a fourth reason for hierarchical diffusion we may hypothesize *an "imitation" process* in which independent entrepreneurs in S-enterprises in smaller centers mimic the actions of enterprises in larger centers (Pedersen, 1970).

As a product becomes standardized, it may become possible to introduce new process innovations. This means that during the standardization stage we may expect a simultaneous relocation of production and adoption of new process innovations in the M-enterprises. Consequently, while the adoption of new advanced product innovations can be expected to take place first in large metropolitan areas, the adoption of new process innovations, with applications mainly in standardized production processes, may simultaneously take place within as well as outside the large metropolitan areas, at least in the case of M-enterprises. This means that we expect the organizational structure of an industry to be one important determinant of the innovation diffusion process.

Applying the reasoning so far to the Swedish case, we may think of the Stockholm region as the metropolis where the imported new innovations are first introduced in Sweden. This should be valid for new innovations in the form of new advanced components and new "process" innovations used by the advanced functions within the enterprises. As to process innovations designed for the standardization of the manufacturing processes, the introduction pattern may be mixed. In some cases process innovations may first be introduced in the Stockholm region, while in other cases introduction may take place outside the Stockholm region. In principle this could be connected with the decentralisation of the manufacturing of a product from the Stockholm region. As the standardization process proceeds, the new innovations are diffused to regions of lower and lower industrial sophistication.

The spatial pattern of entrepreneurial innovation diffusion within Sweden will, particularly in the case of S-enterprises, depend on the spatial location of the nodes in the information networks within Sweden. For natural reasons most information networks in Sweden have the Stockholm region as their central node. The Stockholm region is the region in Sweden with the highest international and interregional accessibility. It is also the location of the headquaters of the vast majority of the large Swedish companies.

Finally, turning to empirical studies in the field, few studies of innovation diffusion have actually focussed upon the spatial dimension. A notable exception is the early work of Mansfield, et.al. (1971) which vindicated the discovery of time lags in the adoption of numerically controlled machine tools between states in the USA. In a later study by Rees, Briggs and Hicks (1985), concerning the spread of major production innovations based on information technology in the United States' machinery industry, adoption patterns with spatial differences of varying significance were found. Specific spatial differentiations were also maintained in a study of the rates of adoption of various techniques in Canadian industry by Martin, et.al. (1979). In the British context, research by Thwaites, Oakey, and Nash (1981) claimed that regional variations in the introduction of varying levels of technical innovation, especially product innovation, occur to the advantage of the South East and to the detriment of the peripheral areas of the country, notably the Nothern region, Scotland and Wales. Thwaites (1983) maintained that his research confirms for the UK a spatial innovation diffusion pattern demonstrating a low interregional rate of diffusion for new product innovations from the prosperous (R&D-rich) regions to the development (R&D-deficient) regions. Later research on Britain by Gibbs and Thwaites (1985) also claimed that new products developed in research units or plants in core areas are usually not transferred to outlying regions and that "most innovations are put into production at the location where they were originally developed". In the case of process innovations Thwaites (1983) asserted that differences in regional rates of adoption were small. Gibbs and Edwards (1985), however, have reported on spatial variations in adoptive behavior closely associated with corporate status, plant size and access to R&D facilities.

12.3 A MODEL FRAMEWORK FOR ANALYSING SPATIAL DIFFUSION

It should be clear from the discussion above that the spatial diffusion of new technologies is related to the spatial location pattern with respect to different activities within an enterprise or an industry and to the spatial process by which the location of production changes as products mature. In the archetypal location pattern, knowledge-oriented activities such as research, and technical and commercial development are located in metropolitan and other agglomerated areas, whereas more standardized activities are located at lower levels in the urban hierarchy. Once information technology (IT) has been defined, this familiar scheme of the spatial product life cycle produces some expectations as regards the spatial diffusion of specific IT-applications.

Let us, however, first give two definitions of entrepreneurial innovations in general. A *dynamic innovation* is any measure taken by an enterprise as part of a dynamic competition strategy that increases the perceived quality of its old product(s) or that introduces totally new products on the market. A *cost innovation*, on the other hand, is any measure taken by an enterprise as part of a price competition strategy that reduces its production costs for given products. A common characteristic for both classes of innovations is that, in the case of IT, they may be achieved both through the adoption of IT-applications in administration and management, in the production process and through the adoption of IT-components in the products. Now, this means, as we will argue below, that in the case of IT, it is possible to make a categorization of applications in terms of dynamic and cost innovations, respectively. Some "process" applications are in fact linked to a dynamic competition strategy, i.e. should be categorized as a dynamic innovation, whereas some "product" applications are linked to reducing the costs of producing a standardized product, i.e. should be categorized as a cost innovation.

As it is a new and dynamic technology, definitions of IT are not very precise. It is, however, possible to make some rough classifications of different types of IT-applications with respect to the dimensions *standardation* and *complexity*. Looking first at IT-applications with reference to the production process, numerically controlled machines (NC), computerized numerically controlled machines (CNC), industrial robots (IRb), automated materials handling (AMH), and production systems with limited manning (PBB) are all examples of IT-applications closely associated with standardized production, i.e. with a price competition strategy. The opposite tendency of differentiated production, emerging from an adaptation to customers´ needs, and thus associated with a dynamic competition strategy, is closely associated with computer-aided design (CAD), computer-aided manufacturing (CAM), and flexible manufacturing systems (FMS). As to the complexity of various applications, there is a ranking from low complexity, e.g. individual NC and CNC machines, to high complexity, e.g. integrated systems including sub-systems of CAD, CAM, PBB, FMS, etc (cf., Karlsson, 1987).

Now, we expect the probability of adopting IT-applications of high complexity, and applications adapted to differentiation to vary with the location of the enterprise (plant) with respect to the level in the urban hierarchy. There are several reasons for this. First, the diffusion of this kind of IT-applications is constrained by limitations on the spatial distribution of technological skills. Second, a high level of intraregional accessibility is a constraint also with respect to the requirements of technical support and the aim of flexible customer adaptation.

Regarding IT-components in the product, the question whether there is a hierarchical spatial diffusion pattern is, again, dependent on our beliefs regarding the standardization and complexity level of the product and its IT-components. A hierarchical diffusion pattern is, accordingly, expected with respect to the use of IT-components in products of relatively high complexity and relatively low standardization.

Finally, as regards IT-applications in administration and management, two points should be made. Some administrative routines, as e.g. accounting and budgeting, are general and the corresponding IT-applications are, likewise, standardized. For this group of IT-applications no hierarchical spatial diffusion pattern is expected. But there are other tasks in administration and management of specific importance to enterprises located at higher levels in the urban hierarchy, e.g. R&D, and strategic planning. The probability of adoption of IT-applications for such purposes is consequently expected to vary with the agglomeration of the location.

Summing up our hypotheses, the spatial diffusion of IT-applications is mainly determined by the spatial distribution of different activities within enterprises and industries. This means that, to the extent that there is a hierarchical pattern in the spatial diffusion of IT-applications, this will partly represent fixed or semi-fixed characteristics internal to the enterprise (plant), e.g. the composition of the labour force with respect to education level, skills, etc.

Diffusion is a concept which implies the presence of the time dimension. Modelling spatial diffusion could, for example, mean that the spatial variation in the year of adoption is the subject at issue. We will, however, analyze the spatial variation as regards the adoption of specific IT-applications in the year of observation. The reason, for one thing, is that a major part of the IT-applications are adopted only by a minor number of plants, which makes it less interesting to discriminate between the adopting plants. Second, the characteristics of the plants refer to the year of observation. It would be dubious to assume that the characteristics also refer to the year in which the specific IT-application was adopted, perhaps ten years earlier.

Assuming, that the probability of enterprise (plant) n adopting the IT-application ($i = 1$) depends on a vector of characteristics (Z) and a vector of unknown parameters (b), we will use the logit formulation as our model framework:

$$P_n (i = 1) = \exp b Z_{1n} / (\exp b Z_{0n} + \exp b Z_{1n})$$

Since the model is binary, the logit and probit approaches are very similar and, as it is the most convenient, the former has been chosen.

Among the characteristics to be tested, those representing the spatial dimension are in focus. More particularly, the urban hierarchy is at issue and will be represented by the size of the region in terms of its population. Among the enterprise characteristics, the size of the enterprise in terms of the number of employees is essential, since the capacity and the incentive to adopt new technologies, as well as the readiness to take risks, are assumed to be related to enterprise size. Since data are reported of the plant level, the affiliation to a multi-plant enterprise is a complementary characteristic in this respect. Other characteristics, such as the number of university engineers, the number of employees with higher education, etc., are assumed to be direct indicators of the ability to adopt new technologies (cf. Johansson and Karlsson, 1986). Since past investment outlays may be a retardation factor in adopting new technologies, and the age distribution of the capital stock is assumed to be related to the age of the plant, the latter is included as one characteristic with possible influence upon the adoption probability. One characteristic, customer dependence, is intended to be a direct indicator of standardization vs. differentiation. It is assumed that the higher the share of sales going to the major customer, the higher is the probability of standardization in production. Finally, the industry - machinery, electrical, or instruments - is included as potentially influencing the adoption probability.

Within this model framework the analysis refers to 11 IT-applications in production, 9 IT-applications in administration and management and finally IT-components in the product. From the discussion above, it may be expected that standardization in production and the complexity of the applied technology are the two dimensions related to the spatial dimension of the diffusion process, as regards process applications. We have given some specific examples with respect to these dimensions and it may seem natural to define all IT-applications similarly to get a complete and exclusive classification.

This task is, however, hard to accomplish. First, the way different IT-applications within an enterprise or a plant are coordinated or integrated cannot be judged from available data, nor is there any information of the number of a specific IT-application, e.g. the number of NC machines. Second, definitions even of individual applications are not very precise; just to take one example, it is not possible to differentiate between 'traditional' and new systems of materials resource planning, MRP1 and MRP2. This means that the character of the IT-applications must be judged somewhat *ad hoc*.

The model outlined above is used in two steps. First, it is used with respect to individual IT-applications. As to the process applications, we aim here at making a distinction between technologies supporting standardized and differentiated production respectively, though the distinction is based on less satisfactory definitions. In a second step, the model is used with respect to the adoption of computer-based systems of varying complexity. Although the coordination and integration of different IT-applications are unknown, a simple classification with regard to the number of different applications is assumed to give a reasonable measure of the complexity of the technological system in use. By defining systems as of low, medium, and high complexity, depending on the number of different applications, the adoption probabilities are estimated by a multinomial formulation of the logit model. Although the definitions are made *ad hoc*, they are based on observations as to the adoption rates of some significant applications, as CAD, CAM, PBB, and FMS.

Observing that some IT-applications, e.g. PBB, and FMS, are less frequent in systems of low complexity, implies inherent complexity even in the first step. Thus, the standardization of production and the complexity of the technology are not separable dimensions, a fact to keep in mind, when interpreting the results.

12.3 DATA AND SOME DESCRIPTION.

Our empirical analysis is based on survey data from 263 plants in 3 industries: (i) Machinery and equipment, (ii) Electrical equipment and components, and (iii) Instruments. Data were collected in 1985/86 by postal questionnaires to enterprises and plants in 3 counties: Stockholm, Värmland and Älvsborg. Different survey designs have been used: in Stockholm a stratified sample by size of plant (> 9 employees) was taken, whereas complete surveys were made in the other two counties (> 4 employees). In Värmland and Älvsborg the number of respondents covers more than 90 % of the relevant population of plants, compared to 33 % in Stockholm. The size distribution of sampled plants is presented Table 12.1.

Table 12.1 Sampled plants in Stockholm, Värmland, and Älvsborg by number of employees. Per cent. (Absolute numbers)

	NUMBER OF EMPLOYEES							
COUNTY	<20	20-49	50-99	100-199	200-499	>500	TOTAL	
STOCKHOLM	11	16	22	17	17	17	100	(120)
Total Population inStockholm	37	30	12	8	7	6	100	(366)
VÄRMLAND	42	21	8	18	5	6	100	(62)
ÄLVSBORG	48	30	6	5	9	2	100	(82)

The larger plants in the total sample are thus concentrated to Stockholm; more than two thirds of plants with more than 50 employees are located in Stockholm, compared to 45 percent of all plants in the sample. Also with respect to plants in different industries the distribution varies between the counties, as shown in Table 12.2. Obviously, the sample in Stockholm is biased with respect to plant size and, though to a lesser degree, with respect to industry. In order to avoid bias in the estimations, plants in Stockholm will therefore be weighted according to their distribution in the total Stockholm population.

Table 12.2 Sampled plants in Stockholm, Värmland and Älvsborg by industry. Per cent. (Absolute numbers).

COUNTY		INDUSTRY		
	MACHINERY	ELECTRICAL	INSTRUMENTS	TOTAL
STOCKHOLM	52	29	19	100 (120)
Total Population in Stockholm	51	34	15	100 (366)
VÄRMLAND	71	19	10	100 (62)
ÄLVSBORG	69	28	3	100 (81)

A crucial question is whether the spatial dimension can be reasonably represented by the 3 counties selected, excluding 22 counties. Actually, within this small number of counties the urban hierarchy in Sweden is fairly well represented. Municipalities in Sweden can be grouped into six 'H-regions' according to their level of intra-regional accessibility. Applying this definition of the urban hierarchy, the sampled plants are distributed as shown in Table 12.3.

Table 12.3 Sampled plants by level in the urban hierarchy

H-REGION	DEFINITION	NUMBER OF MUNICIPALITIES	NUMBER OF PLANTS
1	Stockholm/Södertälje A-region	18	118
2	Other metropolitan areas, Gothenburg A-region	3	24
3	Municipalities with more than 90,000 inhabitants within their central labour market area, 30 km	9	63
4	Municipalities with 27,000-90,000 inhabitants (....) and with more than 300,000 inhabitants within 100 km from the municipal center	11	46
5	Municipalities with 27,000-90,000 inhabitants (....) and with less than 300,000 inhabitants within (....)	1	2
6	Municipalities with less than 27,000 inhabitants within their central labour market area, 30 km	3	10

As can be seen from Table 12.3, all levels in the urban hierarchy are represented in our sample. Some H-regions are, however, represented only by a few plants. For analytical purposes it seems more productive to define regions in the same way as H-regions 1 and 2, i.e. as A-regions, defined as aggregates of municipalities connected by labour force commuting. This definition provides more information about the intraregional accessibility at various locations. In our data 10 A-regions are represented (See Figure 12.1).

As spatial characterization we will use the size of the A-region, given by the total number of inhabitants at the end of 1985. One A-region - A-region 42 - is divided into a central and a peripheral subregion, since the commuting criterion is not valid. Other representations of the spatial dimension, such as measures of interregional accessibility and regional infrastructure, must be left out because of the small sample of regions.

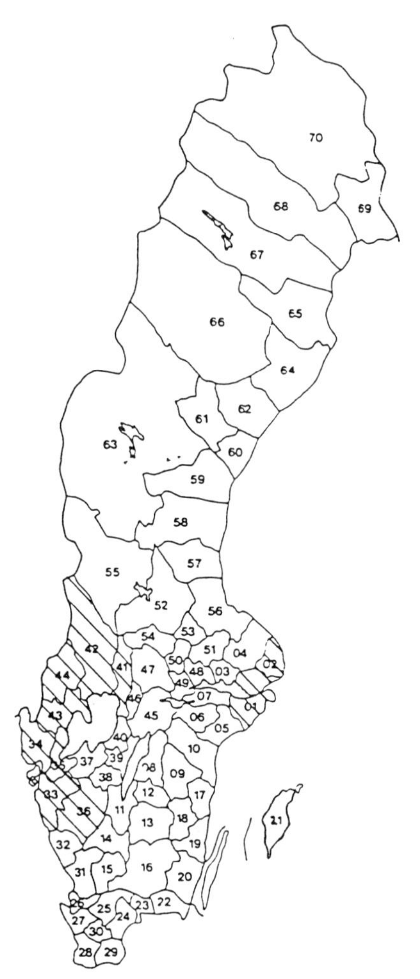

Figure 12.1 A-regions in Sweden. The shaded regions are covered by the surveys.

Among the IT-applications to be studied, the 11 referring to the production process are of major interest and should be specified. In Table 12.4 we report on the first year of adoption, adoption rates with respect to all plants in the sample, and adopting plants in different categories as regards the number of different IT-applications adopted (reflecting 'complexity' according to the discussion above). Further, the applications are reported in order of assumed association with standardized production.

There is no relationship between the present adoption rate and the reported year of introduction (first year of adoption). Not surprisingly, this means that the path and speed of diffusion is varying between different applications. Nor is it surprising that applications adopted only by a small number of plants, e.g. PBB and FMS, are primarily adopted by plants using several different applications. Here it should be noted that 53 % of the adopting plants belong to the first category, i.e. 1-3 applications adopted; 31 % belong to the second category, i.e. 4-6 applications; and 16 % belong to the third category with more than six applications adopted.

Table 12.4 IT-applications in the production process. First year of adoption and adoption aates. Per cent.

	IT-APPLICATION	FIRST YEAR OF ADOPTION	ADOPTION RATE	ADOPTING PLANTS BY NUMBER OF IT-APPLICATIONS ADOPTED:		
				1-3	4-6	>6
1.	Numerical control (NC) and Direct numerical control (DNC)	1965	32.7	33	41	27
2.	Computerized numerical control (CNC)	1970	35.0	37	36	27
3.	Industrial robots (IRb)	1968	16.0	26	34	40
4.	Production with limited manning (PBB)	1974	11.0	7	38	55
5.	Automated materials handling (AMH)	1964	9.1	17	33	50
6.	Materials resource planning (MRP)	1965	38.8	27	47	26
7.	Production planning (PP)	1965	17.5	13	52	35
8.	Computer-aided testing (CAT)	1965	21.7	28	28	44
9.	Computer-aided design (CAD)	1970	32.3	35	34	31
10.	Computer-aided manufacturing (CAM)	1970	32.3	22	48	29
11.	Flexible manufacturing systems (FMS)	1975	9.5	4	32	64

Since adopting and developing a computer-based system is normally a piecemeal process, we should expect to find the initiating plants among those who have adopted several applications. Looking at individual applications, we find that 6 have been introduced (first adopted) by plants using more than 6 applications in the year of observation, that 5 have been introduced by plants using 4-6 applications in the year of observation, and, consequently, that none has been introduced by plants using 1-3 applications in the year of observation. Table 12.5 summarizes the information in terms of the average year of adoption, with regard to the year of introduction. From this table we can conclude that a large (small) number of IT-applications adopted is a good indicator of whether the plant is generally an early (late) adopter.

Table 12.5 Average year of adopting IT-applications by number of applications. Per cent.

AVERAGE YEAR OF ADOPTION	NUMBER OF IT-APPLICATIONS ADOPTED		
	1-3	4-6	>6
Less than 15 years after introduction	20	45	74
More than 15 years after introduction	80	55	26
	100	100	100

Finally, table 12.6 presents a summary description of the characteristics to be included in the models.

Table 12.6 Average characteristics of the sampled plants.

	MEAN	MINIMUM	MAXIMUM
LOCATION CHARACTERISTICS			
Total population in A-region (POP)	807548	42204	1536095
PLANT CHARACTERISTICS			
Total number of employees (EMP)	211	5	8000
Number of university engineers (UNE)	17	0	800
Number of employees with non-technical university education (UNO)	19	0	560
Share of sales to the major customer, per cent (CUST)	22	0	100
Percentage of plants established before 1970 (OLD)	52		
Percentage of plants affiliated to multiplant enterprises (MPF)	62		
Percentage of plants in Electrical equipment and components (ELE)	27		
Percentage of plants in Instruments (INSTR)	12		
Percentage of plants in Machinery and equipment (MACH)	61		

12.5 EMPIRICAL RESULTS

The specification of the binary logit model in Section 3 follows the normal assumption of a function that is linear in parameters and additive in variables. Given such a functional form, the variables included could incorporate more or less complex transformations and interactions of the characteristics presented in Table 12.6. For one reason, there are probably significant interaction effects between location and plant characteristics and among plant characteristics. In the work reported in this paper we have not elaborated on these aspects. The model applied uses the characteristics as presented in Table 12.6, transformed only for computational reasons, and no interaction effects are included.

Since the two characteristics indicating the number of employees with higher education, UNE and UNO, and the total number of employees, EMP are highly correlated, about 0.7 - 0.8, each of these characteristics has been tested separately. The models estimated include all the characteristics in Table 12.6 except MACH, which is used as a reference industry. It appears that total employment, EMP, gives significant estimates (at the 5 % level) for all applications but two, whereas the number of university engineers, UNE, is a significant factor in only two cases. Hence, the estimated models presented in Table 12.7 refer to the specification where total employment is the characteristic included.

Table 12.7 Estimated coefficients of a logit model of adopting specific IT-applications in the production process. (Asymptotic t-statistics).

CHARACTE-RISTIC:	NC,DNC	CNC	IRb	PBB	AMH	MRP
Constant	**-0.574**	**-1.255**	**-3.521**	**-2.376**	**-5.026**	**-1.704**
	(-2.19)	(-5.52)	(-6.60)	(-4.88)	(-6.72)	(-5.48)
POP*10^{-7}	**4.726**	-1.926	**-6.349**	4.599	-5.076	-2.516
	(2.98)	(-1.19)	(-2.52)	(1.66)	(-1.44)	(-1.48)
EMP*10^{-4}	**25.380**	**43.074**	**16.464**	**12.524**	7.111	**32.631**
	(4.31)	(6.03)	(3.47)	(3.31)	(1.79)	(4.32)
CUST	**-0.983**	**0.985**	0.469	**-2.446**	0.648	-0.815
	(-2.25)	(2.35)	(0.67)	(-2.68)	(0.78)	(-1.61)
OLD	-0.294	**-1.188**	**-0.764**	**-1.333**	0.896	0.072
	(-1.49)	(-5.42)	(-2.17)	(-3.46)	(1.86)	(0.32)
MPF	-0.366	-0.124	**2.096**	**0.782**	**1.266**	**1.192**
	(-1.71)	(-0.56)	(4.98)	(2.32)	(2.31)	(5.06)
ELE	**-1.000**	**-1.255**	**0.792**	-0.295	**1.360**	0.237
	(-4.64)	(-5.52)	(2.37)	(-0.88)	(2.68)	(1.01)
INSTR	**-0.654**	**-1.027**	0.516	-0.302	1.149	-0.045
	(-2.17)	(-3.09)	(1.03)	(-0.62)	(1.65)	(-0.13)
2[log L(b) - log (0)]	140.4	178.8	553.0	219.2	679.1	282.0
RHO-SQUARE	0.17	0.21	0.65	0.60	0.80	0.33

Bold type indicates significance at the 5 % level.

Table 12.7 (continued).

CHARACTE-RISTIC:	PP	CAT	CAD	CAM	FMS
Constant	**-3.074** (-7.54)	**-4.746** (-9.66)	**-1.504** (-4.68)	**-1.471** (-4.63)	**-2.968** (-5.07)
POP*10^{-7}	**6.020** (2.66)	**11.546** (4.43)	-3.085 (-1.72)	0.295 (0.17)	**8.223** (2.60)
EMP*10^{-4}	5.332 (1.72)	**21.587** (4.00)	**27.544** (4.58)	**20.398** (3.74)	1.982 (0.87)
CUST	0.043 (0.08)	**1.103** (2.21)	**-1.639** (-2.88)	**-1.531** (-2.75)	**-5.345** (-3.62)
OLD	0.216 (0.82)	0.003 (0.01)	**-0.511** (-2.10)	-0.286 (-1.23)	-0.129 (-0.37
MPF	**0.608** (2.20)	0.429 (1.49)	**0.899** (3.69)	**1.046** (4.38)	**0.738** (2.02)
ELE	0.216 (0.80)	**1.721** (5.82)	**0.527** (2.18)	-0.403 (-1.65)	0.150 (0.43)
INSTR	-0.581 (-1.23)	**1.232** (3.17)	**0.889** (2.72)	-0.311 (-0.89)	-0.768 (-1.18)
2[log L(b) - log L(0)]	408.4	414.8	310.4	293.8	577.6
RHO-SQUARE	0.48	0.50	0.37	0.35	0.68

Bold type indicates significance at the 5 % level.

The IT-applications in Table 12.7 are reported in order of assumed association with standardized production from NC to FMS, as in Table 12.4. If this assumption is correct, we can conclude that the estimated influence of the regional population generally supports our hypothesis; from NC, DNC to MRP population has a significant positive influence in only one case (NC, DNC), while the influence is negative in four cases (CNC, IRb, AMH, and MRP). A reverse pattern applies to IT-applications with assumed low association with standardized production: From Production planning (PP) to FMS regional population has positive, and significant, influence on the adoption probability in three cases (PP, CAT, and FMS), while the influence is negative (although not significant) in one case (CAD).

It should be admitted that the ranking of applications with regard to standardized production is rather poorly underpinned. We have assumed that customer dependence (CUST), is an indicator of standardization. Thus, we should expect the influence of this variable to be positive mainly among NC - MRP and negative among PP - FMS. The result supports this to some extent - with respect to CNC, on one hand, and CAD, CAM, and FMS, on the other - but is not conclusive.

The age of the plant (OLD) appears as a significant characteristic (at the 5 % level) mainly among IT-applications with high standardization ranks. Obviously, CNC, IRb, PBB and CAD are IT-applications more probably adopted in relatively new plants. The affiliation to a multi-plant enterprise (MPF) has the expected positive influence in most cases. Evidently, for some applications, the adoption probability varies by industry, e.g., computer-aided testing (CAT) is an IT-application of relevance primarily for plants producing electrical equipment and components or instruments.

According to the estimates, the number of employees with higher education has generally less influence on adoption probabilities than plant size, as was mentioned above. However, the meaning of this should be qualified. Let us take one specific example - the number of university engineers and its influence on the probability of adopting CAM. First, it should be stated that 80 % of all plants in the sample have less than 10 university engineers employed, and among these we find 65 % of all plants adopting CAM. More specifically, among the 40 % employing no university engineers we find 27 % of all plants adopting CAM, and they are mainly located in Stockholm.

This noteworthy observation does not mean that technical skills and knowledge are irrelevant in Stockholm while relevant elsewhere. It probably means that plants located in Stockholm can have these services provided by the enterprise, if affiliated to a multi-plant enterprise, or can buy them externally. To point out that enterprises involved in diffusing IT or acting as consultants on IT are concentrated to the Stockholm region is, in this context, to point to one important aspect of the urban hierarchy.

Next, Table 12.8 presents the results from estimating models of adopting computer-based systems of varying complexity, as measured by the number of applications adopted. High, medium, and low complexity are defined as more than 6 applications, 4-6 applications, and 1-3 applications, respectively. Three models are reported; model 1 including only the number of university engineers (UNE), and alternative specific constants; model 2 includes UNE and model 3 includes EMP in its place, but they are otherwise similarly specified. The characteristics appear as alternative specific variables with regard to high (H), medium (M), and low (L) complexity, with no applications being the reference alternative.

Table 12.8 Estimated coefficients of logit models of adopting computer-based systems of varying complexity. (Asymptotic t-statistics).

CHARACTERISTIC:	MODEL 1	MODEL 2	MODEL 3
Constant (H)	**-1.795** (-9.22)	**-5.126** (-7.26)	**-5.083** (-7.01)
Constant (M)	**-0.436** (-3.54)	**-1.233** (-4.21)	**-1.221** (-4.14)
Constant (L)	**0.413** (3.92)	0.406 (1.81)	0.386 (1.73)
POP*10^{-7}(H)	-	**11.649** (3.47)	**12.113** (3.35)
POP*10^{-7}(M)	-	**6.697** (3.31)	**6.073** (3.00)
POP*10^{-7}(L)	-	2.521 (1.52)	1.880 (1.15)
UNE*10^{-6}(H)	**2.086** (3.28)	0.449 (1.16)	-
UNE*10^{-6}(M)	-0.026 (-0.30)	-1.219 (-1.52)	-

Table 12.8 (continued)

	MODEL 1	MODEL 2	MODEL 3
UNE*10^{-6}(L)	**-4.501** (-3.25)	**-4.317** (-3.02)	-
EMP*10^{-4}(H)	-	-	**48.575** (4.27)
EMP*10^{-4}(M)	-	-	**34.490** (3.13)
EMP*10^{-4}(L)	-	-	19.755 (1.74)
OLD(H)	-	**0.981** (2.23)	0.595 (1.30)
OLD(M)	-	**-0.676** (-2.75)	**-0.815** (-3.25)
OLD(L)	-	**-1.087** (-5.22)	**-1.137** (-5.48)
MPF(H)	- (4.88)	**2.530** (3.23)	**1.773**
MPF(M)	- (4.36)	**1.132** (2.22)	**0.607**
MPF(L)	- (2.37)	**0.538** (0.31)	0.073
2[log L(b) - log L(0)]	230.5	335.4	352.0
RHO-SQUARE	0.14	0.20	0.21

Bold type indicates significance at the 5 % level.

The estimates of model 1 show that the probability of adopting computer-based systems of high and low complexity is influenced (positively and negatively, respectively) by the number of university engineers employed. The result coincides with the very meaning of complexity. However, the additional characteristics of model 2 dissolve this relation with regard to systems of high complexity. It appears that urban agglomeration, and affiliation to a multi-plant enterprise, are the influential factors in this respect - implying that plants with a large number of university engineers are mainly located in agglomerated areas. That 80 % of all plants in the sample employing more than 10 university engineers are located in the Stockholm region, of course, reflects the regional distribution of plant sizes in the sample. Replacing university engineers with total employment, in model 3, slightly reduces the influence of agglomeration and affiliation to a multi-plant enterprise. The age of the plant (OLD) influences the adoption probabilities, in a way that could be expected, when the piecemal process of developing a computer-based system is taken into account.

The general impression from the results reported in Table 12.8 confirms our expectations; advanced system orientation is a phenomenon mainly found in Stockholm and other agglomerated areas, where the requirements for technical skills and producers' services can be most easily met.

As stated in Chapter 3, we do not expect to find a general hierarchical diffusion pattern with regard to IT-components in the products. Including all relevant characteristics, the results of the two models estimated are reported in Table 12.9.

Table 12.9 Estimated coefficients of logit models of adopting IT-components in the products. (Asymptotic t-statistics).

CHARACTERISTIC:	MODEL 1	MODEL 2
Constant	0.263	0.289
	(1.08)	(1.17)
$POP*10^{-7}$	**-3.302**	**-3.330**
	(-2.25)	(-2.27)
$EMP*10^{-4}$	-	**18.184**
	(3.00)	
$UNE*10^{-6}$	**2.197**	-
	(2.65)	
CUST	**-1.583**	**-1.602**
	(-3.95)	(-3.91)
OLD	**-0.587**	**-0.663**
	(-3.19)	(-3.49)
MPE	**0.652**	**0.557**
	(3.43)	(2.82)
ELE	**0.632**	**0.603**
	(3.26)	(3.10)
INSTR	0.162	0.179
	(0.57)	(0.64)
2[log L(b) - log L(0)]	82.8	85.5
RHO-SQUARE	0.10	0.10

Bold type indicates significance at the 5 % level.

Obviously, controlling for plant size (or number of university engineers), industry, age, and type of plant, the results indicate that the agglomeration factor has a significant negative influence on the probability of adopting IT-components in the products. According to Table 12.9 the typical characteristics of a plant using IT-components in its products are the following: Large size and a number of university engineers employed, newly established at a non-central location, affiliated to a multiplant enterprise and producing electrical equipment sold on a large market (many customers).

The first year of adopting IT-components reported in our data is 1960, and the adoption rate by 1986 is 56 %. Since 'IT-components' refer to simple as well as advanced electronics, a somewhat closer examination is called for. Data include reports on the estimated proportion of the product value attributed to the content of electronics and software. Further, information is given as to whether the software is mainly developed by the enterprise or mainly bought from external sources.

It appears that there is a positive relation between agglomeration, early adoption and the relative value of the IT content. There is yet no relation between agglomeration and the internal development of software. Regressing the IT-value share on the same characteristics as in models 1 and 2 above, it turns out that the agglomeration factor (regional population) and the number of university engineers are the only characteristics that are important in a statistical sense.

Assuming that a high IT-value content reflects a product of high complexity and low standardization, the picture that emerges supports our expectations. That is, advanced use of IT-components is a phenomenon mainly found in plants involved in dynamic competition, located in agglomerated areas and with a high level of technical skills available. The common use of IT-components is, however, mainly found in large branch plants, in peripheral locations, involved in price competition.

Finally, as for IT-applications in administration and management, the results can be briefly summarized. With a model specification similar to that in Table 12.7, the following results were obtained. Controlling for plant size, etc., the agglomeration factor (regional population) has significant influence on adoption probabilities in 7 cases out of 9, the exceptions being accounting and budgeting. The plant size proves to have an influence in all cases but for computerized accounting, while the number of employees with higher education (UNO) is of significant influence only for adopting computer support in distribution/logistics. Being affiliated to a multi- plant enterprise appears to exert an influence on the adoption probabilities in all 9 cases. By and large the results are in line with our expectations.

REFERENCES.

Anderstig, C., and B. Hårsman, 1986, "On Occupation Structure and Location Pattern in the Stockholm Region", *Regional Science and Urban Economics* 16, 97-122.

Berry, B.J.L.1972, "Hierarchical Diffusion: The Basis of Development Filtering and Spread in a System of Growth Centers", in Hansen . Ed., 1972, 108-38.

Brown, L.A., and K.R. Cox, 1971, "Empirical Regularities in the Diffusion of Innovation", *Annals of Association of American Geographer* 61, 551-9.

Chamberlain, N.W., ed., 1969, *Contemporary Economic Issues,* Richard D. Irwin Inc., Homewood Hills, Illinois.

Cohen, Y.S., 1972, "Diffusion of an Innovation in an Urban System: the Spread of planned Regional Shopping Centers in the United States 1949-1968", University of Chicago, Dept. of Geography, *Research Paper No. 140*, Chicago.

Feller, I., 1971, "The Urban Location of U.S. Inventions 1860-1910", *Explorations in Economic History* 8, 285-303.

Gibbs, D.C. and A. Edwards, 1985, "The Diffusion of New Production Innovations in British Industry", in Thwaites and Oakey, eds., 1985, 132-163.

Gibbs, D.C. and A.T. Thwaites, 1985, "The Location and Potential Mobility of Research and Development Activity: A Regional Perspective", Paper delivered at the 25th European Congress of the Regional Science Association, Budapest.

Gillesspie, A., ed., 1983, *Technological Change and Regional Development,* Pion, London.

Goddard, J.B., and D. Morris, 1976, "The Communication Factor in Office", *Progress in Planning* 6, 1-80.

Hansen, N.M., ed., 1972, *Growth Centers in Regional Economic Development*, The Free Press, New York.

Harris, C.D., 1954, "The Market as a Factor in the Localization of Industry in the United States", *Annals of the Association of American Geographers* 44, 315-48.

Hudson, J.C., 1969, "Diffusion in a Central Place System", *Geographical Analysis* 1, 45-8.

Helleiner, F.M. and W. Stöhr, eds., 1974, *Spatial Aspects of the Development Process*, Proceedings of the Commission on Regional Aspects of Development of the International Geographical Union, vol. 2, Allister Typesetting and Graphics, Toronto.

Hägerstrand, T., 1952, *The Propagation of Innovation Waves*, Lund Studies in Geography, Ser. B, Human Geography, No. 4, Gleerup, Lund.

Hägerstrand, T., 1967, *Innovation Diffusion as a Spatial Process* (translated and postscript by A. R. Pred), University of Chicago Press, Chicago.

Jacobs, Jane, 1969, *The Economy of Cities*, Random House, New York.

Johansson, B., and C. Karlsson, 1986, "Industrial Applications of Information Technology: Speed of Introduction and Labour Force Competence", in Nijkamp ed., 1986, 401-28.

Karlsson, C., 1987, "The Introduction of New Technology: The Role of Labour Force Competence in the Introduction of Applications of Information Technology", *Working Paper* 87:1, Research Team for Regional Science, University of Karlstad.

Lasuen, J.R., 1973, "Urbanisation and Development - The Temporal Interaction between Geographical and Sectoral Clusters", *Urban Studies*, 10, 163-88.

Mansfield, E. 1968, *The Economics of Technological Change,* W. W. Norton & Company, Inc., New York.

Mansfield, E., et.al., 1971, *Research and Innovation in the Modern Corporation*, Macmillan, London.

Martin, F., et.al., 1979, *The Inter-regional Diffusion of Innovations in Canada*, Economic Council of Canada, Hull, Quebec

McNeill, W.H., 1963, *The Rise of the West*, University of Chicago Press, Chicago.

Moseley, M.J., 1974, *Growth Centers in Spatial Planning*, Pergamon Press, Oxford.

Nijkamp, P., ed., 1986, *Technological Change, Employment and Spatial Dynamics*, Springer-Verlag, Berlin.

Oakey, R.P., 1979, "Distribution of Significant British Innovations", University of Newcastle, CURDS, Department of Geography, *Discussion Paper No. 25*.

Pedersen, P.O., 1970, "Innovation Diffusion within and between National Urban Systems", *Geographical Analysis* 2, 203-54.

Perloff, H.S. and L. Vingo, eds., 1968, *Issues in Urban Economics*, The Johns Hopkins University Press, Baltimore, Md.

Perroux, F., 1955, "Note sur la Notion de 'Pôle de Croissance'", *Economie Appliquée* 8, 307-20.

Pred, A.R., 1966, *Spatial Dynamics of US Urban Industrial Growth 1800-1914: Intepretive and Theoretical Essays*, The M.I.T. Press, Cambridge, Mass. .

Pred, A.R., 1974, *Major Job-providing Organizations and Systems of Cities*, Association of American Geographers, Washington, D.C.

Rees, J., R. Briggs, and D. Hicks, eds., 1985, "New Technology in the United States' Machinery Industry: Trends and Implications" in Thwaites and Oakey, eds., 1985, 164-194

Schon, D.A.,1963, *Displacement of Concepts*, Tavistock, New York.

Thomas, M.D., 1974, "Structural Change and Regional Industrial Development", in Helleiner and Stöhr, eds., 1974, 39-71.

Thompson, W.R., 1965, *A Preface to Urban Economics*, Johns Hopkins Press, Baltimore.

Thompson, W.R., 1968, "Internal and External Factors in the Development of Urban Economics", in Perloff and Vingo, eds., 1968, 43-62.

Thompson, W.R., 1969, "The Economic Base of Urban Problems", in Chamberlain ed., 1-47.

Thwaites, A.T., 1978, "Technological Change, Mobile Plants, and Regional Development", *Regional Studies* 12, 445-61.

Thwaites, A.T., 1983, "The Employment Implications of Technological Change in a Regional Context", in Gillespie, ed. 36-53.

Thwaites, A.T., and R.P. Oakey, eds., 1985, *The Regional Economic Impact of Technological Change*, Frances Pinter, London.
Thwaites, A.T., R.P. Oakey and P.A. Nash, eds., 1981, *Industrial Innovation and Regional Development*, Final Report to the Department of the Environment, CURDS, University of Newcastle upon Tyne.
Vernon, R., 1966, "International Investment and International Trade in the Product Cycle", *Quarterly Journal of Economics* 80, 190-207.
Vernon, R., 1971, *Sovereignty at Bay*, Penguin, Harmondsworth.

CHAPTER 13

Innovative Behaviour of Industrial Firms: Results From a Dutch Empirical Study

Evert Jan Davelaar and Peter Nijkamp

13.1 PROLOGUE

In recent years innovation research has increasingly been linked to product life cycle analysis. The revival of the 'long waves' debate, the increasing interest in Schumpeterian economic dynamics and the analysis of portfolio cycles reflect the view that new 'technical regimes' are to be regarded as the vehicles of qualitative structural changes and of (quantitatively and qualitatively) more favourable employment conditions.

Clearly, this observation also has important regional dimensions. Regions are to a large extent the 'workfloor' of industrial innovations; they may provide the platform for new activities, but at the same time also embody obstacles to new evolutionary patterns. Consequently, an unequal regional innovation potential will - in the medium and long run - have dramatic impacts on the dispersion of new industrial activities within and between nations, as can already be observed in various countries like the USA, Great Britain, Germany and France (see, among others, Malecki, 1979; Oakey, 1981; and Aydalot, 1984). Thus the regional component of technological dynamism deserves closer attention.

In this context, an important analysis framework for connecting innovative behaviour with economic fluctuations in a spatial setting is offered by the innovation-incubation (I-I) hypothesis (see e.g. Davelaar and Nijkamp, 1987). The I-I hypothesis takes for granted that central (or urban) regions tend to specialize in generating new products (seen from a macro-economic viewpoint), whereas non-central areas tend to specialize in the (often large-scale) production of (mature or standard) commodities that were originally designed and introduced in the centre. Consequently, non-central areas tend to specialize more in process innovations than in product innovations. Clearly, if this I-I hypothesis were true, the result would be a geographical specialization of different types of commodities.

Such geographical patterns have been studied from different angles:

a) *sectoral*. Sector studies analyse the regional distribution of specific sectors that are regarded as important candidates for industrial innovations (e.g. micro-electronics, aircraft, good processing, etc.) (see, among others, De Jong and Lambooy, 1984; Alders and De Ruyter, 1985). By comparing the spatial concentration of such sectors with the spatial pattern of related innovations (e.g., via a location quotient method), the degree of sectoral innovativeness of regions can be gauged.

b) *entrepreneurial*. Entrepreneurial studies deal with the micro features of specific individual firms (e.g., the type of R&D expenditures, the use of venture capital, etc.) (cf. Kleinknecht and Mouwen, 1985; Kok et al., 1985). Also the achievements of such firms (in terms of output) are usually taken into consideration.

c) *regional.* Such studies focus attention on the production environment and the production structure (offered by a given region), which are relevant for favouring industrial innovations. Thus the latter type of approach mainly addresses itself to the external conditions for innovative behaviour (see e.g., Nijkamp, 1987).

The approach adopted in the present paper is a blend of a), b) and c), though with a major emphasis on b). In particular, we will focus attention on the identification of the (relative) importance of specific features of firms with regard to various types of innovations and the regional dimensions thereof. This analysis will be undertaken from the viewpoint of the above-mentioned I-I- hypothesis.

13.2 ANALYSIS FRAMEWORK

The analysis of our I-I hypothesis will be based on a relatively simple geographical demarcation of a country, viz. central, intermediate (or half-way) and peripheral areas. Clearly, more refined geographical subdivisions may be used (see e.g., Nijkamp, 1987), but in that case one sometimes runs the risk of almost empty cells for observations on various types of regions and of firms. Furthermore, this trichotomy also has the advantage of incorporating relevant geographical scale elements in terms of agglomeration and social overhead capital (see Davelaar and Nijkamp, 1986).

Our case study for the Netherlands defines the above-mentioned trichotomy of central, intermediate and peripheral areas on the basis of standard statistical areas (so-called COROP areas) which together make up the above-mentioned 3 types of regions. Thus now we will test the I-I hypothesis that firms in central areas are more specialized in highly qualified product innovations, while firms in intermediate and peripheral areas are more specialized in process innovations.

For a small country like the Netherlands it is, however, often assumed that the regional dimension is not strongly prevailent in industrial innovation patterns. For instance, Pred (1977) has claimed that a densely populated country like the Netherlands may mainly be regarded as a (fairly uniform) urban field. Whether or not such a uniformity is a valid assumption will be further investigated in our case study.

Our case study is based on a postal survey of approx. 1800 industrial enterprises in the Netherlands, for many details see also Kleinknecht, 1987). One of the questions concerned the number of product innovations which were new to the firm and which were introduced in the base year (1983). The regional distribution of these product innovations is presented in Table 13.11 (process innovations exhibit and analogous picture).

Table 13.1 Regional distribution of product innovations (1983) that are new to the firm

	Type of region			
Number of firms	Peripheral	Intermediate	Central	Total
with product innovation	335	404	358	1097
without product innovations	219	296	230	745
total	554	700	588	1842

Addendum: $\chi^2 = 1.6$; degrees of freedom (d.o.f.) = 2.

The above table seems to support the assumption that the regional dimension is lacking in the innovation patterns in the Netherlands, so that the above-mentioned I-I hypothesis is most likely not very plausible. However, before we reach this conclusion, a few relevant remarks have to be make here.

(a) The use of cross-sectional data on regional innovativeness ignores the fact that various different types of innovation may have been considered, each of them being positioned in different phases of a space-time trajectory. For instance, if we have two innovations (A and B) each of them being first generated in a central area but at different time periods, then at time t we may face a situation where innovation A may hardly any longer be adopted in the centre but almost exclusively in intermediate and peripheral areas, whereas innovation B may still be mainly introduced in the centre. This incongruent space-time behaviour is pictured in Figure 1. Thus, in principle, it might be possible that the regular spatial distribution of innovations from Table 13.1 conceals a qualitative difference in the creation and adoption rate of different types of innovation. Thus Table 13.1 measures in fact more the global diffusion pattern of innovations than the creativity pattern (which more closely relates to the I-I- hypothesis).

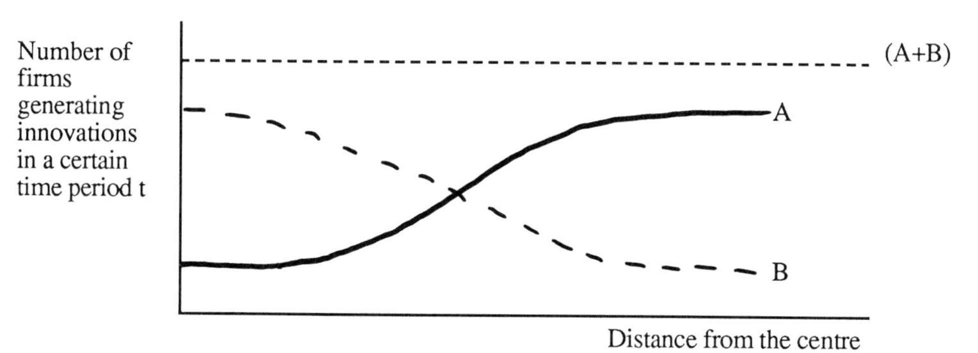

Figure 13.1 Space-time trajectories of two types of innovations

(b) Furthermore, it should also be noted that global innovation data often overlook the difference between product innovations, process innovations and mixed innovations (leaving aside organizational innovations). This is particularly relevant in the framework of our I-I hypothesis, which assumes that the centre tends to specialize in qualitatively high-valued product innovations, while remaining areas may concentrate on qualitatively highly-valued process innovations.

(c) Finally, bivariate methods (like those presented in Table 13.1) are not doubt useful, but also limited in scope, as such two-way tables aggregate over other often relevant categories of explanatory variables of innovative behaviour, such as R&D expenditures, firm size, market structure etc (see also, in general, on this issue, Bishop et al., 1977).

Thus, our conclusion is that a test of the I-I hypothesis has to be done with great care. In order to take account of the foregoing remarks, a distinction has to be made between the *creative* aspect (the invention and introduction of new technologies) and the *adoptive* aspect (the spatial and sectoral diffusion of innovations). In order to avoid a wicked analysis of a cross-sectional dataset, we have made the following subdivision of levels of innovativeness:

(1) *Diffusion of innovations*. At this level it is important to identify the features and driving forces of firms that have introduced (product and/or processes) innovations which are new for the firm concerned, but *not* for the sector the firm belongs to. This will be called the *adoptive* level.

(2) *General introduction of innovations*. At this level the main interest is in whether firms have an innovative attitude, irrespective of the question whether this innovation is new for the firm or for the sector as a whole. This will be called the *generic* level.

(3) *Sectoral introduction of innovations*. This level is concerned with the question whether a certain innovation is entirely new for the sector (and hence also for the firm) or not. This will be referred to here as the *creative* level.

It is evident that the creative level in particular is of great importance for testing the I-I hypothesis, as here we may plausibly assume the validity of the innovation cycle theory that central regions will specialize in product innovations of level 3, while remaining regions will specialize in process innovations of level 3. If instead of Table 13.1 we now construct a new table in which the number of firms which have introduced an innovation that is new to the sector, are considered, we get Table 13.2 (process innovations have a similar pattern).

Table 13.2 Regional distribution of product innovations (1983) that are new to the sector

	Type of region			
Number of firms	Peripheral	Intermediate	Central	Total
with product innovation	445	567	475	1479
without product innovations	99	133	113	345
total	554	700	588	1842

Addendum: $\chi^2 = 0.53$; d.o.f. = 2.

Table 13.2 still seems to contradict our I-I hypothesis. Two remarks are however in order here.

First, Table 13.2 does not provide any insight into the quality, type and socio-economic impact of these innovations, so that the indigenous geographical pattern of these innovations might still differ. And secondly, it would be plausible to assume that, in general, the average quality, type and impact of innovations (in a sector, a firm, or a region) is co-determined by the level of R&D expenditures. Thus, R&D expenditures may provide a key explanation of the *qualitative* pattern of innovations across sectors and regions. This will, however, require the use of a multivariate analysis, in which multiple variables are simultaneously taken into consideration as explanatory variables. The variables selected here are:

(a) The *market* the firm operates on. Here a distinction is made between old and new sectors. Old-line sectors are defined as relatively traditional industrial activities which are mainly found in branches with SBI-codes 20-27.[1] In this respect, we may also introduce notions from product life cycles and innovation cycles (see also Abernathy and Utterback, 1978, and Rothwell and Zegveld, 1985). It is then

[1]SBI-codes refer to the official Dutch Standard Statistical Classification of economic sectors, made by the Central Bureau of Statistics.

assumed that old-line sectors will focus more attention on process innovations, while new-line sectors will specialize more in product innovations.

(b) The *size* of a firm. In this context, a distinction has been made between medium size and large firms (with more than 100 employees) and small firms.

(c) The existence of *external* R&D expenditures. Here the question is whether a firm has an entirely independent R&D division or whether it is (also) dependent on extramural R&D research. It seems plausible to assume the the socio-economic impact of innovations will be higher, as the level of external R&D is higher.

(d) The presence of an *internal* R&D division. Here 3 possibilities exist:
- the firm has its own R&D division;
- the firm has no own R&D division, but has at least R&D activities undertaken by other divisions;
- the firm has no R&D activities.

The foregoing explanatory variables will now be dealt with in the framework of a logit analysis. Separate analyses are carried out for product and process innovations. In both cases, it is not the *number* of innovations that is regarded as relevant, but only at which of the 3 above-mentioned levels a firm has innovated (see also Hansen, 1986).

Out logit model is treated and estimated according to the guidelines of Bishop et al. (1977) (i.e., the sum of total parameters over the various categories of all main and interaction effects is equal to zero). The logit model is particularly suitable for our purposes, as it identifies the relevance of various explanatory (categorical) variables in combination with other variables of the model. The estimation procedure itself will not be treated here. In all cases, the 'optimal' mode appeared to fall in between a model with all the main effects and a model with all first-order interaction effects.

The results are presented in Appendices 1 and 2. The dependent variable is in all cases the firm which is introducing innovations at one of the three above-mentioned levels. In the explanatory part we have used the main effects with respect to the above-mentioned explanatory variables and those with respect to the above-mentioned regional subdivision. In all cases, interaction effects have only been included if they led to a significant improvement of the statistical fit. The results will be discussed in more detail in Section 3.

13.3 RESULTS

13.3.1 Main Effects

The results with regard to the main effects of our logit analysis will be discussed here in a concise manner.

(1) The effect of old and new sectors show a mutually contrasting pattern for product and process innovations. In case of *product* innovations there is a positive relationship between new sectors and the third (creative) level of innovations, whereas in case of *process* innovations the old sectors then become increasingly important. Thus, old sectors are more oriented toward the creation of process innovations, while new sectors are more oriented toward the creation of product innovations. This result supports the above-mentioned *innovation cycle hypothesis*.

(2) If we compare the *adoptive* level with the *creative* level, then we find that, for both product and process innovation, the effect of *external R&D* is greater at the creative level than at the adoptive level. In addition, the importance of external R&D is - for equal levels of innovation - greater for the adoption/creation of *process* innovations than that of *product* innovations.

(3) In general, the importance of *large firms* increases with a rise in the level of innovations. Especially for process innovation the impact of large firms is evident, but with respect to product innovations small firms appear to keep pace with large firms.

However, at the third (creative) level, large firms appear to perform significantly better than small firms, both for product and process innovations. On the other hand, small firms appear to score relatively well with regard to the adoptive (diffusion) level of innovations.

(4) The significance of a firm having its *own R&D division* (and its own R&D activities) appears to be greater at the creative level than at the adoptive level. Thus, internal R&D expenditures are especially important for a firm, if it wants to generate (and *not* to adopt) new innovations. But the role of these R&D expenditures is less relevant for process innovations than for product innovations.

13.3.2 Interaction Effects

In the present subsection the most significant results of our logit analysis for the first-order effects will briefly be discussed.

(1) The *sector* x *firm size* effect shows that *small* firms operating in *old* sectors have a poor performance for all 3 levels and both types of innovation (product and process). *large* firms in *old* sectors, on the other hand, are relatively successful in terms of the adoption of innovations, and, in particular, for process innovations, *Small* firms in *new* sectors score reasonably well for all 3 levels of production innovations, but are relatively weak with regard to process innovations. Finally, *large* firms in *new* sectors appear to be relatively strong at the creative level of product innovations (but weaker at all levels of process innovations.)

(2) The internal *R&D* x *external R&D* effect appears to discriminate only weakly at both the adoptive and the generic level of process innovations. Apparently firms with both internal and external R&D activities do not use both activities for the same purposes (viz. the adoption of process innovations), but for different purposes (viz. either the creation of process innovations or development of product innovations).

(3) The *firm size* x *internal R&D* effect shows that small firms with their own R&D division perform very well at all levels of product innovations, especially at the adoptive and generic level. At the creative level of product innovations, those small firms lag slightly behind compared to large firms with their own R&D division, but score equally in comparison with large firms without an R&D division but with R&D activities elsewhere in the company. Surprisingly however, small firms with their own R&D division perform much worse for all levels of process innovations. Large firms have a much stronger position here, viz. there is a high score for the creative process innovations of large firms with their own R&D division, and a high score for adoptive and generic process innovations of large firms without a special R&D division but with R&D activities elsewhere in the company. Thus, at the creative level of both product and process innovations, large firms with their own R&D division appear to perform well. In terms of product innovations the position of small firms with their own R&D division is also relatively favourable, while in terms of process innovations large firms with R&D activities elsewhere in the company perform well.

13.3.3 Regional Aspects of Product Innovations

Having discussed in general terms various important dimensions of innovative behaviour, we will now turn to a test of the above-mentioned I-I hypothesis by paying attention to the regional dimensions of innovations at the creative level. As this hypothesis assumes a specialization of product innovations in the centre and of process innovations elsewhere, both types of regions will be dealt with successively (in subsection 3.3 and 3.4, respectively), by using again the above-mentioned logit model for multivariate analysis.

For each of the 3 types of region a separate logit model has been estimated. A surprising outcome in all separate model results appeared to be the existence of regional differences in the parameter related to the explanatory variable 'own R&D division', especially with regard to creative product innovations. The general pattern of this parameter appeared to exhibit a geographical pattern indicated by Figure 13.2.

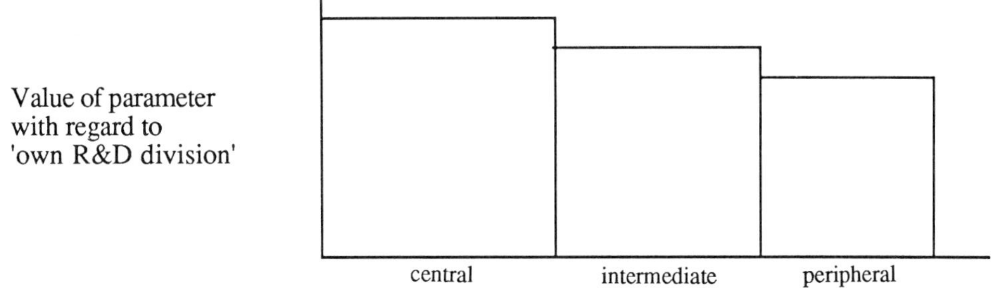

Figure 13.2 Regional differences in the effect of 'own R&D division' on creative product innovations

This picture shows that the presence of an 'own R&D division' in a firm - as mentioned before -, an important condition for product innovations at the third (creative) level - is more significant in the centre than elsewhere. This result supports an I-I hypothesis that the centre is marked by a concentration of creative and *qualitatively high-valued product* innovations, while this degree of specialization declines acording to a distance-decay function from the central zone outward.

The statistical significance of these results has to be considered carefully. The region x own R&D division effects measures the degree to which, in the centre, the parameter related to 'own R&D division' differs from that in the periphery. The results appear to be significant at an 8% level (see also Appendix 1).

Another interaction effect concerns the sector x region dimension. Here it turns out that in the intermediate zones the role of new sectors is more important than in the periphery, and that the periphery scores higher in terms of old sectors. Also in this case the regional effect appears to be significant (8%).

In conclusion, the first part of our I-I hypothesis appears to be confirmed by the results of the logit analysis for product innovations at the creative (third) level.

13.3.4 Regional Aspects of Process Innovations

As mentioned above, the second part of the I-I hypothesis takes for granted that the centre will not have a dominant position as an incubation place for new process innovations.

First we will show some interesting results from the analysis of the region x external R&D effect in our logit analysis. The geographical pattern of the parameter related to 'external R&D' is presented in Figure 13.3.

It has already been shown above that the use of 'external R&D' is an important condition for process innovations at the creative level. This - together with figure 13.3 - supports our hypothesis that intermediate regions (as to a lesser extent peripheral areas) specialize in the creation of *qualitatively* highly-valued process innovations. The results of this analysis appear to be clearly statistically significant (3%) (see also Appendix 2 with respect to variable hulp).

If we add the (significant) region effect to the previous model for main effects, then we may conclude that non-central areas have an overall higher propensity to carry out creative process innovations.

In conclusion: the regional pattern of process innovation is in agreement with the expected pattern emerging from the I-I hypothesis. The intermediate zones appear to score relatively high in terms of creative *high quality* process innovations (followed by the periphery), while the periphery performs reasonably well in terms of propensity to innovate. In both cases, the centre has a much lower performance.

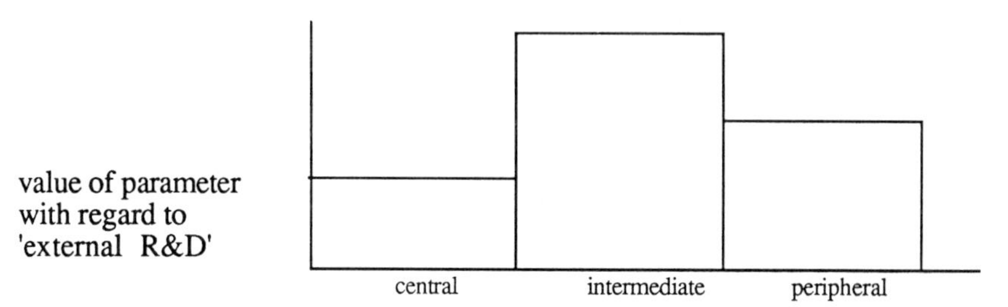

Figure 13.3 Regional differences in the effect of 'external R&D' on creative process innovations

13.4 CONCLUSION

Our analysis has identified - by means of a multivariate logit analysis - important driving forces of innovative behaviour. The role of internal (own) R&D for product innovations appeared to be very important at all 3 levels of innovation (adoptive, generic and creative). However, in the case of process innovations external R&D appeared to be of great importance.

The results for various types of firms (e.g., small vs. large, old-line vs. new-line industries) appeared to vary with the type of innovation (product vs. process) and the level of innovation. For instance, small firms with their own R&D division appear to have excellent results at all 3 levels of product innovations, but poor results at various levels of process innovations. Furthermore, large firms with their own R&D division prove to perform very well at the creative level of both product and process innovations.

These results have clearly important implications for industrial and regional innovation policy. For instance, if one wants to stimulate product innovations at level 3, one has to orient policy measures to a different set of firms as compared to a stimulation of generic innovations. The difference between creation and diffusion (leading to 3 levels of innovation) turns out to be of great importance in the regional dimensions of innovative behaviour. Especially in the context of the I-I hypothesis, the creative level of innvation is of paramount relevance here. Despite the cross-section data, our test has brought to light the existence of a space-time trajectory of innovations which is closely connected with product life cycle theory. Thus the central zone appears to generate more (high quality) creative product innovations than any other zone, while the non-central zone (mainly the intermediate zone followed by the peripheral zone) appears to perform reasonably well in terms of creative (high-quality) process innovations. The peripheral zone as such has a higher overall tendency to introduce process innovations than the central zone. In conclusion, the I-I hypothesis - despite some seemingly paradoxical evidence provided by a superficial glance at overall empirical data on innovations - has some relevance at the level of a small country like the Netherlands.

APPENDIX 1 Estimated Logit Models with Respect to *Product* Innovations

Variable	Level 1		Level 2		Level 3	
	Estim.	Signific.	Estim.	Signific.	Estim.	Signific.
REG1	-0,082	-	-0,0217	-	-0,034	-
REG2	0,12	-	0,1113	-	-0,02	-
BRAN	10,12	1,86	-0,0847	-1,42	-0,2425	-3,4
ERD1	0,144	1,6	0,1833	3,04	0,21	3,13
WERK	10,1576	1,5	0,0497	0,5	-0,1365	-1,96
RDA1	0,8597	6,76	1,27	11,3	0,9689	7,75
RDA2	0,5057	4,16	0,68	6,85	0,7064	6,32
BRAN1 x WERK1	-0,2015	1%	-0,1277	1%		
WERK1 x RDA1	0,08	3%	0,089	6%		
WERK1 x RDA2	-0,2423		-0,2023			
REG1 x RDA1					0,2098	8%
REG2 x RDA1					0,04	-
BRAN1 x REG2					-0,1548	8%
BRAN1 x REG1					0,016	

Legenda

REG1 = effect with respect to central zone
REG 2 = effect with respect to intermediate zone
BRAN1 = effect with respect to 'old line industries'
ERD1 = effect with respect to external R&D
WERK1 = effect with respect to small (w≤100) firms
RDA1 = effect with respect to 'own R&D* department
RDA2 = effect with respect to R&D by other divisions

* With respect to all the interaction effects and the main effects REG1 and REG2, we compared the improvement in the log-likelihood ratio with the change in the degrees of freedom. With respect to the (remaining) main effects, we mentioned the ratio between the estimated coefficient and its standard error (this ratio is asymptotic normally distributed).

APPENDIX 2 Estimated Logit Models with respect to *Process* Innovations

Variable	Level 1		Level 2		Level 3	
	Estim.	Signific.	Estim.	Signific.	Estim.	Signific.
REG1	0,054	-	-0,03	-	-0,1586	6%
REG2	0,0117	-	0,063	-	-0,0462	-
BRAN 1	0,01846	0,34	0,0746	1,32	0,1402	1,71
ERD1	0,1915	2,53	0,3363	4,9	0,3257	3,85
WERK1	-0,066	-0,96	-0,2371	-4,9	-0,3894	-4,49
RDA1	0,1321	1,22	0,3594	3,82	0,6424	4,73
RDA2	0,2734	2,98	0,2839	3,46	0,3212	2,66
ERD1 x RDA1	-0,0587	} 5%	-0,0757	} 5%		
ERD1 x RDA2	-0,2193		-0,1964			
HULP					0,2116	

REFERENCES

Abernathy, W.J. and J.M. Utterback,1978, "Patterns of Industrial Innovation", *Technology Review* 80.

Alders, B.C.M. and P.A.. de Ruijter, 1985, "Kansrijke Economische Activiteiten" in W.T.M. Molle, ed., *Innovatie en Regio*, pp. 33-53,Staatsuitgeverij, Den Haag.

Aydalot, Ph., 1984, "Reversals of Spatial Trends in French Industry since 1974", in J.G. Lambooy, ed., *New Spatial Dynamics and Economic Crisis*, pp. 41-61, Finn-publishers, Tampere.

Bishop, Y.M.M., S.E. Fienberg and P.W. Holland, 1977, *Discrete Multivariate Analysis: Theory and Practice*, MIT Press, Massachusetts.

Davelaar, E.J. and P. Nijkamp, 1986, "De Stad als Broedplaats voor Nieuwe Activiteiten: Theorie en Onderzoek, *Stedebouw en Volkshuisvesting*, No. 2, pp. 61-66.

Davelaar, E.J. and P. Nijkamp, 1987, "The Incubator Hypothesis: Old Wine in New Bottles?", in M.M. Fischer and B.M. Sauberer, eds., AMR-INFO, Vienna, 1987, pp. 198-213.

Jong, M.W. de and J.G. Lambooy, 1984, *De Informatica-sector Centraal, Perspectieven voor de Amsterdamse Binnenstad*, EGI Universiteit van Amsterdam.

Hansen, J.A., 1986, *Innovation Characteristics of Industries in the United States*, paper presented at the conference 'Technologie, Arbeid en Economie', Maastricht, 23-24 October 1986.

Kleinknecht, A., 1987, *Industriële Innovatie in Nederland. Een enquête-onderzoek*, Van Gorcum, Assen.

Kleinknecht, A., and A. Mouwen, 1985, "Regionale Innovatie (R&D): Verhuizing naar de 'Halfwegzone'"?, in W.T.M. Molle, *Innovatie en Regio*, pp. 125-142, Staatsuitgeverij, Den Haag.

Kok, J.A.A.M., G.J.D. Offerman and P.H. Pellenbarg, 1985, *Innovatieve Bedrijven in het Grootstedelijk Milieu*, Geografisch Instituut, R.U.G.

Malecki, E.J.,1981, "Locational Trends in R and D and the Location of Technological Activities, *Regional Studies* 219-234.

Nijkamp, P., 1987, *Revitalisation of Regional Resources*, Research Memorandum, Dept. of Economics, Free University, Amsterdam.

Oakey, R.P., 1981, *High Technology Industry and Industrial Location*, Aldershot.

Pred, A., 1977, *City Systems in Advanced Economies*, Hutchingson, London.

Rotwell, R. and W. Zegveld, 1985, *Reindustrialization and Technology*, Longman, Essex.

CHAPTER 14

Innovating Behaviour of Swiss Industry - Findings and Policy Conclusions

Beat Hotz-Hart

14.1 INTRODUCTION

Swiss industry is facing serious challenge. Like all highly industrialized nations, Switzerland must, in the long run, move away from the development of hardware and focus intensively on software. More specifically, there is a need to produce more complex, less standardized products with a higher share of services and know-how, depending less on raw material, environmental goods and capital.

Confronted with this general trend, strategies which focus only on maintaining stability or exploiting established positions can lead companies as well as economies as a whole into serious crisis. They can cope with the massive changes needed only through innovation. To survive there must be a preparedness to change and develop and a general capability to learn and adapt. The promotion and use of creative potential has become one of the most important tasks for management. For a large part of Swiss industry, economic success can only be achieved through *competition by innovation.*

Therefore, the paper concentrates on the following *questions*: What were the typical problems connected with innovative behaviour in Swiss industry during the last decade and what kind of conclusions for economic policy can be drawn from them? This is dealt with, first, by discussing findings from different studies concerning the innovative behaviour of Swiss industry and, secondly, by suggesting starting points for economic policy promoting innovative development.

14.2 EMPIRICAL FINDINGS ON INNOVATING BEHAVIOUR IN SWITZERLAND

Several findings are a point of some concern about the innovation performance of the Swiss economy. An analysis of *patent applications* shows that in Switzerland not only foreign but also domestic applications have markedly decreased since 1977. The number of applications filed in 1985 amounts to only 58% of that in 1977 and with respect to the maximum registered in 1971 only to 52%. This is absolutely unique amongst eight European countries. The strongest decrease was observed in chemistry, then in electricity/physics, whereas in the field of mechanics the position has remained relatively stable (annual report of the "office fédérale de la propriété intellectuelle" 1985, p.10).

Switzerland's share of the *export of high technology products* of seven OECD countries has decreased remarkably in the last two decades. An analysis of groups of high-tech products shows that a structural weakness of the Swiss economy is obvious.

Table 14.1 Various countries' share of the overall export of high technology products of seven OECD countries, including textile and machine tools (in %).

	1965	1970	1975	1980	1985	1986
USA	33.3	32.6	29.7	28.6	29.5	25.8
Japan	9.2	13.6	14.7	18.1	25.3	26.4
FRG	23.4	23.5	23.6	21.8	18.5	20.5
France	9.0	8.8	10.5	10.5	9.3	9.4
UK	15.5	12.6	12.2	13.3	0.9	10.7
Sweden	2.4	2.7	3.1	2.4	2.4	2.4
Switzerland	7.2	6.2	6.2	5.3	4.1	4.8
Total in Percent	100	100	100	100	100	100

Source: Several OECD statistics and own calculations

14.2.1 Problems of Exchange, Linkage and Transition in the Innovation Process

Innovation can be separated into different phases. The whole innovation *process* from generating an idea via diffusion through to full market success has to be taken into account. For successful innovation the economy has to be strong in each of these phases.

The understanding of innovations as a linear process is fairly widespread. The argument normally begins with the generation of new ideas through research, leading to development, production and marketing and finally diffusion and application.

Innovation is, however, not applied science. A linear model is too much of a simplification and does not take into account some crucial aspects of innovation. *Feedbacks and interactions between the phases and activities* are an inherent part of development processes. Successful innovating needs rapid, accurate, precise and adequate feedbacks and appropriate follow-on actions: Innovation is also a process of learning, and making and exploiting experiences (see Kline and Rosenberg, 1986, p.286).

If the characteristics of the process are taken seriously, the connections and linkages, the exchanges and the transition between the phases become crucial for successful innovation. Innovation and diffusion is, to a large extent, *a problem of interface and bottlenecks*. The speed and the form of communication leading to adjustment and a further development of technologies are crucial for the success of innovation.

In the Swiss system of R&D and innovation there is an *inadequate exchange* and linkage between research on the one hand and technological development on the other, and between research in universities and that in industry. The distribution of expenditure between the public sector with 20% and the private sector with 80% of R&D is unique in an international comparison. Traditionally, *public authorities* have been strongly engaged in basic research, significantly less in applied research and hardly at all in development.

In 1986, expenditure on R&D in Swiss industry per annum was distributed one third to research and two thirds to development. Excluding the chemical industry, the distribution was 10% research and 90% development. By far the greatest amount of expenditure was on development.

In 1986, the chemical (45%), machinery and appliance (23%) and electrical industries (25%) accounted for together 92% of all R&D expenditure of Switzerland (see Vorort, 1987). The chemical industry invested strongly in all phases of research and development, as did the machine industry, but on a significantly lower level. The other industries were in practice exclusively engaged in developing technologies; and even here their activities were weak. Thus *t*he activities of the large majority of companies were concentrated on a small area of development which was rather remote from research.

This may include a *danger* that basic, long term changes in technological development are overlooked or dismissed; and that companies focus too much on short and medium term development. As the OECD experts put it with respect to the watch industry "...though Switzerland has acquired a certain scientific and technical lead (...) it was outstripped in the race to apply the results industrially" (OECD, 1984, p.42). Structural ruptures in the economy are building up; as a result of such behaviour structurally caused crises are more likely.

Similar problems can be identified between the *phases of design and industrial exploitation.* In Swiss industry the difficulties of the transformation of technologies into industrial products and procedures and into technological services are often underestimated, and as a result, this transition is undertaken slowly and hesitantly. As an example: "Progress is hampered by difficulties of communication between machine tool manufacturers and users on the one hand and numerical control suppliers on the other" (OECD, 1984, p.43).

There is a similar *weakness in marketing and the introduction of new products in new markets.* The predominant behaviour is characterised by hesitation and low risk taking. Products are launched only when their market success is more or less certain.

The *reasons* for such behaviour can only be postulated. Industrial institutions which are cartellised and corporative favour pragmatic and short-term oriented behaviour. Other constraints can be found in the training system, the information available, the communication processes and the socio-cultural aspects typical of Switzerland, such as a lack of receptiveness to new things and lack of readiness to take risks.

14.2.2 Risk Aversion: Focussing on Rationalisation and Cost Reduction

Companies are the driving force of economic development. A crucial determinant of company behaviour is the assessment of the future and it is, therefore, connected with insecurity and project evaluation. Decisions are taken on the basis of expected profits and the propensity to take risks. The degree of risk aversion, including the company's attitude towards risks and failures, is of great importance for economic development.

Several studies of innovative behaviour in Swiss companies come to the conclusion that *rationalisation*, i.e. efforts to reduce costs, is predominant, followed by efforts to improve the quality of already existing products. Substantial innovations in products or procedures lie far behind. A study by the Batelle Institute (see Hieronymi et al., 1983) summarizes: there is a *typical Swiss mixture of caution with respect to new basic technologies and an acceptance and open readiness to use technology for rationalisation.* And the OECD registers "a lack of preparedness to innovate rather than an unwillingness to adopt new technologies" (OECD, 1984, p.44).

Focussing only on a strategy of cost reduction can, however, be problematic and dangerous in the long run since, internationally, Swiss companies are likely to face tough competition on costs and fail in the end.

Reasons for such behaviour could be found in surveys of *hindrances of diffusion.* As seen by the companies, these are difficulties in assessing the market properly and realising a profit on investments; reservations about the future of new technologies and an insecurity about the structural change connected with a new technology. Many Swiss industrial companies are *risk averse* with respect to research and development and innovation.

Predominant is a strong wish of entrepreneurs to be independent and a *l*ack of integration and cooperation between companies in the area of R&D. There seems to be a sometimes unfounded sense of security among companies about their own market position, and the particular market niche they operate in. In addition, a lack of training on all levels of the companies seems to be obvious.

Besides these hindrances there are also some stimuli. First of all, the high intensity of competition, particularly for companies operating in an international context, provides a positive stimulus for the main objectives as well as for technology diffusion. The character of Switzerland as a small and open economy, together with the pressure to compete and to perform, supports technology diffusion.

14.2.3 Low Innovative Behaviour in Small and Medium Sized Enterprises

In Switzerland, the size of companies has been found to be an important discriminating factor in innovative behaviour. Concerning the size of companies, surveys show that in Switzerland small and medium sized enterprises are on average less innovative than larger enterprises in comparable situations (see, for example, the survey results of Arend and Stucky, 1984). This might be seen to confirm Schumpeter's hypothesis of a positive relationship between innovative behaviour and monopoly power (see also Mansfield, 1986).

However, the Swiss findings are not fully convincing and need further research. A company survey in 1986 showed that the share of their turnover companies spend on R&D varies according to their size, forming a U-shaped curve: Small companies with less than 50 employees spend 10% of their turnover on R&D. This ratio declines to 3% for those with 1000-1999 employees and reaches its peak of 12% with the companies larger than 5000 employees (Vorort, 1987, p.18). The empirical findings suggest the following explanation:

- Confronted with fast technological change, small and medium sized companies have difficulties in acquiring existing *knowledge and information* adequate to their needs. They have problems in transforming and adjusting such knowledge to their particular purpose without becoming dependent on suppliers of investment goods. The necessity for market research and effective market observation and penetration is often beyond their capabilities.

- A small or medium sized company cannot by itself cover all the phases of an innovation process. Normally they are located in the later phase with insufficient connections and exchanges to the earlier phases.

- They *lack planning and medium-to-long term orientation* in their company policy. SME are guided by habits, and tried and tested methods. They lack information, for example, on new markets and the courage to take risks. The time available is scarce. The demands on the local managerial elite increase and a work overload is likely.

- A further hindrance is the *lack of willingness to cooperate*. Cooperation is well developed in areas such as marketing, sales management and production, but lacking in areas such as the exchange of technologies and experiences or research and development (see Müller, Goldberger, 1986). A mixture of pride and faint-heartedness, which results in retreat and redimensioning, seems to be typical. "Better to go bust independently than to survive in cooperation".

- There is a lack of adequately *qualified personnel* to use and exploit the potential of new technologies (see Nydegger, Oberhänsli, 1983). SME are known to have problems in financing the training of their own personnel. Particularly in the service sector, large companies train in a much more substantial way and small ones take over by pursuing an aggressive recruitment policy.

14.2.4 Sub-Optimal Framework of Rules for Company Foundation

Experience from different countries shows that innovations are often achieved by start-up companies, by the founding of new companies and/or intra-preneurship. They exploit innovations, sometimes they are even the innovator; for example the launching of new companies by university leavers, fellows etc. who try to commercialise inventions made in university laboratories; intra-preneurship, that is spin-offs of large companies by engineers or staff who want to commercialise ideas or experiences outside the established company and become independent.

Both cases can be illustrated in Switzerland. There is room to improve the general conditions for the foundation of new companies and spin-offs in Switzerland. The particular constraints they face are a large set of rules to be fulfilled which are put forward not only by the state but also by interest associations and professional associations within the economic sector. Each attempt to found and develop a new company has to cope successfully with complex rules concerning choice of location, the construction and security of buildings, technical standards, security standards for employees and consumers, labour law, and so on. The question of deregulation has arisen in connection with innovation.

Further reasons for the low degree of innovation by successful start-ups can be seen in the area of information and training. Most entrepreneurs who try to found a company are technicians or engineers. They often lack adequate knowledge in business affairs. Disincentives are caused by the socio-cultural traditions. Because of the basically conservative character of the Swiss, the average risks taken are only small; a failure caused by high risk-taking is negatively sanctioned in social terms. This goes together with the rather conservative assessment of risks by bankers in respect of granting credits.

14.2.5 Increasing Gap of Qualifications Available and Needed in Order to Innovate

Innovative behaviour and diffusion are to a large extent dependent on the situation and development of the factor markets. Problems of financing, such as the acquisition of risk capital, are not dealt with in this paper; however, some crucial aspects of the labour market will be discussed.

According to the employment statistics (census of establishments and enterprises) there is a *substantial shortage of qualified personnel*, particularly of skilled engineers and technicians compared with the needs of the economy. With respect to the number of employees in 1987 (third quarter) this shortage is found in banking and finance (85.8%), insurance (74.2%), electrical engineering (75%) and machinery and appliances (70%). With an average of 60%, the shortage in industry was more substantial than in the service sector. At the same time, there is a surplus of inadequately trained personnel. The Swiss labour market is characterised by structural bottlenecks in specific qualifications particularly with regard to the qualifications needed for the use, service and exploitation of new technologies.

The labour market situation is causing limits to the growth of the Swiss economy and of innovation activities. Bottlenecks in qualified personnel have substantially delayed a more intensive use of computer aided activities in the economy; a modernizing and rejuvenation is needed in order to stay competitive. It could be said that there is *a know-how frontier of economic development.* One of the consequences might even be that R&D is transferred abroad because of the lack of domestic personnel. In 1986 the chemical industry undertook 49% of its R&D abroad and this tendency is on the increase.

Methods of retraining personnel and adjusting qualifications, which were usual up to now, such as "learning by doing" have become insufficient. The economy cannot cope with new technology through qualification on the job only. There has to be an increase in the share of professional training done by schools, which will not be confronted by the immediate pressures of production and profitability. There is a tendency for more of the training to be carried out outside production (see Balmer, Gonon and Straumann, 1986).

At the moment the adequate educational structures required to fulfil the need for the further development of qualifications are lacking. An adequate professional access to new technologies is rare and there are substantial regional differences.

Within the professional training system in Switzerland there is not sufficient mobility and permeability with respect to choice of profession, access to jobs in connection with electronics and career patterns. In traditional industrial regions the one-sided structure of sectors and companies offers only limited and unattractive opportunities for training and retraining and therefore for professional mobility. The results are commuting or migration. Personnel managers say that, on average, only 1 to 2% of the total salary figure of an industrial company is spent on training. This should be contrasted with the 5% considered desirable.

At the moment many different bodies and institutions are taking separate initiatives. Further development is dependent on the supply of training through private institutions (such as private schools, in-house training) and public schools (e.g. in the area of professional training). The situation in the area of professional training has become chaotic and confusing. Certification is unclear, screening does not take place sufficiently. An improved coordination among instructors themselves and between them and the clients is needed.

The adjustment of qualifications is constrained by institutional factors such as interest associations and public authorities which grant the official acknowledgement of apprenticeship and training certificates. Institutional rules have also resulted in rigidity in salary classes within and between sectors and companies. The defence of achieved standards dominates the Swiss labour market and labour costs have become rather fixed.

14.3 CONSEQUENCES OF AN INNOVATION-ORIENTED SWISS ECONOMIC POLICY

There is no separate and clear cut innovation policy because such a policy involves complex coordination processes and requires an assessment of the impact policies and different policy authorities have on economic innovations. Different policies have to build up a stimulating environment for change and a climate-supporting innovation.

The problems and weaknesses with respect to innovation pointed out in the last chapter will be the starting point for the presentation of policy measures. The suggestions made are interdependent. The policy mix has to be consistent, balanced and coordinated.

14.3.1 Transfer Policy: Improving the Exchange between the Phases of Innovation

The concept that technology is transferred in a finalized state from one phase to another is inadequate. Transfer always includes change and adjustment, a further development of the original idea or technology. Therefore, the transfer of technology and knowledge is always a process of learning, and accumulating and using experiences. This process itself produces something new. Both academia and the economy can benefit from such a process. Therefore, transfer policy has to be seen as the *organisation and promotion of a process of communication, learning, and exchange of information and personnel between different phases of innovation and different groups.*

Within Swiss industry the linkage between the phases of innovation, in particular between the early phases of research and their application, has been identified as a particular weakness. State authorities can ease and accelerate the adjustment of structural change by promoting research which can be readily applied. Swiss authorities should widen their traditionally strong activities in support of basic research to cover *development and training which has a closer practical and commercial application.* This effort might be focussed on particular groups (such as SMEs), in particular fields of technology (such as computer integrated manufacturing (CIM)) and promoting international cooperation in R&D. Joint task forces of state and private business should be established.

The success of public programmes depends to a large extent on whether an effective contact with the companies concerned can be achieved. In order to do so, constraints have to be overcome such as lack of information, bureaucratic barriers to applications for support, and managerial problems. Policy authorities have to take the initiative to address potential clients. In Switzerland particular difficulties are caused by the sceptical attitude of Swiss entrepreneurs towards any kind of state involvement which could undermine their autonomy. This is the case even though there is an obvious need for technology transfer. Technology transfer is also a problem of education and political climate.

The conclusions which can be drawn from Swiss experiences are that state intervention should be avoided. Instead of financial support the state should *offer real services* such as training, information and consultation, and help to initiate cooperative arrangements in order to accelerate innovation processes. State authorities should initiate and implement the ideas together with the private economy and withdraw after a certain time. Such an approach is based on the corporate tradition of Swiss policy.

Because empirical findings show that the majority of technology transfer originates from competitors, customers and suppliers, the transfer between companies has to be improved. Well functioning alliances of companies of different size and operating on different levels of technology support innovative development. Networks in the sense of synergetic interaction structures (see Stöhr, 1986) use a Swiss tradition and offer a fair chance for economic success. In this respect, interest associations can play an important role. Transfer means first of all *cooperation.* Second in line is the transfer through recruitment i.e. the *transfer of personnel* in order to transfer ideas. In third place only is the *intervention of state authorities.*

Suggestions have been recently of installing *"technology parks"* in some parts of Switzerland (Zürich, Bern, Lausanne) in order to foster technology transfer particularly through new foundations. Public support can be given with respect to marketing, legal problems and in business affairs. Experience shows that the help of consulting bodies in smoothing all kind of contacts with public authorities is much appreciated.

14.3.2 Innovation Promotion

Generally speaking incentives have to be established for entrepreneurs or companies to take on more risks more quickly. This depends on the general framework of laws and rules (for example, of the fiscal and the patent laws) but also on socio-cultural aspects such as the attitude towards risks and the social esteem of low risk aversion.

In Switzerland there are 13 regionally based innovation consulting institutions. Their main activities involve consultation in the areas of product innovation, process innovation (particularly robotics and handling), materials, and patents and licences. Their help in solving the bureaucratic problems entailed in submitting applications for state support or permission seems to be very effective. Roughly half of the consultation is on behalf of

SME. The consulting institutions can support the assessment of new market developments and marketing (for example, through contacts and fairs). They are most effective when they are closely connected and located with engineering schools.

14.3.3 Promotion of SME to Increase their Innovations

Technology policy oriented towards SME has to supply information, consultation, financial and training services. This necessitates the establishment of an adequate infrastructure, and the promotion of a regional network of relations between companies themselves and the institutions providing these services. Empirical studies show that these services have to be presented and realized in a very pragmatic way which is easily applicable to and useful for the particular problems companies have.

Consultations should not, however, be oriented only towards technology and engineering. There is a danger that such a narrow view is predominant when most consultancy is undertaken only on a very short term basis, which, in reality, seems to be mostly the case. More basic questions should also be tackled, including ecological aspects such as energy saving and a reduction in pollution. Publicly supported consulting offers a fair chance of initiating and pioneering ideas such as qualitative economic growth. Innovation-oriented consulting should go beyond short-term contacts and be involved in a company's development for a longer period of time.

SME have substantial information problems. They not only lack risk capital but also *"risky ideas"*. Therefore, consulting could also help to design attractive policy alternatives. Public authorities, together with interest associations, could help or organise seminars and meetings for SME staff members, which would provide contacts and stimulation and open cooperation. Regular meetings in order to exchange experiences between SME staff might also be useful.

In Switzerland *new forms of cooperation* between SME offer an opportunity to improve company performance and should be taken more into account. In the field of R&D there is some discussion about the foundation of institutes for joint research to fulfil the needs of SME or of associations to promote particular sectors such as a foundation to support SME in the initiating phase particularly with risk capital (Müller, Goldberger, 1986; Boehme, 1986).

Particular efforts on behalf of SME have to be undertaken in the area of *training and retraining* for their staff. This is a field where interest associations supported by state authorities could do more. Some polytechnics already run these kinds of courses quite successfully. The transfer of personnel could be initiated and supported by joint projects between polytechnics and SME and by the organisation of special classes at local schools and training centres.

14.3.4 Deregulation

Particular policy attention has to be given to the identification and removal of constraints on new foundations and intra-preneurship and on the development of SME, particularly in their initiating phase. Such constraints could be eased through specific deregulations: simplifying bureaucratic procedures and rationalising state regulations. In the case of Switzerland single items on such an agenda are the simplification and adjustment of the tax system in a way which supports innovation; rationalising personnel policy through the amendment of labour law; reducing the rules and regulations in the construction and planning law and its administrative procedures, particularly with respect to the permission procedures for new buildings or the conversion of buildings, and amending health and safety regulations without reducing the degree of protection (see Walser, Hunckeler, 1986, p.25).

Deregulation is not only directed towards state authorities and their rules but also and to quite a substantial degree towards the rules of the private interest associations such as standards of market access, procedures for licenses and labour market certificates. The long standing tradition of corporatism and cartels in the Swiss economy has, in this respect, a constraining impact on innovation.

14.3.5 Improving Qualifications and Redesigning Work

To a large extent technology transfer takes place through the *transfer of personnel*. Therefore, the labour market fulfils the important role of transmission in the diffusion process. Training policy is crucial for promoting innovation. The qualifications needed by innovative companies have to be developed.

The know-how limits to growth in the Swiss economy should be tackled through programmes such as an extension of the training schemes; introductory classes to specific areas in the field of high tech (for example, information technology, CIM); offers of consultation with respect to training and career planning; programmes for retraining elderly craftsmen. Education and training has to be improved in a qualitative sense: information-oriented sciences have to be introduced in the curriculum of schools at different levels. In professional schools the share of management training or training in business affairs has to be increased. Public schools such as polytechnics have to extend their supply of further training and retraining in a variety of forms. A change in structure and curriculum of universities and polytechnics is needed. Last but not least, incentives have to be given to invest more in training.

A particular problem is presented by the unpredictability of the need for new qualifications. Therefore, training has to be *a mixture of a specialised, craft-oriented training and a more general training to improve flexibility*. Training activities have to be adjusted periodically to meet structural economic change. To support labour to cope with structural change, its flexibility, mobility, initiative and innovativeness has to be improved. This should be accompanied by the state taking over the social costs caused by economic change.

The need of further training and retraining can arise unexpectedly and has to be fulfilled quickly. The Swiss federal government has recently launched programmes to base general economic management policy on structural types of measures focussed on complementary training to supplement rather than replace existing skills, including a national software school and a national training centre for information management in business. The programmes have served to remedy certain short term deficiencies in the education system fairly rapidly.

To sum up: The purpose of the paper was to stress the characteristics of innovation as a complex process including linkages, exchanges, feedbacks and appropriate follow-on actions. The idea was to make clear the need for a well functioning network of organisations covering all kinds of complementary functions. This could lead to a new look at the division, organisation and control of labour leaving the tradition of Taylorism behind and moving towards what was called more recently "flexible specialisation".

All in all, an economic polity which favours an innovative development needs the coordination of different policy areas and institutions. A consistent policy mix dependent on the ideosyncratic problems of the particular sector of the economy has to be achieved. In Switzerland, firstly, the general framework of laws and rules has to be improved through the reduction of injunctions, hindrances and constraints to innovation and the promotion of new institutional forms of cooperation particularly between SME. Secondly, innovation and technology transfer have to be promoted and investment in human capital and risk capital has to be provided. Policy has to make sure that social hardship does not arise or at least that, in the case of hardship, assistance is available. Because *innovation is first of all a social event in a complex social network.*

REFERENCES

Arend, M., and B. Stucky, 1984, "Zu den Ursachen räumlicher Innovationsdisparitäten in der Schweiz", in E.A. Brugger, ed., *Regionale Innovationsprozesse und Innovationspolitik*, Diessenhofen.

Balmer, K., Ph. Gonom and M. Straumann, 1986, *Innovation und Qualifikation*, Aarau

Boehme, J., 1986, *Innovationsförderung durch Kooperation*, Berlin.

Brooks, H. ,1982, "Toward an Efficient Public Technology Policy. Criteria and Evidence, in H. Giersch, ed., *Emerging Technologies: Consequences for Economic Growth, Structural Change, and Employment*, Tübingen.

Brugger, E, and B. Stucky, 1987, "Regional Economic Structure and Innovative Behaviour in Switzerland", in *Regional Studies*, Vol. 21.3, pp. 241-254.

Freeman, Ch., 1982, *The Economics of Industrial Innovation*, 2nd edition, London.

Hotz-Hart, B., 1987, "Diffusion von Informations- und Kommunikationstechnologien und Regionalentwicklung - Theorie und Thesen am Beispiel der Schweiz", in B. Hotz-Hart, and W. Schmid, eds., *Neue Informationstechnologien und Regionalentwicklung*, Zürich.

Hieronymi, O., A. Gabus, H. Hattemer and E. Sallin, 1983, *La diffusion de nouvelle technologies en Suisse*, Saint Saphorin,

Kline, S.J. and N. Rosenberg, 1986, "An Overview of Innovation", in Landau, R. and N. Rosenberg, eds., *The Positive Sum Strategy, Harnessing Technology for Economic Growth*, Washington D.C.

Landau, R. and N. Rosenberg, eds., *The Positive Sum Strategy, Harnessing Technology for Economic Growth*, Washington D.C.

Mansfield, E., 1986, "Microeconomics of Technological Innovation", in Landau, R. and N. Rosenberg, eds., *The Positive Sum Strategy, Harnessing Technology for Economic Growth*, Washington D.C.

Müller, K. and E. Goldberger, 1986, *Unternehmenskooperation bringt Wettbewerbsvorteile*, Zürich.

Nydegger, A., H. Oberhänsli and R. Harringer, 1983, *Investitionen und Innovationen in kleineren Industriebetrieben*, Diessenhofen.

OECD, 1984, *Economic Surveys - Switzerland 1984/85*, Paris.

Office Fédérale de la Propriété Intelectuelle, ed., 1985, *Annual Report*, Bern.

Schelbert, H., N. Harabj and Ph. Halbherr, eds., 1985, *Schweizerische Volkswirtschaft unter Innovationsdruck*, Bern.

Staudt, E., ed., 1980, *Innovationsförderung und Technologietransfer*, Berlin.

Stöhr, W.B., 1986, "Territorial Innovation Complexes", in Ph. Aydalot, ed., *Milieux innovateurs en Europe*, Groupe de Recherche Européen sur les Milieux Innovateurs (GREMI),, Paris.

Vorort, Schweizerischer Handels- und Industrieverein, 1987, *Forschung und Entwicklung in der schweizerischen Privatwirtschaft 1986*, Zürich.

Walser, R. and U. Hunckeler, 1986, *Die Belastung der Klein- und Mittelbetriebe durch staatliche Regelungen*, Zürich.

Zschaur, E., 1986, "Government Policies for Innovation and Growth", in Landau, R. and N. Rosenberg, eds., *The Positive Sum Strategy, Harnessing Technology for Economic Growth*, Washington D.C.

CHAPTER 15

Infrastructure for Technological Change: Intellectual Property Rights

Derek L. Bosworth and R.A. Wilson

15.1 INTRODUCTION

The patent system is, in certain respects, unique in the context of the law governing monopolies and restrictive trade practices. For various reasons which are explored in more depth below, the government chooses to give a monopoly right to those who produce (and/or own) certain intellectual property. Within this system of protection, the drafting and examination of patent specifications are of paramount importance. As a consequence, the law affords Patent Agents and, in effect, the Patent Office, a privileged position with a view to ensuring that this process is carried out expertly. This paper reports on research undertaken by the authors for the Office of Fair Trading, prior to its ruling (for a more detailed discussion see OFT, 1986; Bosworth and Wilson, 1988).

Section 2 examines the nature of the current system for the protection of intellectual property in the U.K. (which is also typical of other industrialized countries). Section 3 looks at the implications of monopoly power, examining the inter-related goals of the Patent Office and Patent Agents. Section 4 outlines the empirical evidence regarding the monopoly position of Patent Agents, based on two new survey data bases commissioned by the OFT. Finally, Section 5 draws the main conclusions and recommendations based on this study.

15.2 PROTECTION OF INTELLECTUAL PROPERTY

15.2.1 Aims of the Patent System

The traditional stylised view of invention and patent law is the 'reward theory' (Balz, 1975, p.1). In effect, the patent system offers the inventor (or owner of the invention) monopoly rights over their intellectual property in return for the disclosure of information about the invention. The grant of monopoly rights is intended to ensure that the inventor can recoup the associated investment in research and development. Without this protection, imitators could use the invention without incurring these costs and, by implication, would be placed in an advantageous competitive position. The incentive to undertake inventive activity would therefore be seriously reduced in the absence of the patent system and the level of such activity would be sub-optimal from a social welfare point of view.

The patent system is known to fulfil other important roles, one of which is the disclosure of information about the invention to other inventors and to potential users of the invention. The extent to which the inventor wants information to be disclosed will vary from one situation to another, dependent on both the nature of the invention and the objectives of the inventor. The information published in the patent specification might be useful in 'inventing around' the patented idea and, in this instance, disclosure acts as a disincentive to patent. In some other cases, disclosure is a form of informative advertising which will be observed by potential licensees using the regular abstracting and watching services. While patent law stipulates full disclosure of the invention, there is, however, no requirement that the patentee describe the commercially most viable form of the invention. There is clear evidence that a user of patented knowledge must often seek additional information if the inventive idea is to be used successfully (Tookey, 1972).

There are, however, other, less conventional views of the operation of the patent system, which suggest that it fulfils other roles. One such view is known as the 'prospect theory' (Kitch, 1977). In this case, the patent system determines the most adept *potential prospector* in a given area of invention. The patent, in awarding a monopoly right in that area of inventive activity, ensures that inventive resources are used in an efficient manner. In effect, the individual needs to demonstrate the greatest ability to produce the invention and, then, is awarded a monopoly over the production of the invention. This to some degree appears to avoid the common pool problem, where there are competing researchers, only one of which will be awarded monopoly rights (Dasgupta and Stiglitz, 1980; Wright, 1983).

It should be borne in mind that substantial changes in procedure have been brought about by recent patent legislation (e.g. through the *Patents Act, 1977* and the introduction of European patents.) However, the main focus of attention concerns the general principles surrounding the application and renewal stages of a patent's life, and, in the main, we abstract from the detail of the modifications, although reference is made to them where appropriate. Much of the argument developed below applies directly to European patents.

15.2.2 Role of the Patent Office

The Patent Office can control the levels of activity surrounding what it perceives to be its main goals, for a number of reasons: it is a monopolist; it breaks activities down into a large number of detailed categories, and can manipulate the level of activity in each area; it charges separate prices for the services it provides in each of these areas. Examination of the accounts over the post-War period indicates that, on balance, the Patent Office 'broke even' on its patenting activity. This tells us very little about its pricing behaviour *per se* (under different cost and demand conditions, a zero profit may be consistent with either profit maximisation or with output maximisation subject to a zero profit constraint); however, the Patent Office has recorded its, "stated intention to break even over the period 1969-72" (Annual Report, 1970, p.1).

There are a variety of activities going on and the Patent Office does not have to price each and every activity at (average) cost in order to achieve its overall breakeven goal. Indeed, causal empiricism suggests that it is hard to explain the observed structure of fees in this way. In particular, it is difficult to see how, if prices are supposed to reflect underlying average costs: (i) the application and search stage (which involves the bulk of the Patent Office work) can be less expensive to operate than some of the renewal stages; (ii) why later renewal stages should be more expensive than earlier ones (when they all involve broadly the same amount of work)? An obvious response is that this structure of prices is adopted to encourage the application for patents, maximise the throughput of applications, and thereby the use of the patent system and the amount of disclosure. It could also be argued that the Patent Office is raising the levels of renewal fees in an attempt to reduce the life expectancies of patents and, thereby, lower the social welfare costs associated with the award of a patent monopoly. However, it is difficult to sustain the latter argument given the size of the tariffs imposed by the Patent Office. The absolute level of fees therefore seem set to generate the greatest incentive to patent and to invent (*Patents, Designs and Trademarks, 1985*, p.1).

15.2.3 Role of Patent Agents

Under prevailing U.K. law, Registered Patent Agents have a virtual monopoly over the drafting of patent specifications, although the individual inventor (or owner of the invention) may write their own specification and a very limited amount of work is also channelled through Solicitors. The arguments about why this should be the case are well known: a mistake in drawing up the specification, particularly with regard to the nature and limits of the patentee's monopoly claim, can result either in a failure to obtain patent protection or in the subsequent contesting and possible revocation of the monopoly right, with consequent financial losses. Whether Patent Agents fulfil this role adequately or efficiently is an important question. The nature of the service offered and its cost to the patentee have recently come under scrutiny, and it is claimed that both patent Agents and Lawyers "... have restrictive practices which raise the cost of the system to the user" (Nicholson, 1983, p.16). Certainly, the figures quoted are high (*op cit.* p.16).

15.2.4 Broader Issues

Despite the fact that Nicholson (1983, p.1) claims that the procedures necessary for the grant of a patent "... give the impression of an arcane world rather than of a modern technological Britain", the focus of the Report is narrow, effectively limiting itself to relatively minor modifications of the existing system (unlike the work of Taylor and Silberston, 1973). The reason for this narrow scope appears to lie in the author's belief that the U.K. system is part of a broader European and worldwide network which, anyway, could not be dismantled (*op cit.*, p.8) and that the current system works well, at least for the larger companies (*op cit.*, p.1). However, the view of the efficacy of the patent system for larger firms is largely an empirical question. It is true that larger firms may have recourse to employing in-house patenting expertise, but, if supplies of such people are restricted over all, they may be forced to pay higher salaries than otherwise. Likewise, larger firms have to compete for such people in the labour market, and, in doing so, they compete with incomes that could be earnt in private practice.

Nicholson argues that the current system is poorly suited to the needs of the individual inventor and the small business because intellectual property rights are relatively inaccessible, and the mechanisms by which they are secured are complex, costly and time consuming (Nicholson, 1983, p.1). The main concern appears to be with the complexity of the specification and, given the need for expert advice in drafting the specification, with the role of Patent Agents. The complexities of the current specification and of processing a claim (particularly for the small enterprise) lead Nicholson to recommend a two tier patent system (*op cit.*, pp.20-2), although this appears to have been rejected by the government (Cmnd 9712). However, the lower tier of patents would have offered a lower degree of protection for intellectual property, and, would not therefore have been a close substitute for the higher tier, allowing Patent Agents to retain much of their market power. In addition, there are other aspects to this question. Individual inventors and small firms, in particular, may require much more than the drafting services of a Patent Agent. They may require advice on both the technical and commercial merit of the application, assistance in finding help with further technological development, sources of financial help, etc. A Patent Agent may not be the best source of this type of advice and help, and the current, restrictive forms of company organisation within which Patent Agents are forced to operate may not be ideal to expand the range of services on offer.

15.3 MONOPOLY POWER: THE INTERRELATED GOALS AND FEES OF THE PATENT OFFICE AND PATENT AGENTS

15.3.1 Goals and Pricing Behaviour of the Patent Office

The Patent Office is a monopoly supplier of patent protection for inventions. The other forms of protection (such as secrecy) can be looked on as substitute products and their existence will affect the elasticity of the demand curve for patent services. Figure 15.1 shows the cost curve at the application stage to be C (i.e. including a normal per unit profit level of r^o, built into the cost curve). Working on the basis of this cost level, the Patent Office could, in principle, price as a monopolist, setting marginal revenue equal to marginal cost, pricing at P_r* and dealing with Q_r* of applications. In the light of the perceived goals discussed earlier, however, it is more accurate to assume that the Patent Office sets a price of P_q*, with a throughput of applications of Q_q*. The difference between the two goals is associated with Q_q*-Q_r* of applications when costs are at C. However, by means of cross-subsidisation (e.g. from renewals to applications), the Patent Office can treat its cost curve as C' and reduce its price to P_q'. In this manner, a still further Q_q'-Q_q* of patents initially pass through the system. It is an interesting question as to the quality of the additional Q'_q-Q* inventions (both technically and commercially), and the strains that their presence (at least historically) place on the quality of the Patent Office work. However, given both imperfect knowledge and an imperfect capital market for funds to finance expenditures in this area, the excluded group will include those who cannot afford to pay higher fees, irrespective of the quality of the invention. Certainly, in pricing at or below average cost, the patent Office is making a clear decision about casting its net wide.

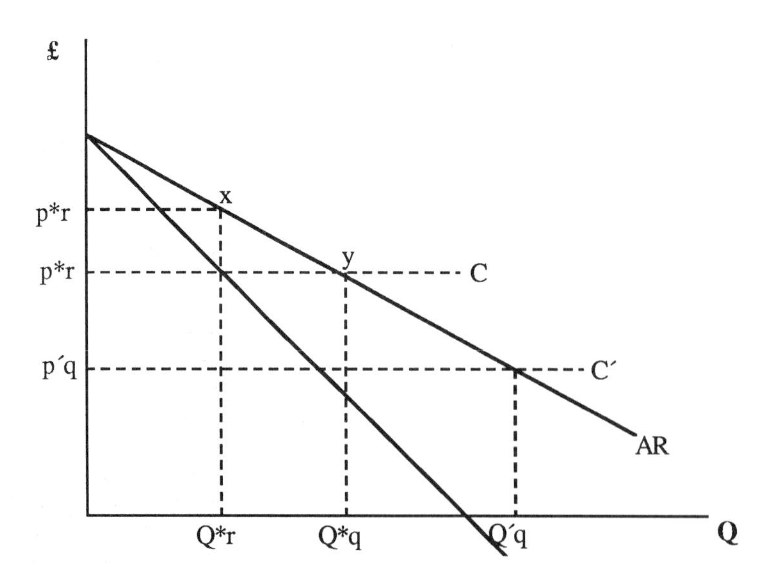

Figure 15.1 Patent office monopoly and discretionary behaviour

Figures 15.2.i and 15.2.ii trace the implication of cross-subsidisation for the renewal of patents. Figure 15.2.i deals with vintage v patents at time t, while Figure 15.2.ii examines the equivalent position for vintage v+1 (i.e. one year older) patents at time t. Typically, because of the reasonably stable numbers of patent grants in each year, the demand curve for vintage v+1 patents at time t lies at a lower level than for vintage v patents in the same year (thus, $D_{v+1,t}$ lies below $D_{v,t}$). Thus, if actual costs can be represented by the average cost curve, C (where, approximately, C = C(v,t) = C(v+1,t)) then renewals at time t would be $Q_r(v,t)^*$ under profit maximising and $Q_q(v,t)^*$ under output maximising behaviour. Likewise, $Q_r(v+1,t)^*$ and $Q_q(v+1,t)^*$ are the equivalent vintage v+1 quantities. However, the Patent Office applies an additional mark-up to costs of renewals of, say, C(v,t)'-C(v,t) to vintage v, and C(v+1,t)'-C(v+1,t) to vintage v+1 patents. Again, assuming output maximising behaviour, the effect is to reduce the levels of activity by $Q_q(v,t)^*-Q_q(v,t)'$ and $Q_q(v+1,t)^*-Q_q(v+1,t)'$ for the two vintages, but particularly the older patents. While this pricing structure is consistent with reducing the allocative welfare losses associated with patent monopoly, the mark-up on renewals appears to be at an insignificant level *vis a vis* the value of the patent monopoly, and, more likely, motivated by the welfare benefits of access and disclosure that arise from the cross-subsidisation of applications.

15.3.2 Goals and Pricing Behaviour of Patent Agents

Evidence drawn from the *Earnings and Professional Standards Survey (EPPS)* provides detailed information about the structure of the Patent Agent profession (see Bosworth and Wilson, 1988). Of the 641 respondents, 6 per cent were employed by the government, 42 per cent by industrial and commercial enterprises, 37 per cent were partners in private practices, and 14 per cent were employees within private practices. We noted above that, although larger firms can employ in-house patenting expertise, in practice, this course of action is limited to groups of companies, rather than large independents. Salaries in large firms and government are likely to be relatively high compared with smaller employers. However, large firms and government will have to compete for such people in the labour market and, in doing so, pay salaries that are sufficiently attractive compared with those than can be earned in private practice. Partners in private practice can be expected to act as profit maximisers, as their salaries will be linked directly to the profits of the firm. The level of profits will depend on the degree of competition in the market for the drafting of patent specifications and this, in turn, depends on the extent to which the supplies of qualified people are restricted by existing practitioners.

15.3.3 Interrelated Goals of the Patent Office and Patent Agents

While the fee structure and goals of the Patent office appear to be set to maximise disclosure and the stimulus to invent, their success in these areas will depend on the degree of monopoly power and on the goals of Patent Agents. Figure 15.3 shows the demand curve for applications for patent protection, D, and the associated marginal revenue curve MR. The analysis focusses on independent Patent Agents (IPAs, i.e. partners in private practice). Patent Office services are treated as an intermediate input. For simplicity, the average cost curve is assumed to be horizontal (i.e. AC=MC). Normal profits (i.e. sufficient to keep Patent Agents in that line of business) are also incorporated within the average cost function. If the Patent office charges the full (unsubsidised) costs of processing applications, the charge per applicant is represented by od'. Under competitive conditions, profit maximising behaviour amongst IPAs would lead to a level of activity Q'_C where the average cost curve, AC', cuts the demand curve, D. Under

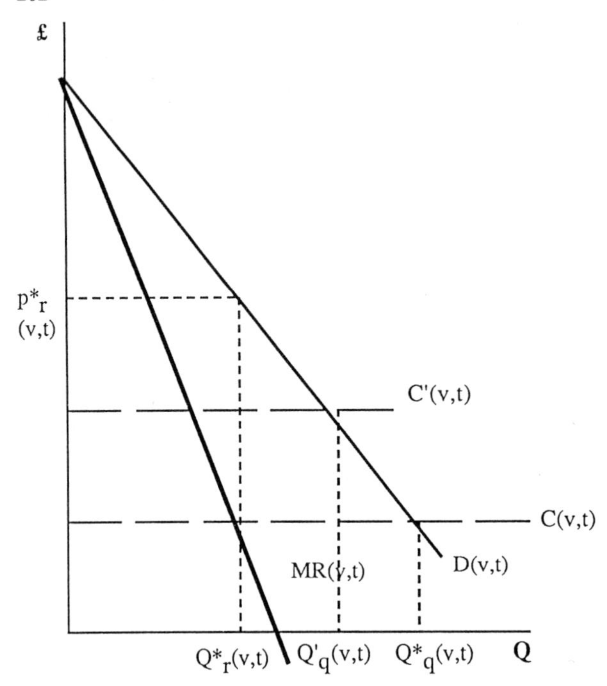

Figure 15.2.i Renewal of vintage v patents at time t

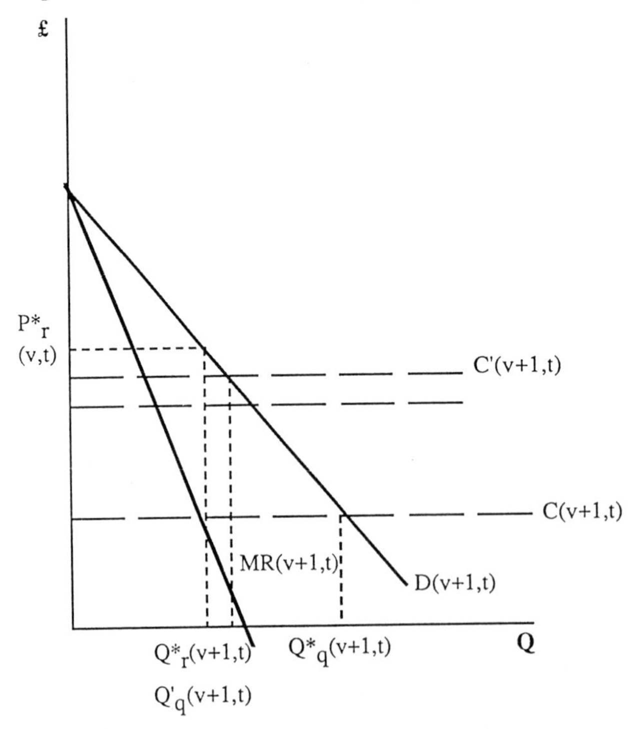

Figure 15.2.ii Renewal of vintage v+1 patents at time t

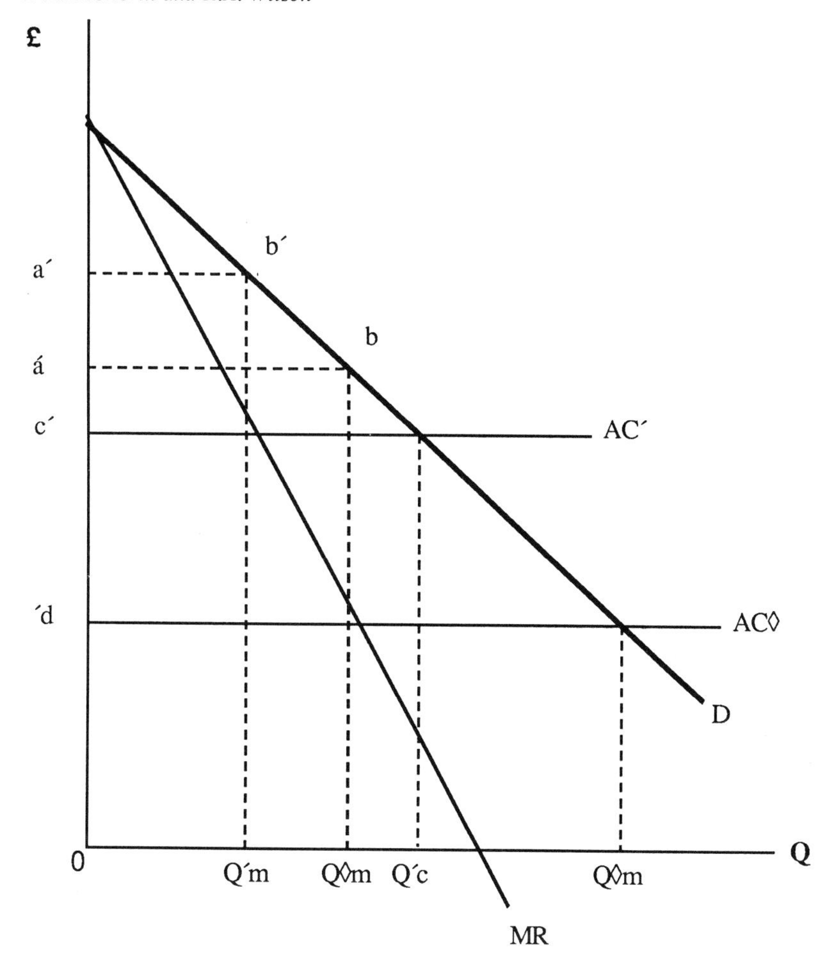

Figure 15.3 Interrelatedness of patent office and patent agent goals

these conditions IPAs earn normal profits. However, IPAs acting in concert, as a monopoly would restrict output to Q'_M and earn abnormal profits a'b'c'd'. If the Patent Office is now assumed to rearrange its fee structure to subsidise applications, then the cost level of the IPA can be represented by A°_C and IPAs make normal profits, while a monopoly industry will move to Q°_M and make abnormal profits of abcd.

There are a number of interesting points to note from this exercise: first that the volume of patent applications is restricted by: the magnitude of Patent Agent costs, their level of normal profits and their degree of monopoly power; second, that the welfare gains to be reaped by the lowering of application costs by the Patent Office are reduced in the presence of monopoly power amongst Patent Agents because only a part of the cost reduction is passed on to the inventor; third, because of the importance of work prior to grant in the Patent Agents' portfolio, the choice of the Patent Office to subsidies the stages of progress through to grant by the revenue from renewals, effectively means that the Patent Agents operate in the most favourable environment possible. This leads to higher profit and income than might be expected under alternative (non-subsidised) Patent Office pricing structures (i.e. abcd>a'b'c'd').

The greater the monopoly power of IPA's, the higher their prices and the greater the residual demand (i.e. that part of the demand for patent applications to the right of point b' or b in Figure 15.3) and the greater the extent of private applications (and, by implication, the lower the propensity to patent). Monopoly power amongst independent Patent Agents can clearly have a major impact on the individual inventor and the small firm sector, which the U.K. government hopes to be a major engine of domestic growth. We have seen that the Patent Office casts its net wide in an attempt to pull in these types of enterprise. With regard to large firms, the consequences appear less clear cut because of the existence of employed Patent Agents. While higher prices in the IPA sector will mean that there will be a tendency to substitute towards in-house patent services, the cost of EPAs is not independent of the profits and earnings to be made in the private sector. The outcome will depend on the ease of substitution between the services of IPAs and EPAs, but also on the mobility of Patent Agents between the two sectors. Equally, restrictions on the supply of Patent Agents in general will tend to raise their incomes in both sectors.

15.4. EMPIRICAL EVIDENCE ABOUT THE PATENT AGENT MONOPOLY

15.4.1 Survey Evidence

There have been few surveys in the patent field in the U.K. The Banks Report (1970) examined a very small sample of patents to ascertain the extent to which they had been used commercially. Taylor and Silberston (1973) report on a survey of 50 major U.K. companies (see particularly pp. 82-3 and 383-404), aimed at the relationship between R&D and patenting, with a view to understanding the underlying costs and benefits of the patent system, when considered against the possibility of replacing it by a 'compulsory licensing system'. Nicholson (1983) does not appear to have collected any survey evidence to test his main hypotheses or to provide support for his main conclusions. The empirical evidence reported below draws on the results of two surveys commissioned by OFT: the Earnings and Professional Practices Survey (*EPPS*) and the Survey of Applicants and Applications (*SAA*).

15.4.2 Structure of the Profession

Other things being equal, we would expect the degree to which IPAs are able to exploit their monopoly position will depend on: (i) the size and structure of the profession; (ii) the market power of their customers: (iii) the experience of their customers, for example, the extent of repeat orders; (iv) they way in which potential customers undertake search and the degree to which the profession undertakes informative advertising. The extent to which Patent Agents possess and exercise market power will be reflected in: (i) the magnitude of charges levied by IPA's for their services; (ii) the success of IPAs in processing patent applications; (iii) customer views on the prices charged and the quality of the service.

The operation of the profession is overseen by the Chartered Institute of Patent Agents. CIPA sets examinations which are a prerequisite for entry to the profession, which now appears to be graduate only. The survey picked up no-one under 35 without a first degree, and indicated that only 16 per cent of the profession as a whole did not have such a qualification. In addition, a wide range of knowledge is required to pass the CIPA examinations, and training takes a further three or four years. Given the magnitude of the potential barriers to entry to the profession that the qualifications represent, the justification for this requires careful scrutiny. Certainly, the way in which entry to the profession is organised appears to be ideal for the operation of restrictions on entry: with degree-only entry and lengthy training periods linked to the limited number of existing practitioners.

The numbers in the profession within the U.K. are small, around 1200. Almost half of the total number operate within the London area and the concentration to London is higher still for those in private practice. The reasons seem likely to be the location of the R&D divisions of companies in London and the South East, coupled with proximity to the Patent Office. It is difficult to draw any hard and fast conclusions about the consequences of this concentration. On the one hand, direct and indirect contact between a considerable proportion of the profession seems likely to be high. In addition, many of the Patent Agents are located physically close to CIPA. On the other hand, for those clients who are willing and able to go to London, the extent of choice is that much larger. If we are willing to accept that concentration in London enhances competition, a natural consequence of this line of reasoning is that it must be reduced in other parts of the country, where Patent Agents are more widely and sparsely distributed. Clearly, however, this is a more important consideration for those clients that are more remote from London and less mobile, particularly individuals and smaller enterprises.

Competition between Patent Agents seems likely to be limited by the segmentation of the profession by class of employment. Thus, the supply to those not in a position to employ 'in-house' services is limited to approximately 621 Patent Agents in private practice: 171 of these 621 are *employees* within private practices and of the remaining 450, 46 per cent are in Central London, 52 percent are in Greater London and 67 per cent are in London of the South East. Many of these individuals belong to a partnership and the number of private practice partnerships is a more crucial indicator of the degree of competition. CIPA evidence shows 168 firms, 50 per cent of which were sole practitioners, 19 per cent were in of size 2-3, 19 per cent were 4-10 and 12 per cent had more than 10 Patent Agents. One practice had 40 Patent Agents (6.4 per cent of the total), the largest 5 had 107 (17 per cent) and the largest 8 had 146 (23.2 per cent). The CIPA figures indicate 629 Patent Agents and, thus, the average number of Patent Agents per practice is just under 4, with the larger of these practices based in London.

15.4.3 Explanation of Earnings

Multivariate regression methods were used to analyse the correlates of earnings, enabling the identification of the separate effects of education and other factors, *ceteris paribus*. Earnings functions have been widely estimated, particularly in the U.S. (Mincer, 1974; Gordon and Williams, 1981; and Wilson, 1987). The methodology enables some light to be thrown on the influences on the earnings of Patent Agents, and the primary aim of the analysis reported in this section is to explain differences in earnings across individuals. However, particular attention is paid to those variables used in explaining earnings which may point to some form of restrictive business practices on the part of Patent Agents. The regressions make use of the individual observations, obtained from the responses to the *EPPS*. However, as the sample covered only Registered Patent Agents, and not the cohort of people with qualifications appropriate to following a course of study that would have enabled them to undertake the CIPA final examination (whether they did so or not), it was not possible to calculate the rate of return to obtaining the CIPA qualification from an earnings function approach.

The overall sample size was sufficiently large to enable a general function to be estimated. The basic model assumes that earnings can be explained by: (i) personal characteristics; (ii) employment characteristics; (iii) pricing behaviour; (iv) attitudinal variables. The inclusion of the latter is entirely *ad hoc*, with no theoretical justification, carried out merely in an attempt to see if there was any link between earnings and attitudes about restrictive practices in the profession.

Table 15.1 Earnings functions: Variables, definitions and expected signs

Category of Variable	Variable Name	Definition	Expected Sign/ Actual Sign	Remarks
Dependent Varible:				
	E	gross earnings	≠	sign column is not relevant for dependent
	F	fringe benefits	≠	variable
Independent Variables:				
(i) Personal Characteristics				
	AGE1	age	+/+	
	AGE2	age squared	-/-	
	EXPA1	experience	+/+	length of time since CIPA qualification
	EXPA2	experience squared	-/-	
	QUAL	level of qualification		1 no formal educational qualifications - set equal to zero
			+/+	2 qualification below degree level
			+/+	3 degree
			+/+	4 post graduate qualificaiton
	SUB	subject of degree	≠	1 chemistry/biochemistry
			≠	2 physics/applied physics/solid state
			≠	3 mechanical engineering
			≠	4 chemical engineering
			≠	5 engineering only
			≠	6 electrical engineering
			≠	7 mathematics only
			≠	8 electronics/electronics communication
			≠	9 metallurgy
			≠	10 natural science
			≠	11 science/international science
			≠	12 not indicated
			≠	13 other specified
	CIPA	member of CIPA	+/+	previously passed CIPA final examination not member set equal to zero
(ii) Employment Characteristics:				
	CLASS	employment class		1 government, nationalised industry, ... - set equal to zero
			≠/≠	2 industry and commerce
			+/+	3 partner in private practice
			≠/-	4 employee in private practice
			≠/-	5 other, not elsewhere specified
	NORPA	number RPAs	+/+	CLASS2 only
	NOSUP	number support staff	+/	in patenting activity
	SIZE1	size of company/group		CLASS2 only/less than 200 employees -
	SIZE2		+/+	CLASS 2 only/200-500 employees
	SIZE3		+/+	CLASS2 only/over 500 employees
	NOPART	number of partners	+/+	CLASS 3 and CLASS 4
	REG	region		1 Central London - set equal to zero
			≠/-	2 Greater London, n.e.s.
			≠/-	3 South East
			-/-	4 Midlands
			-/-	5 Northern England
			-/-	6 UK, n.e.s.
(iii) Pricing Practices:				
	PROP	patenting activity	+/-	proportion of income from patenting
	PRICE1	access to CIPA guidance	-/≠	those with access set equal to zero
	PRICE2	use of CIPA guidance	-/≠	those who follow set equal to zero

Table 15.1 cont.

Category of Variable	Variable Name	Definition	Expected Sign/ Actual Sign	Remarks
(iv) Attitudinal Variables:				
	ATTA1-5	restrict right to practice as Patent Agent, PA, 1977	-/*	1 very desirable .... 5 very undesirable
	ATTB1-5	allow trading as a limited company	+/*	1 very desirable .... 5 very undesirable
	ATTC1-5	allow mixed partnerships	+/*	1 very desirable .... 5 very undesirable
	ATTD1-5	publish CIPA's guidance	+	1 should not be published

Notes: +/- indicates anticipated positive sign/estimated negative sign;
≠/ no *a priori* sign;
/≠ sign not firmly established in regression equation;
* indicates might be expected to differ between classes of Patent Agents (i.e. between independent PAs and employed PAs).

A summary of the explanatory variables tested in the earnings functions are provided in Table 15.1. The table also presents the anticipated signs in the regression estimates. In regressions of this type, individuals with particular characteristics are omitted as the 'base group' or 'standard person', and the coefficients for individuals with other characteristics indicate their position *vis a vis* this base group (i.e. whether earnings are expected to be higher or lower than the standard person). In certain instances, no *a priori* expectations existed about the signs (these are denoted ≠ in the table). In the vast majority of cases, the estimated signs correspond with the anticipated signs (the exceptions, which are quite interesting, are discussed below). Space does not permit a discussion of the sizes of the coefficients, and these are not reported in Table 15.1. However, a more detailed description can be found in Bosworth and Wilson (1988).

The overall fit of the earnings function using the whole sample is broadly in line with earlier studies of this type. A number of influences on earnings are well established in the results, particularly the importance of the regional and employment class dummies. The inclusion of the class dummies improves the performance of the qualification variables, although the coefficients are not as systematic or as significant as human capital theory might suggest. One firmly established result is that earnings increase significantly with experience, although at a decreasing rate (with a turning point at, or just below, age 65). The attitudinal variables show no pervasive, systematic pattern and the associated coefficients are, almost without exception, insignificant. Finally, the variable which reflects the proportion of patenting work appears to indicate that the other activities are, if anything, at least as well paid as patenting.

The results are improved by separating the classes of employment. The relatively low R^2 and F statistics for partners in private practice probably reflect the fact that they are subject to greater random influences on their levels of activity and prosperity and, by implication, these groups will be less susceptible to analysis of this type. The role of experience follows the same pattern throughout, although the rate of growth of earnings with experience differs substantially across classes. The results reveal some interesting differences between classes in the role played by each group of explanatory variables. The important role of formal educational qualifications in CLASS2 is almost entirely absent from the other classes. The regional dummies appear most important in CLASS3, while they play almost no part in the explanation of earnings in CLASS 1. The proportion of patenting activity follows the same pattern in all classes, although the associated coefficient only appears to be significant in CLASS 2. Larger patenting departments tend to pay higher salaries to their employees, larger companies tend to pay higher incomes and partners in larger practices tend to earn more. Some part of the regional effect appears to have been associated with variations in partnership size across regions.

It was always thought that evidence of restrictive practices from the survey would, at best, be rather tenuous. A key indicator is the proportion of patenting activity. In practice, where the variable is significant, it indicates that the other activities open to Patent Agents are at least as lucrative as patenting. This is not proof of an absence of restrictive practices. It might be the case, for example, that these other activities are even more restricted than patenting itself, and, even, that entry to these activities can only be obtained by Patent Agents. However, the survey gives no evidence about such speculations. Neither the patenting activity variable nor the pricing behaviour variables are significant in the explanation of earnings of partners in private practice. Neither the use of CIPA guidelines for charging nor the attitudinal variable (ATTD), concerning the publication of the CIPA guidelines, is significantly related to earnings. Finally, we turn to the other attitudinal variables (ATTA-C). It was thought that, if restrictions are operating, and, thereby, raising earnings, this might be most evident for partners in private practice. This is clearly not the case, if anything, the role of the ATT variables is weakest for this group. There is some evidence of systematic and significant coefficients for employees of private practices. The associated results indicate that those on higher income levels appear to prefer the removal or lessening of restrictions. Some tentative interpretation might be laid on this, for example, relating to the future prospects of entering self employment/ becoming a partner, which might be tenuously related to current restrictions. However, it is impossible to translate this result into any concrete conclusion about restrictive practices.

15.4.4 Rate of Return Calculations

In this section we report on estimates of the rate of return to becoming a Patent Agent and compare these rates with the equivalent values for other professions. Rate of return calculations allow the contribution of education and training to earnings to be assessed. If C_t represents the flow of income if the person does not undertake the investment plus the direct costs of education and training, and B_t is the corresponding stream if the education and training is undertaken, then the internal rate of return, r, is defined by the following expressions:

$$\sum_{t=16}^{65} \frac{B_t - C_t}{(1 + r)^t} = 0 \qquad (1)$$

where the precise definition of B_t and C_t will differ depending on whether it is the private or social rate of return that is being calculated. For simplicity it is assumed here that the decision to undertake the investment is made at age 16 and that the individual retires at age 65. Neither of these assumptions is critical to the results obtained below. In all three years, the typical educational/training profile is assumed to involve full time schooling to age 18, a three or four year university degree course, followed, in the case of Patent Agents, by a minimum of four years part time study (at zero direct cost). The pass rate at the intermediate level is around 75 per cent, and, at the final level, around 40 per cent, although many people resit and a much larger proportion of individuals eventually qualify.

Estimates of the return to becoming a Patent Agent have been made using data from the *EPPS*, together with information from two earlier surveys conducted by CIPA. The main results for Patent Agents are given in Table 15.2. Using the basic data described above, without any further adjustment, the estimated private rate of return to becoming a Patent Agent was 18.5 per cent in 1970, 13.0 per cent in 1981/2 and 15.5 per cent in 1984/85. If a four year degree course is assumed, then the estimate is reduced by about 1 percentage point. Allowance for various effects of self employment tend to reduce the rate of return by about 3/4 of a percentage point, but this is almost exactly offset by adjustments for fringe benefits. However, the additional costs of location in London would also reduce the rate of return by 1 or 2 percentage points. Table 15.2 also presents results for various branches of the profession, with adjustments for on-the-job training, self-employment, and fringe benefits being made, as appropriate. The social rates of return are also presented for the profession as a whole. It is worth adding that many of these other adjustments apply equally to other professional groups and, so, the most accurate basis of comparison, given the currently available estimates of rates of return across professions, is provided by the basic calculations outlined first.

Table 15.2 Results for Patent Agents

%

	1970	1981/2	1984/5
Private rates of return			
Basic result, 3 year degree course	18.5	13.0	15.5[a]
- SE adjustments	18.0	12.5	15.0[a]
- SE & OJT adjustments	17.5	12.0	14.5[a]
- SE, OJT & FB adjustments	18.5	13.5	15.5[a]
- 4 year degree course	17.5	12.5	14.5[a]
Result for sub-groups			
Self-employed, SE & OJT adjustments			
Private practice	n/a	13.0	14.5[a]
Employees, OJT & FB adjustments			
Private practice	n/a	12.5	15.0[a]
Industry and commerce	n/a	15.5	17.0[a]
Government	n/a	13.5	14.5[a]
Social rate of return			
All patent agents	11.5	9.5	11.0[a]

Notes: (a) Adjusted to reflect an average career profile.

Tables 15.3 and 15.4 provide a comparison between the estimates for the various professions, based on as common a methodology as possible (Wilson, 1980, 1983, 1987; Bosworth and Wilson, 1988). It is clear from inspection of the results that the rate of return to becoming a Patent Agent is high compared with most professions. The major decline in the rate of return between 1970 and 1981/2 is common to most other professions, although, if anything, slightly larger in the case of Patent Agents. This may reflect the quite rapid growth in the numbers entering the profession over this period. However, the estimated rate of return was still above average in 1981/2. The increase between that year and 1984/5 is more exceptional, with the figures for most other professions being static. Again, the result is consistent with the various indications of a recent shortage of such people.

Table 15.3 Private rates of return for different professions compared: 1955-1985

%

Profession	c.1955	c.1970	c.1980	c.1985
All graduates	17.5	12.5	9.0	8.5
Chemists	20.0	12.5	9.5	9.0
Physicists	20.0	13.5	11.0	10.0
Engineers	14.5	13.0	10.0	10.5
Architects (b,e)	13.5	13.5	7.0	6.0
Quality surveyors	16.0	15.0	5.5	10.5
General medical practitioners (b,c,d)	21.0	15.5	12.5	n/a
Consultants (b,c,d)	22.0	16.0	12.4	n/a
Other hospital doctors (c,d)	13.0	10.5	7.5	n/a
General dental practitioners (b,d)	35.0	24.5	17.5	n/a
Solicitors (b)	14.0	11.5		n/a
Barristers (b)	10.5	10.5		n/a
Lawyers in industry	17.0		18.0	17.0
Primary school teachers	6.0f	4.0	-ve	n/a
Secondary school teachers (e)	10.5f	8.0	1.5	n/a
Further education teachers (e)	15.0f	11.0	4.0	n/a
University lecturers etc. (e)	20.5	13.0	10.5	n/a
Accountants	16.0	16.0	13.5	n/a
Economists	18.5f	22.5	16.0	n/a
Statisticians	21.0f	22.0	12.0	n/a
Patents agents (b)	n/a	18.5	13.5	15.5

Notes: (a) Assuming a three year degree course of study unless elsewhere specified
(b) Self-employment incomes adjusted for pension provision and capital outlays
(c) Adjusted for hours of work
(d) Assuming a 5 year degree course
(e) Assuming a 4 year degree course
(f) Mid-1960s

The differential compared with all graduates is about 7 percentage points. Compared with most professional scientists and engineers, it is around 5 percentage points. However, the returns are higher in some other occupations such as lawyers in industry and certain medical professions. These are also groups that, it could be argued, have managed to exploit a position of market power to raise their incomes above the level that would have occurred in a more competitive labour market. It is notable that those groups, such as teachers, whose market power is probably not very great, have the lowest rates of return. A pertinent question is what would constitute a significant difference? One way of looking at this is to compute the net present value associated with each career path. Based on a discount rate of 10 per cent, the net present value of the human capital investment would be (approximately) zero for most professions. In contrast, the net present value of becoming a Patent Agent was about £11 thousand in 1984/5 prices, with a net present value of over £13 millions across all Patent Agents.

Table 15.4 Social rates of return for different professions compared: 1955-1985

<table>
<tr><th></th><th></th><th></th><th></th><th>%</th></tr>
<tr><th>Profession</th><th>c.1955</th><th>c.1970</th><th>c.1980</th><th>c.1985</th></tr>
<tr><td>All graduates</td><td>10.5</td><td>7.5</td><td>5.5</td><td>5.5</td></tr>
<tr><td>Chemists</td><td>9.5</td><td>8.0</td><td>6.0</td><td>6.0</td></tr>
<tr><td>Physicists</td><td>10.5</td><td>8.5</td><td>7.0</td><td>6.5</td></tr>
<tr><td>Engineers</td><td>6.5</td><td>7.0</td><td>6.0</td><td>6.0</td></tr>
<tr><td>Architects (b,e)</td><td>6.5</td><td>7.0</td><td>4.5</td><td>4.0</td></tr>
<tr><td>General medical practitioners (b,c,d)</td><td>10.5</td><td>8.0</td><td>7.5</td><td>n/a</td></tr>
<tr><td>Consultants (b,c,d)</td><td>12.0</td><td>9.0</td><td>8.0</td><td>n/a</td></tr>
<tr><td>Other hospital doctors (c,d)</td><td>6.5</td><td>5.0</td><td>4.0</td><td>n/a</td></tr>
<tr><td>General dental practitioners (b,d)</td><td>16.0</td><td>12.0</td><td>9.5</td><td>n/a</td></tr>
<tr><td>Solicitors (b)</td><td>9.5</td><td colspan="2">9.5</td><td>n/a</td></tr>
<tr><td>Barristers (b)</td><td>8.0</td><td colspan="2">8.5</td><td>n/a</td></tr>
<tr><td>Primary school teachers</td><td>4.0f</td><td>2.5</td><td>-ve</td><td>n/a</td></tr>
<tr><td>Secondary school teachers (e)</td><td>6.0f</td><td>4.5</td><td>-ve</td><td>n/a</td></tr>
<tr><td>Further education teachers (e)</td><td>8.0f</td><td>6.0</td><td>2.0</td><td>n/a</td></tr>
<tr><td>University lecturers etc. (e)</td><td>10.0f</td><td>8.0</td><td>6.5</td><td>n/a</td></tr>
<tr><td>Accountants</td><td>9.5</td><td>10.5</td><td>9.0</td><td>n/a</td></tr>
<tr><td>Economists</td><td>11.5f</td><td>22.5</td><td>13.0</td><td>n/a</td></tr>
<tr><td>Statisticians</td><td>12.0f</td><td>13.5</td><td>8.5</td><td>n/a</td></tr>
<tr><td>Patents agents (b)</td><td>n/a</td><td>11.5</td><td>9.5</td><td>11.0</td></tr>
</table>

Notes: (a) Assuming a three year degree course of study unless elsewhere specified and including direct costs of education
(b) Self-employment incomes adjusted for pension provision and capital outlays
(c) Adjusted for hours of work
(d) Assuming a 5 year degree course
(e) Assuming a 4 year degree course
(f) Mid-1960s

There are a number of other issues worth mentioning briefly, some of which are dealt with elsewhere (Bosworth and Wilson, 1988). First, there is the question of the 'transitional gains trap' (Tullock, 1985). Second, no account has been taken of the capital value of the private practices of independent Patent Agents. Third, the discussion has dealt with estimates of the direct value of the Patent Agent monopoly, it has not fully considered the broader externalities associated with the reduction in the dynamic performance of the economy.

15.4.5 Imperfect Information

Evidence from the *SAA* presents a fairly consistent picture which maps certain types of applicants with certain routes by which they seek to obtain patent protection. As a stylised view of the world, the individual appears to go down the private application route, the independent enterprise uses an independent Patent Agent and the enterprise which is part of a group uses an employed Patent Agent. This matching of categories of applicant and route is almost certainly a reflection of the volume of activity of each type of applicant and, thereby, the perceived costs and benefits of each route for each category of applicant.

The survey evidence indicated the much higher volume and different character of the patent work required by the largest of enterprises and groups, with their greater emphasis on foreign activity, the higher number of applications made over time and the greater progress that they tend to make through the various stages of the patenting process.

Some sectors of the market for patenting services possess a number of the essential characteristics of Akerlof's market for 'lemons'. The inherent characteristics of inventive and patenting activity seem likely to mean that certain groups (i.e. individuals and small enterprises) may possess highly imperfect information, in this instance, about the 'product' they are selling *vis a vis* other products available on the market, whereas Patent Agents, with their wealth of experience about inventions, will often have a clearer idea of the novelty, patentability and commercial potential of a given idea. In addition, a considerable proportion of the applicants who use IPAs make highly infrequent applications, often only one. In the light of the 'lemons' analogy, these appear to be ideal conditions for claiming what appears to be a reasonable ('standard') fee, but, in practice, for providing a low quality of service. Akerlof's theory suggests that, in a market with imperfect and asymmetrical information, infrequent purchases and no guarantees of product quality for the customer, 'bad drives out good' (Akerlof, 1970). Thus, there is a tendency for the quality of the goods supplied to fall and, taken to its natural extreme, the market for the product or service in question can collapse (i.e. the supply and demand curves do not intersect).

The question immediately arises as to whether it is in the Patent Agent's interest to exploit this position when a successful application would produce additional revenues through renewal and other related activities. Unlike the Patent Office, however, the charges that Patent Agents seem likely to be able to levy for renewal activities appear, *a priori*, to be relatively low. At this stage of the patenting process, the Patent Agent has little monopoly power because the degree of substitution with private activity becomes higher (renewals involve little skill and the individual can successfully carry out this activity). This is reinforced by the natural enthusiasm of the patentee to pursue the desire for patent protection, which may be misplaced either because of the inadequacy of the invention *per se* or simply that the application is premature. In all probability, Patent Agents follow the line of the Patent Office and claim no innate ability to judge the worth of an invention placed before them. Thus, they cannot be held responsible if an application fails. If Patent Agents had some broader role within the infrastructure for the inducement and exploitation of inventive activity (i.e. which evolve with the removal of the entity restrictions), this potential problem might be avoided.

The results indicate that individuals and small enterprises were less likely to seek advice prior to applying. In addition, the chances were still higher that individuals and small enterprises would make a private application, even where they had previously sought advice. No more than 18 per cent of any of the categories of applicant 'shopped around', and the activity was largely confined to the smallest size group. Given the lack of information about Patent Agents, this seems to suggest high marginal costs of search caused by the fact that hard information is difficult to come by. It should be added that this result is probably typical of many services, but seems to be particularly important in a situation where repeat demands are infrequent and the information network is patchy.

Competition effectively requires that the buyer has accurate knowledge of the products offered by all of the producers in the market. Advertising is one means by which such information is transmitted, bearing in mind that advertising can be qualitatively different. The fact that information is available in the telephone directory suggests that a certain amount of advertising takes place. In practice, this is limited to the name, address and telephone number of the Patent Agent. While this hardly constitutes persuasive advertising, neither is it informative. In addition, no other forms of advertising are allowed by the profession Thus the infrequent applicant has no readily available information about specialisms, charges, etc., let alone the degree to which Patent Agents are successful in obtaining patent protection for the inventors who approach them. In a situation of this type the marginal costs of information about the supplier of services is likely to be high and, therefore, will tend to restrict the extent of search. Location/ proximity is an important factor in the final choice. While there appears to be little problem in finding an IPA, the lack of an extensive or methodical search raises the issue of whether the applicant has obtained the most appropriate IPA at the least cost.

15.4.6 Charges for IPA Services and Related Issues

One other feature is the low weight given by customers to cost as a basis for choosing between IPAs. This may reflect that the demand for patent services is not responsive to the price charged; however, a more likely interpretation is that the expected differences in cost between IPAs are lower than the additional costs of finding out about these differences. This may be the result of using common guidelines for charging, reducing differences between IPAs, and the lack of readily available information about the prices and nature of the services being offered by different agents.

The earlier discussion showed that, in general, the individual appeared to demand less in the way of services from the IPA than independents, and independents less than groups. We would expect to find a distribution of charges that reflects these differing requirements. Certainly, the most important range for individual applications is that of less than £500 (51.6 per cent), but the ranges £501-1000 and £2501-5000 are also important (16.1 and 19.4 per cent respectively). The independents are more evenly spread across the various charging bands. The greater importance of the higher charge ranges amonst independents appears to reflect the greater success of this category in pushing their applications through to the later stages of patenting and their somewhat greater demand for more extensive international protection. The distribution of charges looks perverse in the case of the group enterprises, where we would have anticipated a significantly higher proportion in the high charge ranges. One explanation is that group enterprises tend to use the patent system for disclosure rather than for protection.

Of those whose applications failed (or were waiting) at the first hurdle 59.5 per cent were charged £1000 or less, but 19.1 per cent claim to have been charged over £2500. While there is a general tendency for those in force to have been charged more than those which have failed, nevertheless, 10.5 per cent of the patents in force appear to have been charged less than £500. One possible explanation for these results is associated with the fact that some IPAs charge on a stage-by-stage basis, while others charge for the application as a whole. A total of 16 per cent indicated that they paid for the application as a whole, but this differed between sizes of enterprise (21.4, 20.1, 14.3 and 5.4 respectively for the smallest through to the largest) and by type of enterprise (21.4 for the individual, 16.0 for the independent and 12.5 for the group). Given the differing probabilities of success, these results appear surprising.

A related question concerns the extent to which different routes to obtaining a patent are differently successful. If we take the proportions of patents in force as a measure of success when compared with the overall levels of activity by each route, IPAs appear to do well. They have 50.0 per cent of the patents in force, compared with only 8.7 per cent for private applications and 37.0 per cent for EPAs. On the other hand, if we take patents in force as a percentage of total activity as the measure of success, IPA's have 11.6 per cent in force, while PA's have 5.3 and EPA's have 13.6. Alternatively, if we examine the failures as a percentage of the total of activity, the resulting percentages are 61.1 for IPAs, 65.8 for PAs and 68.0 for EPAs (the latter reflecting the use of the patent system as a means of disclosure). The primary problem with drawing any hard and fast conclusions about this concerns the quality of the material that enters each of the routes and the different uses to which the patent system is put (i.e. protection *versus* disclosure).

The *SAA* asks a number of questions about the degree to which applicants are satisfied with the services and charges of IPAs, on a scale of 1 to 5 (1 is very dissatisfied and 5 is very satisfied). In the case of IPA services up to the stage of grant, the scores are less than 4 in the case of individuals and enterprises with less than 100 employees, but over 4 for the larger two sizes. Individuals, rather than independents or groups, register the lowest score. The opinion of the quality of services after grant is slightly lower in all sizes and types of enterprise, and is less than 4 in enterprises up to 500 employees. There is a fairly even distribution amongst the dissatisfied and the very dissatisfied across applications at all stages of the patenting process, including those in force (although absolute numbers in each of the 10 categories distinguished are necessarily small). There is a greater level of dissatisfaction with regard to IPA charges and the overall costs of gaining patent protection. In the cases of individuals and enterprises with less than 100 employees, the scores are below 3, and all of the scores both by firm size and type are less than 4.

15.5. CONCLUSIONS

Each element of the evidence taken by itself might not be conclusive, but taken as a whole, is consistent with the existence of some degree of monopoly power amongst IPAs. The results indicate that the current system appears least satisfactory for individual applicants and small enterprises. The picture painted by the preceeding discussion is, therefore, very much consistent with the arguments of the Nicholson Report. In particular, it seems fairly certain that the magnitude of the charges restrict output in the cases of individuals and small firms. However, the evidence outlined above is also indicative of the possible existence of monopoly power which may affect all users and, thereby, the operation of the patent system as a whole. Thus, the problem is broader than suggested in the Nicholson Report.

The empirical evidence presented above indicates that the Patent Office attempts to establish social goals that are inconsistent with and can be frustrated by profit maximising behaviour. The evidence with regard to the special position of Patent Agents (set out under the Patent Act), the role of CIPA, the size and structure of the profession and the level of earnings and rate of return to qualifying as a Patent Agent all tend to point to profit maximising behaviour in a situation where the decision makers have some degree of monopoly power. Thus, the discussion has demonstrated that, insofar as U.K. Patent Agents have monopoly power and maximise their profits/earnings: (i) the goals and behaviour of the Patent Office produce the optimal environment for maximising the Patent Agents' profits and incomes; (ii) the separate identity and behaviour of Patent Agents can, unintentionally, frustrate the achievement of the optimal social welfare outcome. While these static welfare effects may be highly significant, the dynamic effects may be more important still, not only affecting the current use of the system amongst existing inventors, but also reducing the incentive to invest and affecting the dynamic performance of the British economy.

Our main conclusion, however, is not that the Patent Office should necessarily rethink its fee structure, which it has presumably adopted in an attempt to maximise social welfare, but that the market for IPA services should be made as competitive as possible, thereby reducing prices and monopoly profits, and increasing the level of activity. Since first outlining this finding, the Office of Fair Trading have removed the Patent Agent monopoly. Increased competition seems likely to lead to welfare improvements, and it will be interesting to chart the changes in behaviour and prices after the OFT ruling.

It is an interesting question whether Patent Agent charges would ever be sufficiently low to achieve the social welfare optimum even under competitive conditions; after all, Patent Agents have to make a living. If, as seems likely, this proves to be the case, an alternative solution may be required. This might involve regulating the charges for the services currently provided by Patent Agents, perhaps by providing subsidised services to applicants from within the Patent Office at the expense of somewhat higher future renewal fees.

REFERENCES

Akerlof, G., 1970, "The Market for Lemons: Qualitative Uncertainty and the Market Mechanism". *Quarterly Journal of Economics* 84, 488-500.

Balz, M.W., 1975, *Invention and Innovation Under Soviet Law*, Lexington Books, Massachusetts.

Banks, M.A.L. (Chairman), 1970, *The British Patent System*, Report of the Committee to Examine the Patent System and Patent Law, Cmnd 4407, HMSO, London.

Bosworth, D.L. and R.A. Wilson, 1988, *Infrastructure for the Protection of Intellectual Property: Patent Agents and Patent Monopolies*, Research Report, Institute for Employment Research, University of Warwick, Coventry.

Dasgupta, P. and J. Stiglitz, 1980, "Uncertainty Industrial Structure and the Speed of R&D", *Bell Journal of Economics* 11, 1-28.

Kitch, E.W., 1977, "The Nature and Function of the Patent System", *Journal of Law and Economics* 20, 2, 265-90.

Marplan, 1985, *Report on a Survey of the Earnings and Professional Practices of Patent Agents*, Prepared for the Office of Fair Trading, Marplan, London.

Mincer, J., 1974, *Schooling, Experience and Earnings*, National Bureau of Economic Research, Human Behaviour and Social Institutions Series, NBER, New York.

Nicholson, R.B., 1983, *Intellectual Property Rights and Innovation*, Cmnd 9117, HMSO, London.

Office of Fair Trading, 1986, *Review of Restrictions on the Patent Agents' Profession*, OFT, London.

Patents, Designs and Trademarks, Annual Report of the Controller General of Patents, Designs and Trademarks, HMSO (annual), London.

Sample Surveys, 1986, *Survey of Applicants and Applications*, Survey for the Office of Fair Trading, London.

Taylor, C.T. and Z.A. Silberston, 1973, *Economic Impact of Patents*, Cambridge U.P., Cambridge.

Tookey, G.W., 1972, *Restrictive Business Practices: an Analysis of the WIPO Model Laws for the Developing Countries from the Point of View of the Export Interests of the Developing Countries*, United Nations, New York.

Tullock, G., 1975, "The Transitional Gains Gap", *Bell Journal of Economics* 9, 2, 671-8.

Williams, G. and A. Gordon, 1981, "Perceived Earnings Functions and ex ante Rates of Return to Post Compulsory Education in England", *Higher Education* 10, 199-227.

Wilson, R.A., 1980, "The Rate of Return to Becoming a Qualified Scientist or Engineer in Great Britain, 1966-76", *Scottish Journal of Political Economy* 21, 41-62.

Wilson, R.A., 1983, "Rates of Return: Some Further Results", *Scottish Journal of Political Economy* 30, 2, 114-27.

Wilson, R.A., 1987, , "The Determinants of the Earnings of Professional Engineers in Great Britain 1981", *Applied Economics* 19, 983-94.

Wright, B.D., 1983, "The Economics of Invention Incentives: Patents, Prizes and Research Contracts", *American Economic Review* 73, 691-707.

PART III

THE POLITICAL ECONOMY OF STRUCTURAL CHANGE

CHAPTER 16

High-Technology Location and Worker Mobility in the U.S.

Henry W. Herzog Jr. and Alan M. Schlottmann

16.1 INTRODUCTION

In recent years, nearly all of the job creation in the U.S. manufacturing sector can be attributed to growth in high-technology industries.[1] The magnitude of this "technological renaissance" has not only spurred competition among states and metropolitan areas for high-technology employment, but also rekindled academic interest in the spatial dimensions of innovation process. However, amid this interest in high technology firms and centers, technological competition, and ensuing local development issues, very little information has been developed to date on the specialized labor resources (for both production and research and development) that generate the observed geographic patterns of interest.

What recent information that has been obtained from surveys and case studies of high technology firms (and their agglomeration in centers) indicates that such firms are drawn both to specialized resources such as labor skills and education, and to factors that help attract and maintain a skilled work force.[2] In fact, although such surveys and case studies show high technology firms to be "footloose" in terms of the more traditional location factors (such as market access and transportation), they also provide evidence that these firms are "dominated" in their location decision by their ability to obtain and retain individuals with specific technical, scientific and engineering occupations.

In previous work by the authors [Herzog, Schlottmann and Johnson (1986)], it was demonstrated that one can discover much about the location determinants of high-technology industry (as well as its R&D arm) by examining the location (migration) decision of workers with high technology occupations.[3] In that work, high-technology workers were shown to be more mobile overall than "other" workers, and to place greater value on transportation access when considering relocation. In addition, high-technology

---

[1]Based upon employment in total manufacturing and Standard Industrial Classification (SIC) codes 28, 35, 36, 37 and 38 for the period 1975-1984.

[2]For example, see the excellent staff study conducted for the Joint Economic Committe by Robert Premus (1982).

[3]Bradbury and Malecki apparently agree when they state that "mobile professional and technical workers not only serve to define what high-tech is, but they also largely determine the location of high-tech activities and therefore the development of places", [Bradburn and Malecki (1987), p. 1)]. On the other hand, the location decisions of high-technology firms and workers are, of course, interdependent. For a general treatment of this labor market simultaneity as it relates to migration, see Muth (1971).

migrants were distinguished from other migrants by their preference for large metropolitan areas. This latter tendency has also been noted by Clark (1981) and Glasmeier (1986), and is quite likely related to the alternative employment needs of dual career (professional) couples.[4]

This study will address the metropolitan dimensions of high-technology industry location directly by examining the migration decision of high-technology workers employed within metropolitan statistical areas (MSAs). In addition to the location factors often considered important for the attraction and retention of a skilled work force (to include transportation access), the study will examine the locational response of these workers to: (1) urban high-technology employment "centers" (represented by employment concentration or "intensity"), and (2) the distribution of such employment between central cities and suburbs.

The correspondence between high-technology industries and occupations is provided in the following section, which concludes with an occupational definition of the high-technology work force. The affinity between high-technology industry employment and metropolitan location is examined and compared to the location pattern for other manufacturing in Section 3. In addition, Section 3 examines location factors believed important for the attraction of high-technology industry, and by extension, for the attraction and retention of a highly skilled work force. Based upon these factors, a model of the 1975-1980 migration decision of high-technology workers is developed in Section 4. Logit estimates of this model are then presented and discussed. Conclusions of the study follow in Section 5.

16.2 DEFINING HIGH-TECHNOLOGY: INDUSTRIES VS. OCCUPATIONS[5]

According to Premus (1982), high technology industries consist of heterogeneous collections of firms that share several attributes.

> "First, the firms are labor-intensive rather than capital-intensive in their production processes, employing a higher percentage of technicians, engineers and scientists than other manufacturing companies. Second, the industries are science-based in that they thrive on the application of advances in science to the marketplace in the form of new products and production methods. Third, R&D inputs are much more important to the continued successful operation of high technology firms than in the case for the other manufacturing industries." [Premus (1982), p. 4][6]

No agreement has yet been reached as to which manufacturing industries satisfy the above criteria; on the other hand, there does appear to be widespread agreement that high-technology industries, however defined, can be found within Standard Industrial Classification (SIC) codes 28, 35, 36, 37, and 38.[7]

High-technology industries in the present study were selected to closely match industries delineated by Doody and Munzer (1981) in their study of high-technology

[4]For an excellent discussion of dual career couples and geographic mobility, see Bradburn and Malecki (1987).

[5]Much of this section draws upon the authors' earlier study. See Herzog, Schlottmann and Johnson (1986).

[6]For comparison of innovation rates and R&D intensity among manufacturing industries classified as either high or low-technology, see Rees (1979), Table 1. For temporal dimensions of associated structural change, see Andersson and Haag (1987).

[7]Four-digit SIC code definitions of high-technology and defense-related high-technology industries are provided in Armington (1986) and Markusen (1986), respectively. In addition, see the survey by McQuaid and Langridge (1984).

manufacturing employment distribution across 24 states. Under this definition, high-technology industries include drugs (SIC 283), ordnance and accessories (SIC 348), electronic computing equipment (SIC 3573), office and accounting machines (remainder of SIC 357), electrical and electronic machinery, equipment and supplies (SIC 36), aircraft, guided missiles, space vehicles and parts (SIC's 372 and 376), and scientific instruments (SIC 381). Notice that this definition permits truly high-technology industries such as electronic computing equipment (SIC 3573) to be distinguished from machinery (SIC 35).

Although there appears to be some agreement on the definition, or delineation, of high-technology industry, little information is available at present on the occupational composition of the high-technology work force (hereafter high-technology workers). Studies of this industry imply that such workers are certainly concentrated throughout the ranks of technicians, scientists and engineers, but little additional help is available to these researchers in the literature.

Our approach to the delineation of high-technology occupations (workers) followed two steps. In the first step, occupation listings and descriptions were examined, and a tentative list of high-technology occupations was extracted. Industry-by-occupation employment percentages were then determined in the second step utilizing both the Doody and Munzer (1981) delineation of high-technology industries (see above) and occupations defined in step one. Occupations were then deleted from this list if their employment shares in various high technology industries were below the equivalent shares computed relative to total manufacturing employment (for a significant number of high-technology industries).

This two-step process led to the following definition of high technology workers (specific examples): computer specialists (programmers); engineers (chemical, electrical, mechanical, metallurgical); mathematical specialists (mathematicians, statisticians); life and physical scientists (space, biological and marine scientists, chemists, geologists, physicists); operations and systems analysts; and engineering and science technicians (biological, chemical and mathematical technicians, and electrical, industrial and mechanical engineering technicians).[8]

An industry-by-occupation matrix for these six high-technology occupations and the seven high-technology industries defined above is presented in Table 16.1. For each high-technology industry listed in the first column of the table, numbers shown under each occupation and total denote percentage of industry employment in 1978.[9] Equivalent percentages are also listed at the bottom of Table 16.1 for both total manufacturing and all industries.

[8]With the exception of operations and systems analysts, these occupations match those employed by Riche, Hecker, and Burgan (1983) in their study of the high-technology industry. They note that "most workers in these technology-oriented occupations are directly involved in developing or applying new technologies. Their work requires in-depth knowledge of theories and principles of science, engineering, and mathematics underlying technology - a knowledge which distinguishes them from computer operators, computer service technicians and other high-tech machinery repairers, or workers in a wide range of occupations who use word processing machines, computers or other high-technology products, but rarely have - or need - such in-depth knowledge. Workers in these technology-oriented occupations generally need specialized post-high school education in some field to technology - ranging from an associate degree or its equivalent to a doctorate - education with a thorough high school preparation in science and mathematics as a prerequisite" [Riche, Hecker, and Burgan (1983), pp. 54, 55].

[9]Note that this 1978 high-technology industry-by-occupation classification falls within the interval over which high-technology worker mobility is observed in this study, namely 1975-1980.

Table 16 1: High technology industry/occupation distribution in 1978
(Percent distribution of industry employment by occupation)[a]

	High Technology Occupations						
High Technology Industries[b]	Computer Specialists	Engineers	Mathematical Specialists	Life and Physical Scientists	Operations and Systems Analysts	Engineering and Science Technicians	Total
Drugs	1.11	1.75	.15	7.32	.48	6.43	17.24
Ordnance and Accessories	2.38	12.01	.27	.80	1.27	4.68	21.41
Electronic Computing Equipment	13.69	9.20	.14	.37	.99	7.14	31.53
Office and Accounting Machines	3.34	4.82	.08	.29	.77	3.79	13.09
Electrical and Electronic Machinery, Equipment and Supplies	1.04	6.96	.05	.24	.60	4.84	13.73
Aircraft, Guided Missiles, Space Vehicles and Parts	1.70	13.85	.17	.38	1.40	4.44	21.94
Scientific Instruments	1.06	7.28	.05	.50	.81	5.64	15.34
All manufacturing	.66	2.81	.03	.44	.33	2.24	6.51
All industries	.40	1.23	.05	.30	.13	1.03	3.14

[a]Reprinted by permission of the *Journal of Regional Science*. See Herzog, Schlottman and Johnson (1986), Table 1. Percentages were derived from the U.S. industry-by-occupation matrix. See Table 1 of U.S. Department of Labor (1981)

[b]These industries were selected to closely match the high-technology industries examined by Doody and Munzer (1981). For corresponding Standard Industrial Classification (SIC) codes, see the text.

Notice in the last column of Table 16.1 that workers in the six high-technology occupations are highly concentrated in each of the seven manufacturing industries utilized in the Doody and Munzer study (1981). This concentration ranges from 13.1 percent in office and accounting machinery to 31.5 percent in electronic computing equipment, and significantly exceeds the industry norms reported at the bottom of Table 16.1. Also note in this table that, with few exceptions, employment concentrations in each of the high-technology occupations delineated above exceed industry norms.

16.3 HIGH-TECHNOLOGY INDUSTRY LOCATION

16.3.1 Metropolitan Dimensions

Prior to discussing the multivariate analysis of high-technology worker migration (Section 4 below), it is instructive to first examine the spatial distribution of high-technology *industry* across urban and rural territory.

Table 16.2 indicates the 1980 percent distribution of high-technology industry employment (see Table 16.1) across metropolitan and nonmetropolitan areas, and for the former, between central cities and metropolitan territory outside central cities (suburbs). Equivalent distributions are also shown for manufacturing other than high-technology, and for nonmanufacturing industry. Geographic areas in Table 16.2 are those delineated in 1980 Census microdata [U.S. Bureau of the Census(1983)].

Table 16.2: Industry location in 1980.
(Percent distribution of employment by geographic area)[a]

	Industries		
	High-Technology[b]	Other Manufacturing	Non-Manufacturing
Inside Metropolitan Areas:			
Cental City	28.1	26.7	33.1
Outside Central City	57.7	45.0	41.4
Total	85.8	71.7	74.5
Outside Metropolitan Areas:	14.2	28.3	25.5
Percent of Total Employment	4.8	19.5	75.7

[a]Compiled by the authors from the Public Use "B" Sample of the 1980 Census [U.S. Bureau of the Census (1983)].

[b]Included industries are those listed in Table 16.1. For corresponding Standard Industrial Classification (SIC) codes, see the text.

Notice first in Table 16.2 that high-technology jobs (4.8 percent of U.S. employment) are highly concentrated in metropolitan areas in 1980, and significantly exceed such concentration of other manufacturing and nonmanufacturing industries (85.8, 71.7 and 74.5 percent respectively). In addition, high-technology employment is highly concentrated *within* metropolitan areas, 57.7 percent of total U.S. employment (67.2 percent of metropolitan employment) being sited outside central cities. Comparable percentages are 45.0 (62.8) for other manufacturing and 41.4 (55.6) for nonmanufacturing. Finally, only 14.2 percent of high-technology employment was located outside metropolitan territory in 1980.

Reasons for this concentration of high-technology industry within metropolitan areas in the first instance, and suburban components of these areas in the second most probably derive from the continuing needs of such firms for technicians, scientists and engineers (high-technology workers), individuals known to prefer metropolitan to more remote work/residence locations. These preferences for metropolitan areas appear to be quite strong, high-technology firms often finding it difficult to attract highly trained workers to rural locations [Clark (1981) and Glasmeier (1986)].

It is interesting to note that high-technology concentration or "intensity" (high-technology industry employment relative to total employment) varies considerably across metropolitan areas. The mean of this intensity within the 233 metropolitan statistical areas included in the study is 5.4 percent, and varies from a low of .2 percent to a high of 28.4 percent. Based upon this measure of industrial concentration, Nashua, New Hampshire (a portion of Boston's Route 128 complex) and San Jose, California (Silicon Valley) are the two most high-technology intensive metropolitan areas in the U.S. (26.7 and 28.4 percent respectively).

This index of high-technology industry employment concentration will be utilized on several occasions within the multivariate analysis below. Although not representing in a strict sense the presence of a high-technology center or complex (an employment agglomeration subject to areal definition), this index (a continuous variable) most certainly does represent employment opportunity, and perhaps diversity as well, for high-technology workers.

16.3.2 Determinants

Aside from their observed preference for metropolitan over nonmetropolitan territory, and for a suburban location within metropolitan areas, high-technology firms must of course choose from a great number of such alternatives. Location factors upon which such choices are based have recently been compiled from a survey of high-technology firms conducted by the Joint Economic Committee during 1981 and 1982 [hereafter the Premus study (1982)].[10]

The survey form adopted in this study asked high-technology respondents to twice rate various area characteristics as to significance in their location decision, first with respect to choice of region within the U.S. and second with respect to sites (such as metropolitan areas) within such regions.[11] When choosing among regions, factors rated highest (by rank) were labor skills and availability (1), labor costs (2), taxes (3), academic institutions (4), cost of living (5) and transportation (6). Note that each of these six location factors considered most important by high-technology companies in choosing locations among regions relate either directly, or indirectly, to the procurement and retention of a highly skilled work force. Thus, although labor availability directly affects the number of technicians, engineers and scientists accessible locally, factors such as taxes, proximity to academic institutions, living costs and transportation indirectly affect this same accessibility through their influence on worker mobility, and thus labor supply. In a like manner, cultural amenities (10) and climate (11) should also affect the ability of high-technology firms to attract and maintain a highly skilled work force.

---

[10]For a comparison of the Premus study with other survey results, see Rees and Stafford (1986).

[11]Questionnaires were mailed to 1,750 high-technology companies selected from: (1) members of the American Electronics Association, the leading high-technology trade association in the U.S.; and (2) approximately 400 companies in the Route 128 area of Boston. Based upon 691 responses, the typical respondent was a young, small, high-technology firm with fewer than 400 employees operating in a single plant. The region/site feature of the questionnaire permits location choice to follow a two-stage decision process. In this respect, see McMillan (1965).

When choosing among sites (metropolitan areas) within a region, labor skills and availability were again ranked as the most important location determinant of high-technology firms in the Joint Economic Committee sample. Other highly rated factors that indirectly affect, through migration and labor supply, the availability of high-technology workers were taxes (2), good transportation for people (5), proximity to good schools (7), and proximity to recreational and cultural opportunities (8).[12]

Thus, eight (five) of the more important region (site) location determinants of high-technology companies relate, either directly or indirectly, to the ability of these firms to attract and retain a highly skilled work force. It is this aggregate dominance of location factors related, directly and indirectly, to the *supply* of highly skilled workers to high-technology (to include a metropolitan location) that provides both the justification and point of departure for an analysis of location decisions of high-technology workers themselves.

16.4 HIGH-TECHNOLOGY WORKER MOBILITY

16.4.1 Data

Microdata required for a multivariate analysis of high-technology worker mobility were obtained from the one percent Public Use "B" Sample of the 1980 Census (1983). This microdata was processed to extract householders aged 24-60 in 1980 (19-55 in 1975) who were members of the labor force in 1980 and reported a high-technology occupation (see Table 16.1) in the same year.[13] Excluded from these high-technology workers were individuals attending college in 1975 and/or 1980, members of the armed forces in 1975 and/or 1980, and inmates of institutions in 1980.

In addition to this information for individual high-technology workers (which identifies personal characteristics such as race, sex, age and education), data were also obtained on the 233 metropolitan statistical areas (MSAs) in which the sample workers lived in 1975. These metropolitan area characteristics, many chosen to represent the region (site) location determinants identified above, were obtained, with several exceptions, from Boyer and Savageau (1981). All other MSA characteristics were either derived from Census microdata or obtained directly from published Census documents.

16.4.2 1975-1980 Migration Decision

Considerable information is available in the literature on both the determinants of migration and the socioeconomic characteristics of migrants. Surveys of this literature are provided by Greenwood (1975, 1985), Ritchey (1976) and Shaw (1975). However, little information relevant to the migration decisions of workers disaggregated by specific occupations is available.[14]

---

[12]This ranking is similar to others based upon surveys of general manufacturing such as the plant site surveys of McGraw-Hill. The importance of taxes, for instance, is indicated in analyses of urban manufacturing location. Although it is difficult to compare surveys (for example, those with an intra-regional versus an interregional perspective), the high-technology respondents in the Premus (1982) study apparently place greater emphasis on labor skills and availability than do other manufacturing firms.

[13]A householder is a person, or one of the persons, in whose name a dwelling was owned or rented. See U.S. Bureau of the Census (1983). Occupations were not reported for 1975, the outset of the five-year migration interval employed in the study. For an analysis of the extent to which occupation (and industry) "switching" occurs over such intervals, see Schlottmann and Herzog (1984).

[14]For comparisons of this decision among major occupation groups, see Herzog and Schlottmann (1984) and Ladinsky (1967)

In order to model this behavior for high-technology workers residing within metropolitan areas in 1975, the decision to either remain within the MSA or relocate over the ensuing five-year period was related to both personal and metropolitan area characteristics. All personal characteristics other than age and education were represented by binary variables set to unity (vs. zero) in accordance with their names (i.e., Female = 1 if a worker's sex is female). It is expected that 1975-1980 migration of high-technology workers is augmented by increased education, by personal unemployment in 1975, and by prior geographic mobility.[15] On the other hand, such 1975-1980 mobility should be reduced somewhat among females, nonwhites and the disabled. In addition, mobility is expected to decrease with increased age and family responsibility (marriage and children), and when one's spouse is employed (dual career couples). Finally, geographic mobility is expected to vary among the high-technology occupations listed in Table 16.1.

If high-technology workers truly prefer suburban residences (and work sites) as discussed above, then less 1975-1980 mobility is expected for such workers residing outside the central cities of MSAs in 1975. A binary variable representing a suburban residence will be employed to test this hypothesis (under ceteris paribus conditions).

Turning now to metropolitan area characteristics, high-technology workers are expected to reveal through their migration decisions (i.e., "voting with their feet") preferences for various dimensions of both work and life quality within MSAs. In this respect, 1975-1980 outmigration of high-technology workers should be diminished somewhat as high-technology intensity or employment share (high-technology industry employment relative to total MSA employment) increases. As discussed above, this index of high-technology centers or complexes represents local employment opportunity and, most likely, diversity as well. It is also expected, based upon information in Table 16.2, that the distribution of high-technology industry employment between central cities and suburbs of metropolitan areas should influence 1985-1980 migration, such mobility probably declining as employment opportunity within suburbs is enhanced.

In addition to general economic conditions (total employment growth and per-capita income), city scale (population), and the MSA crime rate, ten metropolitan area characteristics were chosen to represent, as closely as possible, those high-technology industry location determinants that also affect worker mobility. Considered in Section 3.2 above, these factors include climate, cost of living, taxes, educational quality, and accessibility to cultural amenities, recreation, and transportation. Variables representing climate and accessibility of metropolitan dwellers to cultural amenities, recreation, and transportation were obtained from Boyer and Savageau (1981), and were set equal to their scores developed for each of these dimensions of urban life quality. Since higher scores are associated with "better" climate and accessibility, 1975-1980 outmigration from MSAs should be reduced accordingly.[16]

It is instructive to note that transportation accessibility in each MSA is determined by Boyer and Savageau on the bases of airline flights, passenger train departures, interstate highways, and mass transit and freeway mileage. Thus, their transportation score measures both intra- and extra-MSA transportation accessibility, a subject that we will address in more detail below.

Local costs of living and taxes are represented by three variables: home prices, effective property-tax rates (percents) and the percent of household income required for state and local sales and income taxes. Educational quality within metropolitan areas is

---

[15]Prior mobility was determined on the basis of reported states of birth and residence in 1975.

[16]Boyer and Savageau's (1981) climate factor (score) is related to the presence of distinct seasons, heating- and cooling-degree days, relative humidity and extreme temperatures. Cultural amenities (the "arts" in Boyer and Savageau's study) represent accessibility to cultural facilities to include symphony orchestras, opera companies, dance companies, theatres, public television, fine arts radio, and museums. Finally, the recreation factor is determined in a similar fashion, and is related to the number of public golf courses, bowling alleys, public parks, movie theatres, neighborhood bars, zoos, aquariums, etc. (each adjusted by population) within the MSA.

also represented by three variables: expenditure per student and student-teacher ratios to measure such quality in grades one through twelve, and a binary variable set to unity (vs. zero) for MSAs receiving an A, AA, or AAA rating for higher education options by Boyer and Savageau (1981). It is expected that higher costs of housing and local tax rates will promote outmigration of high-technology workers, while quality education will reduce such mobility ceteris paribus.[17]

These metropolitan area characteristics were matched to personal characteristics of high-technology workers based upon MSA of residence in 1975. Of the 5,389 high-technology workers described in Section 4.1 that were metropolitan residents in 1975, 949 (176 per thousand "at risk") relocated between 1975 and 1980, the majority moving to another metropolitan area. This high rate of mobility is consistent with similar rates observed for both professional and technical occupations (Herzog and Schlottmann (1984) and Ladinsky (1967) and scientists and engineers [National Science Foundation (1980a, 1980b)]. Binary logit estimates of the determinants of this migration are provided in Table 16.3.

For each of the 5,389 high-technology workers "at-risk" to 1975-1980 migration, the dependent variable in the logit equation was set to unity for migrants (vs. zero for nonmigrants). Asymptotic t-values and significance levels for each coefficient estimate are provided in the right-most column of Table 16.3.

Note first among significant personal characteristics that high-technology workers, like other workers, demonstrate both age and education selectivities of migration (negative age and positive education coefficients). In addition, 1975-1980 mobility is reduced somewhat among female and/or nonwhite workers, by a working spouse, and by school-age children. On the other hand, high-technology worker mobility is augmented both by unemployment (at the outset of the migration interval) and by prior migration. Notice also that 1975-1980 migration likelihood varies systematically by high-technology worker occupation, computer specialists, engineers, and scientists being more likely to relocate over the five year period than were mathematical specialists, operations and systems analysts, and engineering and science technicians (the omitted occupation).

No evidence is provided in Table 16.3 that high-technology worker mobility is responsive either to residence location within a metropolitan area or to the likelihood of suburban employment (high-technology employment location). On the other hand, such mobility is significantly diminished (albeit at a declining rate) as high-technology intensity increases within MSA's.[18] Thus, as the relative magnitude and diversity of *alternative* high-technology employment increases (perhaps within high-technology centers), workers are less likely to relocate, ceteris paribus.[19]

[17]The authors have recently ranked metropolitan areas based upon 1975-1980 outmigration response to Boyer and Savageau's scores. See Herzog and Schlottmann (1986).

[18]Based upon the quadratic term for employment share in Table 16.3, increases in high-technology intensity beyond 16.8% provide no additional restraint upon outmigration. Metropolitan areas at this level of high-technology intensity in 1980 include Bridgeport, Conn., Lowell, Mass.-N.H., Mansfield, Ohio, and Norwalk, Conn.

[19]Results equivalent to those in Table 16.3 were also obtained when employment share was replaced by total high-technology employment within each MSA.

Table 16.3: Determinants of 1975-1980 migration of high-technology workers: Binary logit estimates[a] (for metropolitan residences in 1975)

Variable	Coefficient	Asymptotic t-value
Constant	1.1561	1.498
Personal Characteristics		
Female	-.4498	-2.583***
Nonwhite	-.3218	-1.813*
Age	-.0798	-17.065***
Education	.0659	3.3833***
Married	-.0127	-0.108
Employed spouse	-.3069	-2.283**
School-age children	-.3811	-4.164***
Disability	-.1700	-0.785
Unemployed in 1975	.6703	3.035***
Prior migrant	.8295	9.917***
Occupation:[b]		
Computer specialist	.4387	3.033***
Engineer	.2433	2.503**
Mathematical specialist	.0826	0.283
Life or physical scientist	.3520	2.250**
Operations or systems analyst	.0998	0.405
Metropolitan Residence		
Reside outside central city in 1975	.0176	0.184
Metropolitan Area Characteristics:[c]		
High-technology intensity (%)		
Employment share	-.0565	-2.557**
Square of employment share (10^2)	.1677	1.989**
High-technology employment location		
Percent outside central city (10^3)	.9294	0.516
Climate[d] (10^3)	-.2501	-0.480
Crime (10^3)	.1928	1.272
Home prices[d] (10^3)	.0065	1.651*
Local taxes (rates)[d]		
Property (10^2)	.1808	3.351***
Sales and income	.3604	0.207
Educational quality[d]		
Expenditure per student (10^3)	-.3920	-1.743*
Student/teacher ratio	-.0131	-0.502
Higher education options	-.2447	-1.767*
Accessibility to[d]		
Cultural amenities (10^3)	.0072	0.490
Recreation (10^3)	.0791	0.493
Transportation (10^3)	-.0442	-2.385**
Economic conditions		
Employment growth (1975-1980)	-.0071	-1.765*
Per-capita income (10^3)	-.0216	-0.224
City scale		
Population (10^6)	-.1247	-1.301
Population squared (10^6)	.0137	1.266

[a]All variables are defined in the text. High-technology workers are listed by occupation in Table 16.1. The analysis was based upon 5,389 observations, and the log likelihood ratio test statistic was significant at the one percent level.

[b]The excluded occupation is engineering and science technicians.

[c]These variables pertain to the metropolitan statistical area (MSA) of 1975 residence, and assume 1975 values except where noted.

[d]Listed among the more important location determinants of high-technology firms that relate, either directly or indirectly, to the retention of a highly skilled professional work force. See Section I3.2 above and Premus (1982)

* t-test significant at 10%
** t-test significant at 5%
*** t-test significant at 1%

Metropolitan area characteristics followed by the superscript "d" in Table 16.3 represent high-technology industry location factors (see Section 3.2) that relate, either directly or indirectly, to the retention of a highly skilled work force. Notice in this respect that high-technology workers are (significantly) less likely to relocate between 1975 and 1980 in the face of lower home prices and property taxes, quality education (to include higher education), and higher levels of intra- and extra-MSA transportation accessibility. With the exception of 1975-1980 employment growth (which retards outmigration), all other metropolitan area characteristics in Table 16.4 to include city scale (population) are insignificant determinants of high-technology worker mobility. Finally, based upon the signs and significance levels of coefficients on high-technology employment share and MSA population(as well as their squared terms), the preference of high-technology workers for larger metropolitan areas [Herzog, Schlottmann and Johnson (1986)] likely derives from concomitant increases in employment opportunity and diversity rather than increased city size per se.

An experiment was conducted with the model just described to provide insights on two issues or questions pertinent to this study: (1) To what extent are metropolitan areas more "attractive" to high-technology workers who both work and reside outside the central city?; and (2) Given the significance of transportation access to high technology workers [documented above and in Herzog, Schlottmann and Johnson (1986)], does such importance derive from transportation linkages within, without, or both within and without MSAs? Based upon a reestimation of the model shown in Table 16.3 with several interactive terms: (1) High-technology workers residing in the suburbs surprisingly find metropolitan areas no more attractive as employment opportunity within these suburbs is increased; and (2) The importance of transportation access to high-technology workers apparently derives from extra-MSA transportation linkages rather than intra-MSA access (namely commuting).[20]

---

[20]Interactive terms were formed from three variables listed in Table 3, namely RESIDE (a binary variable set to unity for a 1975 residence outside the central city), PCTSUB (the percent of MSA high-technology industry employment located outside the central city), and TRANS (combined intra- and extra-MSA transportation accessibility. Based upon these variables, interactive terms defined as RESIDE*PCTSUB and RESIDE*PCTSUB*TRANS were added to the high-technology worker migration model.

All signs and significance levels of the logit estimates listed in Table 3 were maintained. Denoting coefficient estimates on RESIDE, PCTSUB and TRANS by α_1, α_2 and α_3 respectively, and those on the two interactive terms by β_1 and β_2 respectively, the following sign pattern is indicated (at the 10 percent level of significance): $\alpha_1 = 0$, $\alpha_2 = 0$, $\alpha_2 < 0$, $\beta_1 = 0$ and $\beta_2 = 0$. Representing partial deriviates of migration likelihood (with respect to any of the above variables) by $M'(\cdot)$, and recognizing that partial derivatives are simple transformations of logit estimates, note that:

(1) $M'(\text{PCTSUB}) = \alpha_2 + \beta_1 \cdot \text{RESIDE} + \beta_2 \cdot \text{RESIDE*TRANS}$

(2) $M'(\text{TRANS}) = \alpha_3 + \beta_2 \cdot \text{RESIDE*PCTSUB}$

The expectation that workers find metropolitan areas more attractive when they both reside and (become more likely to) work there requires that $\beta_1 < 0$ in equation (1). In addition, the expectation that workers

To place these results in perspective, it might be useful to conclude with a summary of findings from Herzog, Schlottmann and Johnson (1986), which examined the location (migration) decisions of workers with high-technology occupations *relative* to other members of the labor force. Within a multivariate setting, high-technology workers demonstrate significantly greater responses in migration than do other workers to both age and the presence of school-age children. In addition, high-technology workers are influenced in their migration decision by intra- and extra-MSA transportation accessibility at their present residence, while other workers apparently are not. However, once the decision to move has been made, high-technology workers act no differently from other workers when considering metropolitan characteristics (exclusive of city scale) at alternative destinations. In addition, high-technology workers were found to migrate with greater likelihood than other workers, ceteris paribus, such differential mobility most probably stemming from occupation-specific human capital.

16.5 CONCLUSIONS

It has been our contention throughout this study that one can discover much about the location determinants of high-technology industry by examining the migration decisions of workers with high-technology occupations. Evidence for this contention was based upon the aggregate dominance of firm location factors related both directly and indirectly to the procurement and retention of a highly skilled work force. Given the applicability of this supply-side approach, high-technology occupations were delineated on the basis of both judgement and industry concentration. The resulting high-technology work force (termed high-technology workers) consists of computer specialists, engineers, mathematical specialists, life and physical scientists, operations and systems analysts, and engineering and science technicians.

High-technology industry employment is concentrated in metropolitan areas to an extent far greater than that of other manufacturing industry, and within such areas, is disproportionately concentrated in territory outside of central cities (i.e. suburbs). Reasons for this concentration of high-technology firms within metropolitan areas in the first instance, and suburban components of these areas in the second, must probably derive from the continuing needs of such firms for technicians, scientists and engineers (high-technology workers), individuals known to prefer metropolitan to more remote work locations.

Within a multivariate setting, high-technology workers were allowed to "vote with their feet", thereby revealing through their migration decision preferences for various dimensions of urban life quality. With respect to the migration decision, high-technology workers are less likely to relocate in the face of lower home prices and property taxes, quality education (to include higher education), and higher levels of transportation accessibility, factors also considered important for the attraction of high-technology industry to metropolitan areas. Such mobility also varied systematically by occupation, computer specialists, engineers, and scientists being more likely to relocate, all else held constant, than mathematical specialists, operations and systems analysts, and engineering and science technicians.

Worker mobility was shown to be significantly diminished within metropolitan areas characterized by higher levels of high-technology employment concentration, or "intensity". Thus, as the relative magnitude and diversity of alternative employment increases (perhaps within high-technology centers or complexes), worker outmigration is diminished. It is interesting to note in this regard that the preference for larger metropolitan areas often attributed to high-technology workers most probably derives

---

consider intrametrapolitan transportation access important (relative to access without the MSA) requires that $\beta_2 > 0$ in equations (1) and (2).

from the increased employment opportunity and diversity that accompanies urban growth, rather than from city size per se. On the other hand, no evidence was found that worker mobility is responsive to either residence location within metropolitan areas or to high-technology employment opportunities within the suburbs (even for workers residing there). Finally, transportation accessibility was found to be an important determinant of worker mobility, this importance deriving primarily from transportation linkages outside of metropolitan areas rather than from those within.

REFERENCES

Andersson, Å.E. and G. Haag, 1987, "Structural Change and Technological Development", *RICE Symposium*, Karlstad.

Armington, C., 1986, "The Changing Geography of High-Technology Business", in Rees, J., *Technology, Regions, and Policy*, Rowman and Littlefield, Totowa, 75-93.

Boyer, R. and D. Savageau, 1981, *Places Rated Almanac*, Rand McNally and Company, Chicago.

Bradbury, S.L. and E.J. Malecki, 1987, "Dual Career Couples and Geographic Mobility", *Southern Regional Science Association Meeting*, Columbus.

Clark, G. 1981, "The Employment Relation and the Spatial Division of Labor: A Hypothesis", *Annals of the Association of American Geographers*, 412-424.

Doody, E.J. and H.B. Munzer, 1981, *High Technology Employment in Massachusetts and Selected States*, Massachusetts Division of Employment Security, Boston.

Glasmeier, A.K., 1986, "High-tech Industries and the Regional Division of Labor", *Industrial Relations* 25, 197-211.

Greenwood, M.J., 1975, "Research on Internal Migration in the United States: A Survey", *Journal of Economic Literature* 13, 397-433.

Greenwood, M.J., 1985, "Human Migration: Theory, Models, and Empirical Studies", *Journal of Regional Science* 25, 521-544.

Herzog, H.W. Jr. and A.M. Schlottmann, 1984, "Labor Force Mobility in the United States: Migration, Unemployment, and Remigration", *International Regional Science Review* 9, 43-58.

Herzog, H.W. Jr. and A.M. Schlottmann, 1986, "The Metro Rating Game: What can be Learned from the Recent Migrants", *Growth and Change* 17, 37-50.

Herzog, H.W. Jr., A.M. Schlottmann and D.L. Johnson, 1986, "High-Technology Jobs and Worker Mobility", *Journal of Regional Science* 26, 445-459.

Ladinsky, J., 1967, The Geographic Mobility of Professional and Technical Manpower, *Journal of Human Resources* 2, 474-494.

Markusen, A.R., 1986, "Defense Spending and the Geography of High-tech Industries", in Rees, J. *Technology, Regions, and Policy*, Rowman and Littlefield, Totowa, 94-119.

McMillan, T.E. Jr., 1965, "Why Manufacturers Choose Plant Locations vs. Determinants of Plant Location", *Land Economics* 41. 232-238.

McQuaid, R.W. and R.J. Langridge, 1984, "Defining High-Technology Industries", *British Section Meeting of the Regional Science Association*, Canterbury.

Muth, R.F., 1971, "Migration: Chicken or Egg?", *Southern Economic Journal* 37, 295-306.

National Science Foundation, 1980a, *Employment Patterns of Academic Scientists and Engineers*, 1973-78, G.P.O., Washington, D.C.

National Science Foundation, 1980b, *Science and Engineering Personnel: A National Overview*, G.P.O., Washington, D.C.

Premus, R., 1982, *Location of High Technology Firms and Regional Economic Development*, G.P.O., Washington, D.C.

Rees, J., 1979, "Technological Change and Regional Shifts in American Manufacturing", *The Professional Geographer* 31, 45-54.

Rees, J. and H.A. Stafford, 1986, "Theories of Regional Growth and Industrial Location: Their Relevance for Understanding High-Technology Complexes", in Rees, J., *Technology, Regions, and Policy*, Rowman and Littlefield, Totowa, 23-50.

Riche, R.W., D.E. Hecker and J.U. Burgan, 1983, "High Technology Today and Tomorrow: A Small Slice of the Employment Pie", *Monthly Labor Review* 106, 50-58.

Ritchey, P.N., 1976, "Explanations of Migration", *Annual Review of Sociology* 2, 363-404.

Schlotmann, A.M. and H.W. Herzog Jr., 1984, "Career and Geographic Mobility Interactions: Implications for the Age Selectivity of Migration", *Journal of Human Resources* 19, 72-86.

Shaw, R.P., 1975, *Migration Theory and Fact*, Regional Science Research Institute, Philadelphia.

U.S. Bureau of the Census, 1983, *Census of Population and Housing, 1980: Public-Use Microdata Samples Technical Documentation*, G.P.O., Washington, D.C.

U.S. Department of Labor, Bureau of Labor Statistics, 1981, *The National Industry-Occupation Employment Matrix, 1970, 1978 and Projected 1990*, G.P.O., Washington, D.C.

CHAPTER 17

Economic Expansion and Establishment Growth on the Periphery

Michael Wiseman

17.1 INTRODUCTION

Consider the following hypothetical policy problem. A local government agency is involved in training and placing persons in private sector jobs. In order better to identify potential job openings and to match skills training with skill requirements, the agency decides to survey employers. A quick review of someone's old statistics text reveals what most suspected: it is unnecessary to survey all employers to meet the agency's objectives. But, as usual, the textbooks fail to help much when the student, now far distant from the foot of his or her professor, contemplates actually applying the theory. Which employers should be queried?

Agency personnel divide quickly into two camps. On one side are the "simplicists" who argue for the convenience of a random sample of employers. The opposing group, the "proportionists", question this strategy. Because all local employers do not have the same size workforce, a random sample of employers is not the same as a random sample of area jobs. Such a survey would oversample small establishments, and without reweighting the results would be disproportionately influenced by the needs and experiences of small operations. This problem could be corrected if each observation were weighted by employment size, but the efficiency of the sample could be increased by stratification by size of employer. "Concentrate on the big employers", the proportionists assert, "that's where the jobs are".

The simplicists try a new tack. "It is our experience, they argue, "that we are most successful in placing new employees with employers who are expanding their workforces. No only do such employers fill more job openings than are generated by normal turnover, but expansion often creates precisely those skills shortages that our agency is charged with meeting. And it is commonly claimed that small employers create most new jobs. (Birch, 1979) Therefore, while some reweighting may be necessary, we do better to oversample small employers, because that will improve our sample when considered in relation to the likely distribution of employment expansion."

Not to be daunted, the proportionists invoke Authority: "Gibrat's law", one member of the camp intones, "is inconsistent with your contention. That law states clearly that employment *growth* is unrelated to firm *size*. Therefore sampling various employers in proportion to their workforces represents the best strategy for collecting information on requirements associated with employment increase".

Convinced they have won the day, the proportionists prepare to stratify. But a last champion of the simplicists is unwilling to give up. "Wait", she says, "how dare you cite Gibrat's Law!" Consider the following:

- "We are concerned here with employment in our local area. In most of the literature Gibrat's Law is applied to firms, not establishments.[1] Since we are a peripheral region, many of our employers are branches of national and even multinational firms headquartered elsewhere and coordinating production here with activities of fraternal establishments all over the world. Even if Gibrat's law should apply to the firms taken as a unit, why should it apply to individual establishments?

- "Furthermore, most empirical work on Gibrat's law has focussed on manufacturing.[2] While manufacturing is important in our economy, other sectors, in particular services, are even more dynamic. How do we know that this "Law" applies in the service sector?

- "Much of the empirical work utilizes samples restricted to firms which are in business and continue in business, albeit with fluctuating employment. (Hall, 1987; Lang and Leonard, 1987) Yet I believe growth here is attributable more to business births than expansions. New firms tend to be small ones; sampling small firms might help us understand what skills new establishments will require.

- "The literature that is available often includes results that are inconsistent with Gibrat's law: Hall (1987), Lang and Leonard (1987), and Evans (1987b) all report significant inverse relationships between firm size and rates of growth.

- "And finally, there is the problem of theory: There is little in the neoclassical theory of the firm to explain why we should ever observe a stable size *distribution* in a single industry in which firms share a common technology of the type usually assumed in our textbooks? While models of firm behavior with properties that satisfy Gibrat's law exist, the restrictions required to obtain it are substantial."[3]

Unfortunately, the proportionists had a telling response to this barrage of literature. They pointed out that while it may be true that a proportional sample would be inefficient as an instrument for collecting data from growing firms, even the simplicists were coming to acknowledge the complexity of the problem. How should such a survey be conducted? The simplicists, acutely aware of the shortage of information on the distribution of employment levels and change by establishment size at the regional level,[4] admitted that they did not know.

This paper looks at the growth-size relationship in precisely the context described in the parable above. I use a new data set to study the decomposition of growth by establishment size in a dynamic peripheral economy. The study seeks answers to the following questions:

- What is the relation between size and jobs increase among employers in a growing regional economy?

---

[1]See the review in Scherer (1980), pp. 145-150.

[2]See Evans (1987a) and Hall (1987) and the references cited therein. Leonard (1987) uses data for establishments in a variety of non-manufacturing industries but fails to consider the definitional complications introduced by expanding analysis from manufacturing to, for example, retail trade.

[3]See Jovanic (1982) and the discussion of Jovanic's model in Evans (1986), Appendix A.

[4]There are exceptions. See, for example, Fox, Mayo, Miles, Bott, and Price (1987), Jackson (1985), and Teitz, Glasmeier, and Svensson (1981).

- What combination of establishment expansion, birth, and death produces the observed evolution of the industrial structure of employment?
- What are the implications for the placing of new employees and skills training?

It is my intention to use results for the area which I investigate to motivate consideration of extension of such research to regions outside of the U.S. As will become apparent, some of the results are quite surprising; it would be useful to know the extent to which the characteristics of establishment expansion in the context of overall regional growth that are observed here carry over to other institutional, economic, and social environments.

The paper is organized as follows. Section 2 provides background on the region and describes the data. The composition of employment expansion is described in Section 3. Section 4 studies the relation between employer size and growth. The research is summarized in Section 5.

17.2 THE DATA

The Sacramento Economy. The "dynamic peripheral economy" studied here is that of the Sacramento, California, Metropolitan Area. For reasons that will soon be apparent, the Sacramento Metropolitan Area is considered to have one of the strongest regional economies in the United States.[5] While the area was selected for this work primarily because of data availability, at least two other features make it an attractive object for regional study. One is the importance of "high technology" employment in the region's industrial base: Sacramento has been the beneficiary of much spinoff of microelectronics device fabrication from the congested "Silicon Valley" area that is only 125 miles away (Hoerter and Wiseman, 1986). A second is the importance of services in recent regional expansion. The 1980's have seen the Sacramento economy evolve to become the financial and services center for all of north-central California. As a result, the area presents an opportunity to study development in a sector that plays an important role in central city employment (Stanback, et al., 1981). In addition to manufacturing and services, the area's economy depends on employment in government (Sacramento is the capital of California), defense (Sacramento County includes two major U.S. Air Force Bases), and agriculture (the area's firth largest employer is the California Almond Growers' Exchange[6]).

The metropolitan area includes Sacramento, Yolo, El Dorado, and Placer Counties (see Figure 17.1). Total population of the area was about 1.3 million in 1986, 70% of which was in Sacramento County. Between 1979 and 1986 wage and salary employment in the metropolitan area grew at an average rate of more than 3% per year; most of this growth was accumulated following the end of the business recession of 1981-82. By 1986 area employers reported 526,000 jobs.[7]

Figure 17.2 illustrates Sacramento metropolitan employment growth rates by sector. The rapid growth of manufacturing employment is an exceptional feature of the area; this is one of the few labor markets in the United States in which the share of manufacturing employment in total jobs has stayed constant or increased in recent years. The burgeoning economy has lured a large number of in-migrants. However, the in-migration of labor has not been sufficient to offset the impact of the economy's growth on unemployment. Since 1982 the unemployment rate has declined by around 5 percentage points; it is now under 6%.

---

[5]For example, Chase Econometrics, Inc., a national economic forecasting firm, recently pronounced the economic outlook for the Sacramento Metropolitan Area "one of the brightest...for any region in the nation" (1986, p.5).

[6](Sacramento) *Business Journal*, 18 May 1987.

[7]Numbers reported in this paragraph are from California Employment Development Department (1986) plus unpublished data supplied by James McClelland, EDD Regional Specialist.

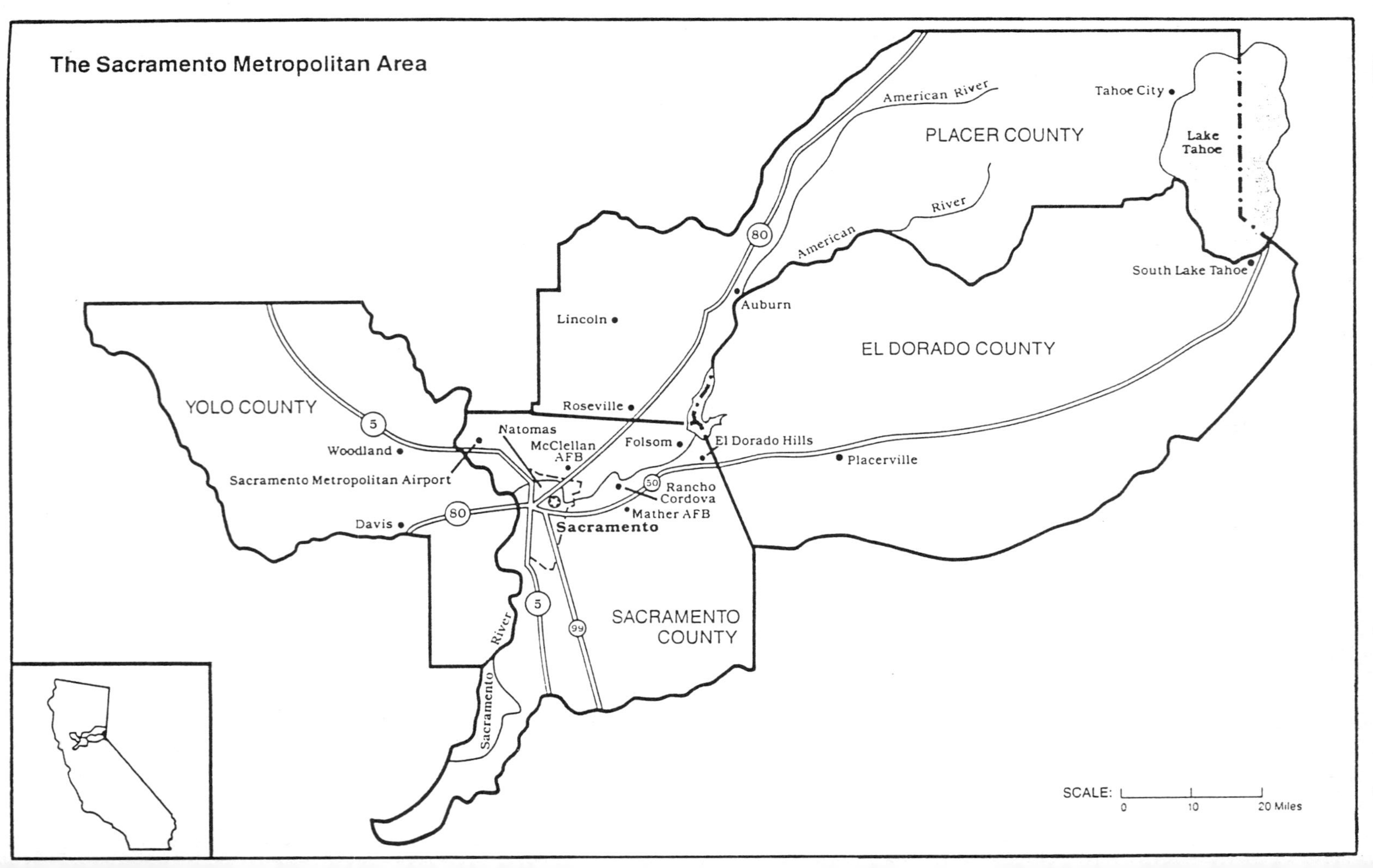
The Sacramento Metropolitan Area
American River
PLACER COUNTY
Tahoe City
Lake Tahoe
River
American
South Lake Tahoe
Auburn
Lincoln
EL DORADO COUNTY
YOLO COUNTY
Roseville
Natomas
McClellan AFB
Folsom
El Dorado Hills
Placerville
Woodland
Sacramento Metropolitan Airport
Rancho Cordova
Mather AFB
Davis
Sacramento
River
Sacramento
SACRAMENTO COUNTY
80
50
5
99
SCALE: 0 10 20 Miles

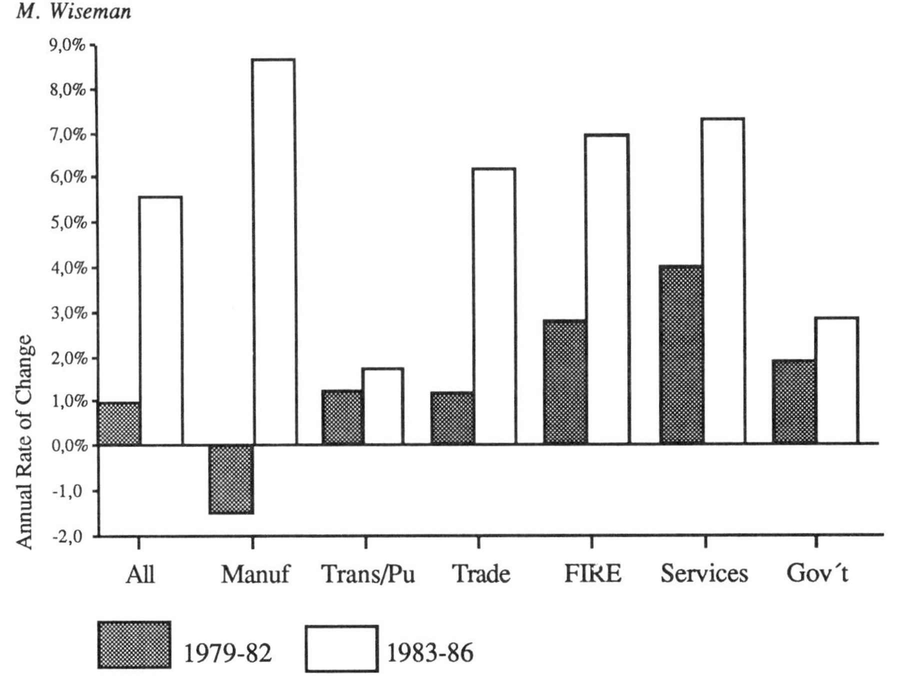

Figure 17.2 Growth rates by sector, Sacramento metropolitan area

Note: "All" is all wage and salary employment; "Manuf." is manufacturing; "Trans/PU" is transportation and public utilities; "Trade" is wholesale and retail trade; "FIRE" is finance, insurance and real estate; "Services" is personal and business services; and "Gov't" is all government employment.

The Data. As is well known, data on firms of the quality available in the Scandinavian countries are not collected, or at least are not publicly available, in the United States. This is especially true at the regional level, and the problem has been exacerbated in recent years by cutbacks in government statistical programs. However, attention to the structure of employment expansion has grown recently in California because of the planning requirements of a new state employment initiative for recipients of low-income assistance, the Greater Avenues for Independence (GAIN) program. In conjunction with planning for GAIN, the state's Employment Development Department has made available certain employer data to county planning agencies under strict confidentiality protocols. This paper utilizes a set of such data and abides by the protocols.[8]

The data cover all employers in the Sacramento MSA who pay at least one employee $1,500 or more in any quarter of the year. The data are collected as part of the administration of the national unemployment insurance program; this system is the basis for most federal employment and earnings statistics. Compliance with reporting requirements is very high because the program has been in operation for a long time and criminal penalties are imposed for errors and omissions. The largest excluded employment category is self-employed workers; this leads to understatement of the number of establishments in such industries as management consulting, legal services, and farming.

---

[8]In particular, except in instances in which publicly-available information may be quoted, it is not possible to identify any particular establishment from the data reported in this paper. For similar work done for an entire state, see Connor, et al. (1985).

The data, which cover employment by month, are collected quarterly. In principle, reports are filed on an establishment, rather than a firm basis. "Establishment" means a single operating facility including production, management, and support personnel. Firms with multiple production or distribution facilities thus file multiple reports. In practice, some firms in retail trade and various financial services are allowed to file one report that covers an entire county. For example, the Bank of Sacramento, with multiple branch offices, appears as four establishments on the tape - one for each county.

Because of changes in data maintenance procedures, it was possible to obtain information for only three years, 1984, 1985 and 1986. For these years I rely on average monthly employment reported in just the third calendar quarter. Employment contraction and expansion will be identified by year-to-year changes in average third quarter employment. If an establishment reports no employment in any year it is considered dead then; establishments which appear in 1985 or 1986 after a year with no employment report are treated as being born.

The most serious problem posed by these data arises as the product of enterprise transfers. If a printing and publishing plant in these data is sold, the original unemployment insurance account is closed and a new one, with a new number, is established. When such a change appears in reports of employment by establishment, the event is superficially identical to a sequential establishment "death" and "birth". Unless such sales are eliminated, the data will exaggerate the actual extent of genuine job loss and creation in the area. The original records maintained by the state included information suitable for identifying such transactions and merging the establishment records that in fact relate to a single enterprise. While it was not possible to undertake a complete clean-up of all of the numbers for this paper, all establishment death/birth sequences in which establishments were "born" at exceptional employment levels or deaths were matched by a subsequent birth of an establishment in the same four-digit standard industrial classification class were examined and, where appropriate, the data were adjusted to reflect the actual status of each establishment as an economic organization. I found, in making these adjustments, that such transfers did not pose a significant problem. Instead, the most serious distortions in the data were generated by failures of large, multi-establishment firms to report correctly employment by county. In particular, in manufacturing some large "births" simply reflected a reporting adjustment in which a firm with a Sacramento area establishment began, often well after the establishment had begun to employ workers in the area, to report such employees as working in Sacramento County instead of, for example, the Silicon Valley.

The original file created for this research includes 38,320 separate establishment records identified by the 4-digit Standard Industrial Classification code. For tractability and to assure the confidentiality of the individual establishment data, I work here with just three industrial subclassifications. These are *Printing and Publishing*, what I will term *High Technology Manufacturing*, and *Business Services*. These groups are defined in the following way. "Printing and Publishing" includes all establishments with an S.I.C. code in the two-digit category 27. Just what constitutes "High Technology Manufacturing" is a matter of considerable controversy. Here I include all firms in S.I.C. categories 283 (Drugs), 349 (Ordnance and Accessories), 357 (Office Computing and Accounting Machines), 376 (Guided Missiles and Space Vehicles and Parts), 379 (Miscellaneous Transportation Equipment) and all firms in the S.I.C. two-digit categories 36 (Electrical and Electronic Machinery) and 38 (Instruments). "Business Services" is defined here to include all establishments in S.I.C. categories 73 (Business Services) and 893 (Accounting, Auditing, and Bookkeeping). Establishment characteristics in the High Technology category are of interest because of the importance attributed to microprocessor-related manufacture in new industrial growth.[9] Establishment characteristics in Business Services are of interest because of the role of such employment in modern

---

[9]The high-technology category is defined in the same way in Browne (1983).

central city development. Printing and publishing is included because this industry also serves centralized management operations and because the establishments in the category are relatively homogenous.

Table 17.1 presents a first view of the data. These figures confirm the trends evident in Figure 17.2: Growth is very rapid. The timing appears to differ somewhat across industries, with printing and publishing and business services growing more rapidly in 1984-85 than in the subsequent year and high technology employment showing greatest growth (11%!) in 1985-86.

Table 17.1 Data overview

Industry	Printing and publishing			High Technology Industry			Business Services		
Year	1984	1985	1985	1984	1985	1986	1984	1985	1986
Establishments	229	224	262	114	116	107	1755	1965	2080
Employment	4473	5227	5486	8657	9647	10774	19640	23160	24993
Average employment per established	19.5	21.4	20.9	76.6	92.8	100.7	11.2	11.8	12.0
Employment standard Deviation/Mean	5.5	5.2	5.5	4.5	4.1	4.4	3.3	3.4	3.3
Employment growth, Annual rate		15.6%	4.8%		10.8%	11.0%		16.5%	7.6%

Source: Calculations by author from unpublished employment data furnished by the California Employment Development Department.

As might be expected, the smallest establishments are in Business Services, the largest in the high tech manufacturing category. But all three categories contain small and large operating facilities. Perhaps most striking is the diversity of establishment types within what are, by the standards of the literature, relatively homogeneous categories. Some examples of this heterogeneity help to provide a feel for the data. High technology employment includes giant firms like Aeroject Strategic Propulsion Company, a manufacturer of solid rocket motors, and Intel Corporation, a manufacturer of computer components.[10] At the same time the category includes establishments producing household refrigerators and freezers and residential lighting fixtures. Printing and publishing includes everything from newspapers such as the *Sacramento Bee* with 1,300 local employees to general commercial printers with less than 10. Business services cover everything from Management and Public Relations to Window Cleaning. Obviously, care must be taken in treating any of these groupings as collections of similar organizations, with similar technologies, which compete in the same markets for output.

17.3 BIRTHS, DEATHS, EXPANSIONS, AND CONTRACTIONS IN THE PROCESS OF ECONOMIC DEVELOPMENT

Table 17.1 shows that employment in each of these categories grew rapidly in the years covered by the data. The next step is to look at the distribution of employment changes across establishments. This is summarized in Table 17.2.

[10]These names and employment figures and those reported elsewhere in this paragraph are not from the confidential data. They are reported in the (Sacramento) *Business Journal*, 5 January 1987, p.41.

Table 17.2 Establishment changes, 1984-1986

Industry	Printing and Publishing			High Technology Industry			Business Services		
Year	1984	1985	1986	1984	1985	1986	1984	1985	1986
From year (t-1) to Year (t), establishments:									
Born	n.a.	42	48	n.a.	25	17	n.a.	512	477
Expanding employment	n.a.	106	94	n.a.	44	50	n.a.	730	712
With no employment change	n.a.	35	34	n.a.	12	10	n.a.	296	357
Reducing employment	n.a.	61	86	n.a.	35	30	n.a.	427	534
Which die	n.a.	27	30	n.a.	23	26	n.a.	302	362
Total establishments with Employment in year (t)	229	244	262	114	116	107	1755	1965	2080

Source: See Table 17.1

What Table 17.2 says may be explained by example. Consider the changes in Printing and Publishing between 1984 and 1985. Of the 229 establishments in this category in the third quarter of 1983 (see the last entry in the first column of the table), 27 ceased to employ anyone during the subsequent twelve months (next-to-last entry, second column). This loss of employers was offset by the births of 42 new ones (first entry, second column). Of the 202 1984 employers which did not die, by 1985 106 had increased employment, 61 had reduced employment, and only 35 reported no change. The slow-up in the rate of employment growth by 1985-86 (see Table 17.1) is associated with an increase in deaths and the number of establishments contracting employment.

As has been repeatedly discovered using similar data for other locations, these data indicate that even the strongest aggregate growth is the product of a complex process of establishment birth, expansion, decline, and, in many instances, demise. Thus the 11% growth of employment in high technology employment in 1985-86 was accomplished despite the fact that almost one half of the 116 establishments with employment in the third quarter of 1985 reduced employment. Indeed, 23% ceased employing at all. In 1986 18% of printing and publishing, 16% of high technology manufacturing, and 23% of business services establishments were newborn.

Table 17.2 describes the distribution of experience across establishments. But this does not tell us about the numbers of jobs involved in each group. To do this, it is necessary to disaggregate changes in employment by source. This is undertaken in Table 17.3. Like Table 17.2, table 17.3 is best explained by working through an industry. Consider Business Services, the last of the three sets of columns in the table. Between 1984 and 1985 employment in BS increased by 3,520 jobs (bottom line). This was the net result of 6,652 new jobs in new establishments and establishments expanding employment (lines 3 and 5) and 3,132 job losses attributable to the demise or contraction of other establishments (lines 8 and 10).

In some instances these data are affected by large single firm changes. For example, one major closing in 1985 significantly affects the relative size of gains versus losses and the share of employment loss attributable to establishment death for high technology industry between 1984 and 1985. Overall the impression gained from looking at employment is consistent with the message of table 17.2: Even when employment is growing, many jobs are "lost" through establishment death or contraction each year. As a rule of thumb for printing and publishing and business services, it is reasonable to say that about half of all jobs gained come from establishment births and about half of all jobs lost are the result of establishment demise. For high technology manufacturing it appears that a much larger share of employment expansion is attributable to growth, rather than birth of firms than is the case for printing and publishing or business services.

Table 17.3 Sources of employment growth, 1984-86

	Printing and Publishing			High Technology Industry			Business Services		
	1984	1985	1986	1984	1985	1986	1984	1985	1986
Total employment	4473	5227	5486	8657	9647	10774	19640	23160	24993
Jobs gained		994	670		1750	1523		6652	6438
From establishment births									
Number		528	195		158	157		2647	3075
Proportion of gain		0.53	0.29		0.09	0.10		0.40	0.48
Establishment expansion									
Number		466	475		1592	1356		4005	3362
Proportion of gain		0.47	0.71		0.91	0.90		0.60	0.52
Jobs lost		-239	-411		-760	-386		-3132	-4605
From establishment deaths									
Number		-116	-200		-570	-157		-1438	-2278
Proportion of loss		0.48	0.49		0.75	0.41		0.47	0.49
Establishment Contraction									
Number		-123	-211		-190	-229		-1649	-2327
Proportion of loss		0.52	0.51		0.25	0.59		0.53	0.51
Net increase		755	259		990	1127		3520	1833

Source: See Table 17.1

17.4 ECONOMIC EXPANSION AND THE SIZE OF NEW ESTABLISHMENTS

New jobs in the Sacramento Metropolitan Area show up in small establishments if they are created by establishment birth and in large establishments if they are created by establishment expansion. Table 17.4 shows average establishment size for jobs created through establishment birth and expansion as well as the size of establishments associated with employment loss. Consider the data for 1985 for printing and publishing. Table 17.1 shows that in 1985 the average establishment in printing and publishing employed 21.4 people, up from 19.5 the previous year. Table 17.4 shows that the jobs created between 1984 and 1985 by establishment birth were in establishments of well below average size (12.6 employees compared to a group average of 21.4), but the average job gained through expansion was located among very large establishments (average size was 310 employees). The same is true for high technology industry and business services. For jobs lost, the picture is mixed. For printing and publishing and high technology manufacturing, the jobs lost through demise tend to be in establishments with below-average employment, while the jobs in printing and publication which were lost through contraction were located in establishments with slightly above-average employment in 1985 and below-average employment in 1986. For business services on average in both years the jobs lost both through demise and contraction were located in establishments of above-average size.

Table 17.4 The Environment of new jobs, 1985-86

Industry	Printing and Publishing		High Technology		Business Services	
Year	1985	1986	1985	1986	1985	1986
Average size of establishment						
In year t, jobs born since t-1	12,6	4.1	6.3	9.7	5.2	6.4
In year t-1, jobs gained through expansion since t-1	309.9	423.6	860.6	1711.9	93.0	72.7
In year t, all new jobs	280.1	335.4	1034.2	1534.3	19.2	90.4
In year t-1, jobs lost due to establishment demise since t-1	4.3	6.7	28.4	14.6	13.1	14.6
In year t-1, jobs lost through establishment contraction since t-1	24.4	17.3	105.9	117.4	65.3	149.3
In year t, all jobs	21.4	20.9	92.8	100.7	11.8	12.0

Source: See Table 17.1

Table 17.5 considers the distribution of change by establishment size. The data are sorted in two ways. One is by rank, so that establishments are ordered by size and then characteristics for the bottom 50% are compared to the same characteristics for the upper 50%. The second starts with the same ordering but combines in a "smaller establishment" category all enterprises which cumulatively account for 50% of the industry's employment. The table is read as follows. As was presented in Table 17.3, in 1984-85 printing and publishing employed increased by 755 jobs, the net result of 994 new jobs and 239 job declines. Over half of all expansions were attributable to births. 6.9% of job expansions were accumulated by the smallest 50% of firms, while 39.9% were attributable to the largest 50%. Differentiation by cumulative employment share presents a different picture: 35.8% of all expansion employment was attributable to the smallest firms accounting for half of printing and publishing employment in 1984. The large firms accounting for the other half produced only 11% of job expansions. Roughly the same picture, with opposite signs, is presented for the distribution of employment contraction. Note that in this table employment loss through demise is included with contractions, while births are listed separately from expansions since by definition all firms born between 1984 and 1985 had zero employment in 1984.

For business services the general picture is similar to printing and publishing. Again, the bottom half of the firm size distribution accounts for only a very small proportion of expansions and an insignificant share of net increase. But when the distribution is considered on the basis of cumulative employment, the smaller firms which together account for half of all employment in the industry account for a disproportionate share of both expansions and contractions. Clearly 1985-86 was a period of significant change for the industry; all net employment change is attributable to establishment births. In high technology industry the picture is different: In both periods, the very largest firms were responsible for a disproportionate share of the net industry increase in employment.

Table 17.5: Employment change among small and large establishments

	Printing and Publishing				High Technology Industry				Business Services			
	1984-85		1985-86		1984-85		1985-86		1984-85		1985-86	
	No.	Share of increase	No.	Share of increase	No.	Share of increase	No.	Share of increase	No.	Share of increase	No.	Share of increase
Job increases	944		670		1750		1513		4005		6438	
All births	528	53.1%	195	29.2%	158	9.0%	157	10.4%	2647	66.1%	3,075	47.8%
Expansion, Bottom 50%*	69	6.9%	67	10.0%	73	4.1%	62	4.1%	507	12.7%	535	8.3%
Expansion, Top 50%	397	39.9%	408	60.9%	1543	88.0%	1293	85.5%	3498	87.3%	2827	43.9%
Expansion, Smaller	356	35.8%	263	39.3%	837	47.8%	778	51.4%	2419	60.4%	2204	34.2%
Expansion, Larger	110	11.0%	211	31.5%	755	43.0%	578	38.2%	1586	39.6%	1158	18.0%
Job Decreases	-239		-411		-760		-386		-3132		-4605	
Contraction, Bottom 50%	-70	29.4%	-92	22.4%	-42	5.5%	-30	7.9%	-466	14.9%	-562	12.2%
Contraction, Top 50%	-169	70.6%	-319	77.6%	-718	94.5%	-356	92.1%	-2666	85.1%	-4043	87.8%
Contraction, Smaller	-227	94.8%	-383	93.0%	-760	100.0%	-386	100.0%	-1849	59.0%	-2766	60.1%
Contraction, Larger	-12	5.2%	-29	7.0%	0	0.0%	0	0.0%	-1283	41.0%	1839	39.9%
Net change between years	755		259		990		1127		3520		1833	
Births	528	70.0%	195	75.3%	158	15.9%	157	13.9%	2647	75.2%	3075	167.8%
Net change, Bottom 50%	-1		-26		31		32		41		-27	
Net change, Top 50%	228		89		825		937		832		-1216	
Net change, Smaller	129		-119		77		392		571		-562	
Net change, Larger	97		183		755		578		303		-681	

Source: See Table 17.1
* "Bottom" and "Top" refer to lower and upper 50% of establishments ranked by employment.
** "Smaller" and "Larger" refer to lower and upper 50% of establishments ranked by cumulative employment.

The results summarized in Table 17.5 are most effectively illustrated by a Lorenz-type illustration. In Figures 17.3a, 17.3b and 17.3c I have plotted the distribution of employment by firm size for the three industry groupings. For each figure, establishments were first ranked by size in 1984 from the smallest to the largest and employment, employment increase, and employment contraction cumulated down the list. When plotted the cumulation of employment, expressed as a percent of total employment in 1984, yields the familiar Lorenz curve. In Figure 17.3a, this is the curve marked "employment" for printing and publishing. The curve reveals an important feature of the distribution of employment in the industry; the smallest 60% of establishments account for only about 10% of area employment in the industry. The "Employment Increase" and "Employment Decrease" lines respectively show the share of total industry employment expansion and employment decrease between 1984 and 1985. The fact that these lines are everywhere above the Lorenz curve for employment tells us something important: both job expansions and job contractions are more uniformly distributed across the establishment size distribution than is employment. The fact that contractions are the most evenly distributed says that, in the absence of a countervailing influence over time, average establishment size would increase substantially. The countervailing influence is, of course, employment generated by new establishment births. Table 17.4 says that new employers in this industry are relatively small, and this serves to stabilize the size distribution overall.

The greater equality in the distribution of employment change than in employment is also evident in high technology manufacturing and business services, but the difference between the distribution of change and the distribution of employment size is less striking in those industries. The results for high technology employment are exceptionally influenced by one large expansion and a significant demise.

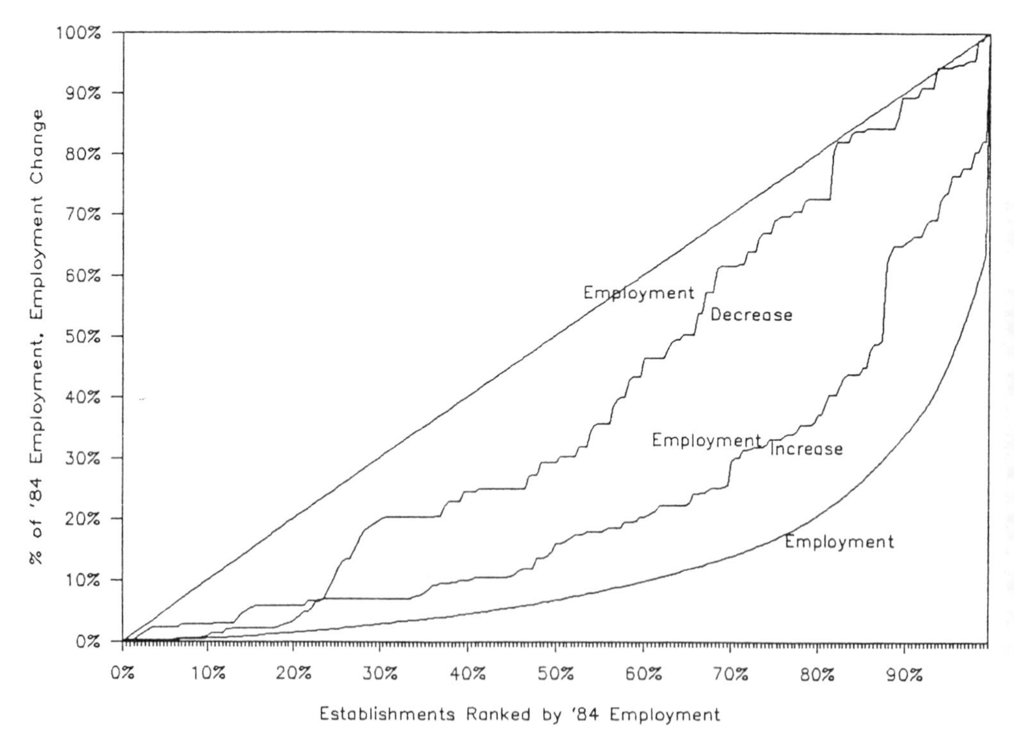

Figure 17.3.a Employment and employment change distribution, 1984-85 Printing and publishing

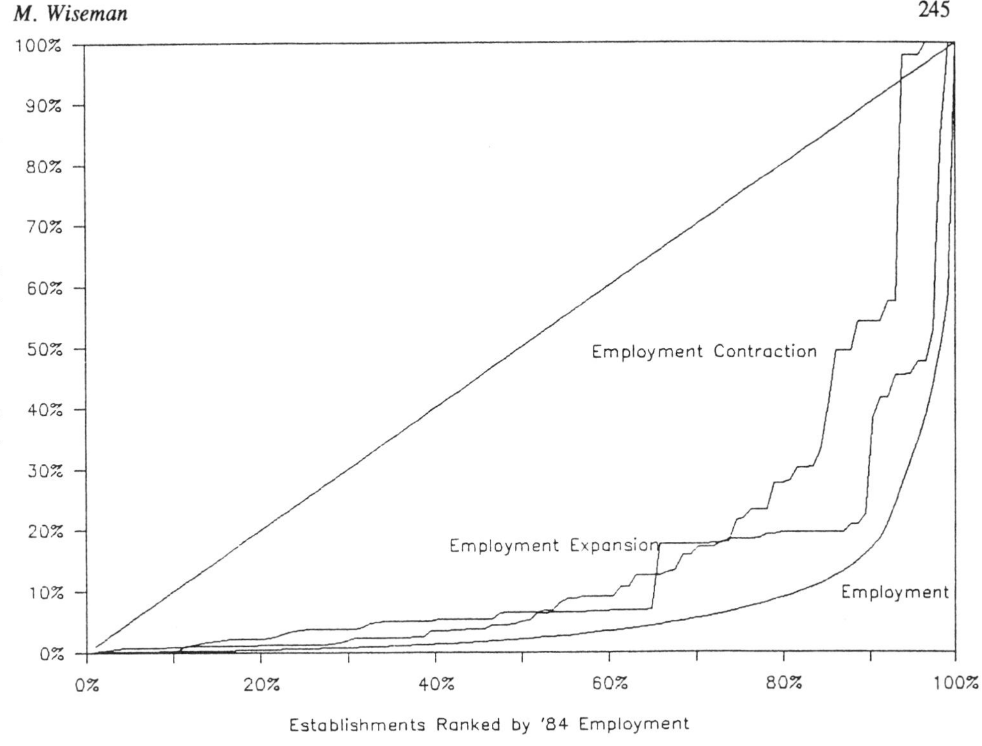

Figure 17.3.b Employment and employment change distribution, 1984-85
High technology manufacturing

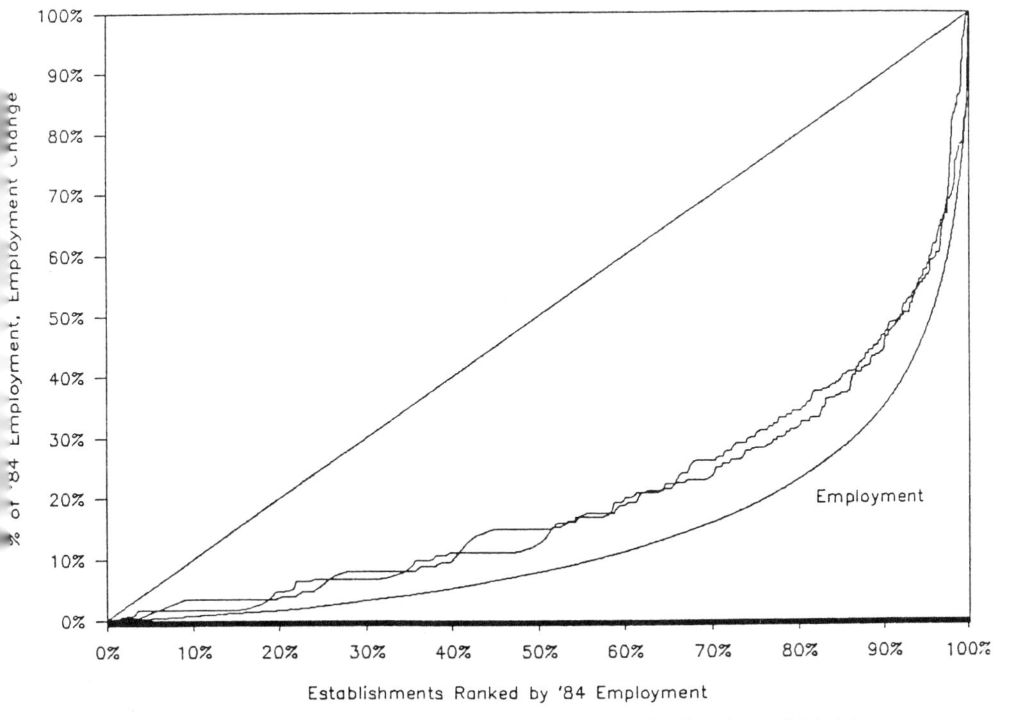

Figure 17.3.c Employment and employment change distribution, 1984-85
Business services

The greater equality in the distribution of employment change than in employment is evident also in high technology manufacturing and business services, but the difference between the distribution of change and the distribution of employment size is less striking in those industries. The results for high technology employment are exceptionally influenced by one large expansion and a significant demise.

The greater equality across establishments of job demise compared to employment size implies something important: the hazard of job loss because of establishment demise or contraction is substantially greater in smaller than larger enterprises. Table 17.6 illustrates the risk for employment in firms ranked on both of the two criteria introduced above. These rates are substantial: jobs in the 50% of firms that are smallest have on average about a 25% chance of disappearing. Because of the unevenness of the size distribution, of course, the share of total employment in firms in the lower half of the size distribution is small. But the next-to-last line of the table shows that the 50% of all employees located in smaller establishments have a 9-10% chance of job loss in printing and publishing, a 10-20% chance of job loss in high technology manufacturing, and a 19-25% chance of job loss in business services.

Table 17.6 Employment at risk: Probability of job loss by establishment

	Printing and Publishing		High Technology		Business Services	
	1984	1985	1984	1985	1984	1985
Proportion of jobs lost through establishment demise:						
Lower 50%, by rank	0.17	0.15	0.18	0.16	0.22	0.23
Upper 50%, by rank	0.02	0.03	0.06	0.01	0.06	0.09
Lower 50% of employment	0.05	0.08	0.13	0.03	0.10	0.13
Upper 50% of employment	0.00	0.00	0.00	0.00	0.05	0.07
Proportion of jobs lost through establishment contractions:						
Lower 50%, by rank	0.07	0.15	0.07	0.07	0.08	0.09
Upper 50%, by rank	0.02	0.01	0.02	0.02	0.08	0.10
Lower 50% of employment	0.05	0.01	0.05	0.05	0.09	0.11
Upper 50% of employment	0.01	0.01	0.00	0.00	0.08	0.09
Combined hazard, job loss						
Lower 50%, by rank	0.24	0.30	0.26	0.23	0.30	0.32
Upper 50%, by rank	0.04	0.04	0.08	0.04	0.15	0.19
Lower 50% of employment	0.10	0.09	0.18	0.09	0.19	0.24
Upper 50% of employment	0.01	0.01	0.00	0.00	0.13	0.16

Source: Se Table 17.1

17.5 CORRELATION OF SIZE WITH EMPLOYMENT EXPANSION

Finally, these data illustrate a problem with standard "tests" of Gibrat's Law. Such studies frequently confine attention to firms which survive over one or more periods. It is clear from these data that such a restriction differentially effects large and small firms; the result is the selection bias that bedevils so many econometric exercises. In addition, it is important not to confuse rates of change with quantities. Even if small establishments grow faster than large ones, the shortrun consequence for the distribution of actual numbers of jobs created may be inconsequential. An establishment with 5 employees that doubles in size creates only 5 jobs; a lethargic firm with 1000 employees bests this by growing at only a 1% rate.

Space does not permit detailed investigation of these issues here. However, I have calculated the simple correlation between growth rates and size for 1984-85 and 1985-86. Each sample is restricted to those establishments with positive employment in each of the two years over which change is measured. In addition I have calculated for those establishments alive throughout the sample the correlation between growth rate in 1985-86 and growth rate in 1984-85. Gibrat's Law could be rejected by a significant relationship between size and growth, although the same outcome could be the product of a random walk growth process in which outliers on both ends of the distribution tend to regress toward the mean of distribution. If growth "innovations" are random and disturbances are serially uncorrelated, then we would expect growth rates in adjoining years to be negatively correlated (Lang and Leonard, 1987).

Table 17.7 reveals, in every instance, an inverse correlation between initial size and subsequent rate of change. If we view these data as a sample of some universe of growth trials, the inverse size-growth relationship is statistically significant in 3 of 6 tests. In all cases the two year's growth rates are negatively correlated, but only for business services is the inverse relationship substantial and statistically significant. The negative correlations do not for the most part point to numerically important relationships: even for business services the estimated elasticity of growth rates with respect to size is only around .10.

Evans (1987a) shows that for manufacturing firms rates of growth diminish with age. My data do not contain satisfactory information on establishment age. However, my data do permit investigation of whether establishments born in 1985 grow exceptionally rapidly (given initial size) over the period 1985-86. This appears to be the case for all three industries, although the relationship is less certain for business services than for printing and publishing and high technology employment. One implication of this result is that newly-arrived establishments are a good target for placing newly-trained workers.

Table 17.7 Correlation of rate of employment change with size

Industry	Printing and publishing			High Technology Industry			Business Services		
	Employment 1984	Employment 1985	Rate of employment Increase 1984-85	Employment 1984	Employment 1985	Rate of employment Increase 1984-85	Employment 1984	Employment 1985	Rate of employment Increase 1984-85
Rate of employment increase, 1984-85	-.07			-.03			-.15*		
Rate of employment increase, 1985-86		-.15*	-.01		-1.9	.02		-2.0*	-1.5*

Source: See Table 17.1. Correlations marked with asterisk meet 5% criterion of statistical significance or better.

17.5 SUMMARY

Where does all this leave the issue of outreach for training and placement policy? To this point the answers are hardly conclusive. However, some things are clear.

- The Sacramento data are consistent with the proposition that growth rates decline with establishment size. But the relation is weak, and growth rates should not be confused with absolute numbers of new jobs created.

- Persons studying the relationship between size and employment change face a serious selection bias problem if the establishment or firm samples employed include only firms with continuous employment histories. While, on average, smaller establishments in Sacramento may have higher growth rates than do large ones, the probability that any one establishment, or any job in an establishment, will "die" is much greater for small than larger firms.

- Except for the very largest firms, it may be unwise to provide training tailored to the needs of a specific employer. The high rates of job loss evident in these data suggest that for many establishments any new employee is probably at significant risk of job loss. The best training is training he or she can take away from the establishment.

Finally, it would appear that these dynamic processes of establishment growth and decline would be a fruitful topic for international comparative study. Obviously the rate of employment contraction and demise evident in these data contributes to maintenance of a high rate of what is commonly termed frictional unemployment. Is this a common feature of area economies undergoing rapid change?

REFERENCES

Birch, D.F., 1979, *The Job Generation Process*, MIT Program on Neighborhood and Regional Change, Cambridge, Ma.

Browne, L.E., 1983, "High Technology and Business Services", *New England Economic Review*, July/August, 5-17.

California Employment Development Department, 1986, *Annual Planning Information, Sacramento Metropolitan Statistical Area 1986-87*, Employment Development Department, State of California Health and Welfare Agency, Sacramento.

Chase Econometrics: The Short-Term Outlook for the Sacramento Metropolitan Statistical Area. Bala Cycwyd, Pennsylvania: Chase Econometrics, 1986.

Connor, J., S. Heeringa and J. Jackson, 1985, *Measuring and Understanding Economic Change in Michigan*, The University of Michigan Institute for Social Research, Ann Arbor.

Evans, D.S.: Tests of Alternative Theories of Firm Growth, New York University C.V. Starr Center for Applied Economics, Economics Research Report 86-31, December 1986.

Evans, D.S., 1987a, "The Relationship Between Firm Growth, Size, and Age: Estimates for 100 Manufacturing Industries, *Journal of Industrial Economics*, 35(4)

Evans, D.S., 1987b, "Tests of Alternative Theories of Firm Growth", *Journal of Political Economy*, 95(4)

Fox, W.F., J.W. Mayo, M.S. Miles, S.E. Bots, and P.A. Price, 1987, *Entries and Exits of Firms in the Tennessee Economy, 1980-85*, Center for Business and Economic Research, College of Business Administration, University of Tennessee, Knoxville, Tennessee.

Hall, B.H., 1987, "The Relationship Between Firm Size and Firm Growth in the U.S. Manufacturing Sector", *Journal of Industrial Economics*, 35(4).

Hoerter, D., and M. Wiseman, 1986, *Metropolitan Development in the San Francisco Bay Area*, Institute for Urban and Regional Development, Discussion Paper 453, Berkeley, Ca.

Jackson, J.E., 1985, *Economic Change in Michigan: 1978-84*, Working Paper.

Jovanovic, B., 1982, "Selection and Evolution of Industry", *Econometrica*, 50(3), 649-670.

Lang, K., and J.S. Leonard, 1987, "In the Wrong Place at the Wrong Time: The Extent of Frictional and Structural Employment" in Lang, K. and J. Leonard, (eds.), *Unemployment and the Structure of Labor Markets*, Basil Blackwell, Inc., New York, 141-163.

Scherer, F.M., 1980, *Industrial Market Structure and Economic Performance*, Second Edition, Houghton Mifflin Company, Boston.

Stanback, T.M. Jr., P.J. Bearse, T.J. Noyelle and R.A. Karasek, 1981, *Services: The New Economy*, Rowman & Allanheld, Totowa, New Jersey.

Teitz, M.B. A. Glasmeier and D. Svensson, 1981, *Small Business and Employment Growth in California*, Working Paper No. 348, University of California Institute of Urban and Regional Development, Berkeley, Ca.

CHAPTER 18

The Economic, Industrial, and Regional Consequences of Defense-led Innovation

Ann Markusen

18.1 INTRODUCTION

"High tech" and "competitiveness" have become buzz words for the American economy's hopes and problems in the 1980s. The character of major economic issues appears to have changed. No longer do inflation, unemployment or even investment or growth dominate the public agenda. In their place, we have the deficit, the trade crisis, and the troubling question "Can America Compete?".[1] In this paper, I argue that a major contributor to these altered policy preoccupations is the emerging phenomenon of defense-led innovation in the U.S. economy.[2]

On the one hand, the sheer size and multiple products of defense-related research and development appear to have moderated the dynamic of creative destruction mapped out by Schumpeter and periodized by Mensch. By artificially inducing a continual stream of fundamental innovations, the product of Cold War strategy, the government effort may have freed innovations from the Menschian trap where they are only adopted during troughs in long waves. This in turn may have had the effect of obviating a severe recession in the present decade.

On the other hand, the defense-led effort has had a number of unintended consequences of rather major proportions which have created new problems for the economy. It has, of course, been extremely costly, adding substantially to the deficit problem. Second, it has funneled resources, especially in engineering and research, to a select set of defense-related firms and sectors, away from more commercially-oriented industries. As a result, the performance of the latter has declined, especially since firms in other countries, particularly Japan and Germany, have successfully commercialized many of the U.S. defense-originated inventions. This has enlarged both the trade and budget deficits. Third, it may have contributed to the slowdown in productivity growth, because the relatively fast-growing defense-related manufacturing sectors are highly labor-intensive.

A number of wider socioeconomic consequences are also argued to be connected to the rise of defense-led innovation. The accelerated rate of worker displacement in manufacturing may be an unintended product of the priority placed in the defense effort

---

[1]This is the title of Robert Lawrence's book, 1984, and the cover story of a recent *Business Week* special issue. Critics of Reagan economic policies add others to the list, like the "disappearing middle" of the income distribution and the crisis of domestic manufacturing. See Bluestone and Harrison, 1982; Harrison et al., 1986; and Zysman and Cohen, 1987.

[2]Portions of this paper are drawn from several previous publications (Markusen 1986, 1987a and 1987b)

on automating warfare, with its associated spin-offs in CAD/CAM systems, robotics and numerically-controlled machine tools. This, plus the swelling of the scientific and technical workforce in defense-oriented sectors, may be an important contributor to the apparent worsening of the income distribution in the U.S. Finally, the rise of aerospace as the prototypical defense-bred high tech industry in the U.S. has induced a massive shifting of manufacturing capacity, population and infrastructure toward what I call the "defense perimeter" of the country.

All of these phenomena, while present throughout the postwar period, have accelerated in the 1980s. In each case, I make the theoretical argument for the relationship with defense-led innovation and marshall what empirical evidence is at hand. The paper does not offer definitive proof of the existence or the strength of each of these relationships, but does show circumstantial correlative evidence. The policy implications is that before entrusting a competitiveness agenda to an enlarged military research program, as the current Strategic Defense Initiative does, we ought to require hard research results on the nature of the relationship between defense-led innovation and the evolution of the economy.

18.2 THE RISE OF DEFENSE-LED INNOVATION

Since the 1870s, we have have experienced periodic crises in capitalist economies. Whether one prefers the Keynesian underconsumptionist or the Marxist overproductionist theories of this phenomenon, it is clear that entire economies can experience prolonged periods of underutilization of capacity and widespread associated unemployment, in the face of demonstrable human need.

There have been two ways out of this morass, neither mutually exclusive. One has involved intervention by the state to guarantee a market for commodities and to reformulate the relationship between capital and labor in a manner which guarantees profitability. This was the Keynesian revolution. In the past decade, the limits of Keynesian approaches appear to have been reached with the emergence of inflation as a permanent tendency and the acceleration of international market integration which renders national policy efforts less potent.

The other has been a more anarchic, private sector initiative - the generation of a new set of innovations which render existing capacity and products lines obsolete, forcing massive restructuring and calling forth large investment commitments. In each major downturn over the past century and a half, a series of fundamental production-altering inventions were commodified, forming the basis of a new growth dynamic: railroads in the first half of the 19th century, steel and mechanical machinery in the latter half, autos and electrical machinery entering into the new century, petrochemicals and electronics since the 1930s (Hall, 1982; Mensch, 1979). It has been argued that major innovations emerge only in periods of severe crisis because in more prosperous times, entrepreneurs are preoccupied with the more routine tasks of supervising production and marketing their product.

In the most recent era, 1979 to present, national economic policymaking appears to have turned from Keynesian remedies (although one could consider military spending in Reagan proportions just that) toward permanent innovation as the route back to industrial supremacy. This path has been found rather accidentally, a byproduct of defense policy in the postwar period. Rather than waiting for hard times to propel basic innovation forward, U.S. policymakers are increasingly talking about the advisability of pursuing innovation in a sustained way, to germinate "sunrise" industries that will permit us to stay ahead of our industrial and third world competitors.

Permanent innovation is a product of defense policy. Historically, wars have been won with quantity not marksmanship. But cold war rivalry is an entirely different matter. Cold War has been prosecuted by institutionalized, continual innovation in weaponry and delivery systems. The goal of military research and development spending is to render the product of the opposition ineffective by designing new, improved weaponry. This sets in motion a process of rapid depreciation with sequential rounds of newly created technologies. Generations of fighter planes are retired, bins of nerve gas destroyed, missile silos abandoned.

This innovative dynamic does not arise from the conventional crisis of capital accumulation. Defense-based industry responds to a unique client and sports a competitive structure unlike that of industries producing for commercial markets. Demand is fairly much guaranteed by the persistence of the cold war and by the fact that there is one major buyer - the United States Government - who cares more about performance qualities of the products and timeliness of delivery than cost per se. Variations in the level and composition of demand do occur, but without clear links to business cycles.[3] Defense spending *is* subject to swings in geopolitical climate, as the Vietnam build-up and aftermath demonstrate, but these are less apt to affect the underlying innovative impulse than they do in purely commercial sectors.[4] Whatever the fashion in Pentagon budgets, research and development activities aimed at new, more sophisticated instruments of defense and destruction continue to be underwritten.

It appears that constant infusions of defense-led innovation have succeeded in ameliorating the tendency toward crisis in the American economy in the current decade. Without arms research, procurement and exports, plus the commercial spin-offs which have emerged, it is likely that the U.S. would have experienced a much deeper and prolonged recession, beginning in the late 1970s. However, this palliative has its costs. Indeed, a central theme of this paper is that a new set of contradictions has accompanied defense-led innovation, creating new and confusing political pressures. Chief among these, although not explored here, is the massive and growing U.S. deficit. For unlike private entrepreneurship, where the inventive costs are borne by individuals and corporations, the bill for defense-led innovation is footed by the American taxpayer. Since 1980 alone, the Pentagon budget has increased by more than 50%, outlays for weapons research by more than 80%. Thus in the 1980s, the rapid escalation in defense spending has foiled all attempts to balance the budget, despite the stated commitments by both the President and the Congress.

Yet despite its cost, military-led innovation has increasingly been championed as the United States' new competitive strategy. Until the present era, military research was conceived of as a strategic necessity, not as research program for basic industrial innovation. Yet in the most recent defense buildup, which will amount to a $2.5-trillion infrastructure program in the 1980s, the stated rationale has begun to include economic performance as a second, even co-equal, goal. The Strategic Computing Program, for instance, has as its goal "increasing national security and economic strength". The program, funded at $600 million for 5 years, contains an explicit commitment to industrial development, including the intent to stimulate new computer concepts, encourage industrial funding, generate new computer-based applications, and pioneer new techniques in software development and methodology. While the Pentagon has for years been subsidizing the adoption of automation technologies such as CAD/CAM systems and numerically controlled machine tools, this recent thrust brings commercial concerns farther back into the very structure of research programs themselves.

Indeed, the Strategic Defense Initiative places the major responsibility for U.S. industrial innovation formally in the hands of the military, which has informally played that role in any case. Up to World War II, the federal government accounted for only about 15% of research and development funding in the U.S., but the level had increased

---

[3]Though shifts in levels and growth rates of defense spending will, simply because of their sheer size, ameliorate or exacerbate peaks and troughs.

[4]Though even here, competition for resources within the defense sector can cut into funds for innovated research. An example is the ascendancy of combat priorities over research during the Vietnam war.

to above 60% by the 1960s and has not declined since. In computers, for example, the military and federal government accounted for virtually all of the market for computers in the decade following the war, an era which served as the entry point for all of the major computing firms. As I have argued elsewhere, the Pentagon has de facto directed an industrial policy which has encompassed targetted research and development, incentives for technological innovation, guaranteed procurements, a "Buy America" policy, a plant closing policy (Office of Economic Adjustment), lifetime job and income security (including retraining for retired officers), and community impact aid (Markusen, 1986a).

It is difficult to pinpoint the true political drives behind this melding of strategic and economic concerns. Department of Defense officials claim that promoting commercial spin-offs is the only way to make high levels of strategically mandated defense spending palatable to Congress. Defense-sponsored researchers and computer scientists express skepticism both that strategic applications like Star Wars are at all feasible or that this is a likely route to commercial payoffs, but claim this type of initiative is the only way to get badly needed dollars into basic research. Industrial policy advocates, at least those banking on "high tech" strategies, claim that the industrial applications are actually the point, but that this type of intervention into the marketplace is only possible when disguised as strategically necessary spending for defense.

Ironically, in the vast majority of discussions about the strategic computing initiative, the enemy to be overcome is not the Soviet but the Japanese. No one is contending that the Soviets are in any way equipped to compete with the U.S. on this type of high tech frontier. Indeed, it is widely acknowledged that they are far behind in basic computing capability. The motivating fear seems to be that efforts like the new Japanese 5th generation computer project will undermine the U.S. role as historic progenitor of new computer breakthroughs.

The implicit belief is that defense research efforts promise, having delivered in the past, commercially applicable innovations which ensure the nation's rather euphemistically invoked "economic strength". They will do so by maintaining the position of U.S. firms on the leading edge of an increasingly permanent wave of innovation, one set in motion by military concerns to begin with. But this developmental dynamic will work only if the U.S. is best poised to take advantage of its results. We return to this question in section 5.

18.3 DEFENSE-LED RECOMPOSITION OF AMERICAN INDUSTRY

Defense-led innovation has been narrowly channelled into a relatively small number of large and heavily defense-dependent firms, operating in a select set of primarily manufacturing industries. The sectors which fall into this category include military construction, aircraft, ammunition, ordnance, tanks, shipbuilding, missiles, communications equipment, electronics, explosives, scientific instruments, industrial trucks, and nonferrous metals. The firms who are heavily military-oriented operate in markets which differ markedly from the commercial markets in which other firms compete.

First of all, the nature of demand is almost entirely different from other manufacturing sectors. Military goods are sold mainly to a single buyer, the federal government, who is thus a monopsonist. But unlike private sector monopsonists this customer does not use its market power to depress the price of the product (and therefore the profits of sellers) in the market transaction. Performance and timeliness and dominant selection criteria for the Department of Defense; demand is price inelastic and may be zero during wartime. Cost insensitivity has been institutionalized into the cost-plus contract, where overruns in excess of 100 percent are common in larger weapons and equipment procurement programs (Baldwin, 1967, pp. 80-89; Gansler, 1980, pp. 82-92; Melman, 1984, pp. 27-39).

On the supply side, these industries are highly oligopolistic. A small number of corporations do the bulk of Defense Department business (Adams, 1981, pp. 33-41). The top 100 companies account for 70 percent of total business, the top 25 for 50 percent, and the top 5 for 20 percent (Gansler, 1980, pp. 35-45). Concentration ratios at the sectoral level are high and have been increasing over time (Baldwin, 1967, pp. 62-734). At the level of individual products, firms generally compete with only one or two others for the initial contract for a new system and after that, exercise a virtual monopoly. Subcontracting is considerably more competitive, but in this arena too, concentration has been growing rapidly, in part because the larger prime contractors have increasingly chosen to produce "in -house" than to buy from others. Often subcontractors have survived by occupying market niches and becoming monopolists of certain highly specific parts (Gansler, 1980, pp. 128-44).

The nature of the military product and its production process is also quite distinct from commercial manufacturing. The units of output of assembled weapons, equipment, and transportation systems are often (and increasingly) few in number and highly sophisticated. The product embodies constant innovation, requiring large numbers of scientific and technical personnel engaged in design, prototype building, and testing. Skilled labor for assembly (e.g. machinists) is also a large cost component. In contrast, materials account for a relatively small proportion of total costs. Often immense in size (e.g., submarines, military transport planes, missiles), these products tend to be produced in very large-scale facilities. Batch, rather than assembly line, production predominates, so that defense-oriented plants are more like huge craft shops than standardized mass production factories.

The defense build-up has been a major boon to these firms and sectors, whether measured by employment or output (Table 18.1). From 1977 to 1982, job gains were in excess of 30% for many of these sectors. All sectors in the table except small arms will enjoy an increase of 30% or more, emanating directly and indirectly from the Department of Defense from 1982 to 1987. Computers, a sector in which successful commercialization had pushed the military share down to a modest 7% by 1982, will be the single greatest beneficiary of new defense spending, slated for an increase of 141% in Defense Department demand in just five years, which will increase its defense share to 12%.

The heavily military-dependent industries are more or less synonymous with high tech industry.[5] Using a very broad definition of high tech, we found that 47% of all high tech employment occurs in defense-dependent industries. If we had applied a more stringent definition, the share would undoubtedly be higher. Whether measured by R&D share of sales or by engineering share of the workforce, defense-dependent industries rank higher than most commercially-oriented sectors. Indeed, the most "high tech" of any industry is Missiles (SIC 376), with more than 40% of its workforce in the engineering and technical occupations (Markusen, Hall and Glasmeier, 1986).

The heart of this military hardware complex is the aerospace industry, which, in addition to producing the big ticket items like aircraft and missiles, is a big consumer of TV and radio communications equipment, scientific equipment, semiconductors and other electronic devices. As a whole, this group will experience a growth in defense-oriented output of about 60%.

These figures understate the impact of defense-led innovation on American industrial recomposition. The rise of bellicosity in the world in general, and the lifting of limitations by the Reagan administration on the trade of first generation technologies and on defense research aimed at international markets, has created an additional round of impacts from a boom in the arms trade. The significance of the arms trade is not new. The aerospace industry alone has been responsible for the largest U.S. trade surpluses of manufactured goods of any single industry in the postwar period (Boretsky, 1975:81). But the U.S. accounted for about 45% of total world arms transfers by the early 1980s, compared with

---

[5]Tirman, 1985, has made the case for this connection.

20% in 1970 (Klare, 1984). More than 840,000 domestic jobs are currently generated by foreign military sales (Gansler, 1980:208-9).

Table 18.1 Department of defense shares of selected industry output, 1979-87

SIC code		1979	Defense share of output 1982	1987	Defence output growth 1982/87
		(%)	(%)	(%)	(%)
3795	Tanks and Tank components	78.1	93.8	95.0	47.2
3483	Ammunition	95.1	90.9	93.2	55.6
3489	Ordnance	85.1	79.7	81.2	35.3
3761	Complete missiles	71.0	67.5	79.4	64.4
3731	Shipbuilding and repair	47.9	61.7	62.1	24.1
3662	Radio and TV communication	44.8	58.0	62.5	54.2
3724, 64	Aircraft and missile engines	42.3	53.5	56.1	32.9
3728, 69	Aircraft and missile equipment	43.4	41.2	44.2	34.9
3721	Aircraft	35.0	40.4	46.1	58.7
383	Optical instruments	21.6	28.0	30.7	38.0
3811	Engineering instruments	23.5	27.7	33.6	59.9
3676-9	Electronic components	12.0	17.0	19.8	49.3
3674	Semiconductors	9.5	12.5	12.5	51.4
3361	Aluminium foundries (castings)	7.9	9.1	11.2	58.5
3334	Aluminium production	5.8	7.5	9.0	51.4
3469	Metal stampings	5.8	7.3	9.1	60.3
3571-3	Electron tubes	8.3	7.3	11.5	105.3
3573	Computers	3.6	7.1	12.7	141.0
345	Screw machine products	5.6	6.9	8.6	57.5
3462	Iron and steel forgings	7.9	6.9	7.6	31.4
3451	Machine tools - cutting	6.1	6.2	7.5	54.4
3544-5	Special dies and tools	4.9	6.0	7.5	45.4
3499	Fabricated metal products	5.0	5.6	6.8	53.4
281	Chemicals	5.5	5.6	7.1	53.3
3312	Blast furnace steel mills	4.5	5.6	6.7	45.6
3313	Electrometallurgical products	4.9	5.4	6.3	26.7
3542	Machine tools forming	5.0	4.8	6.3	70.0
332	Iron and steel foundries	3.9	4.5	5.2	45.1

Source: David Henry "Defense Spending: A Growth Market for Industry", *U.S. Industrial Outlook*, 1983: XXXIX-XLVII.

As the military-related sectors have flourished under the present administration, the more commercially-oriented manufacturing sectors have languished. The heavier, capital goods industries are particularly in distress. Partly, this is a result of the shift in federal budget priorities toward military procurement and away from social spending and infrastructure provision, with their adverse multiplier effects on commercial sectors. Partly, it is a function of growing imbalance in public commitments of funds for research; compare, for instance, the $30-billion Star Wars program with an endangered $15 million the steel industry is requesting for research on innovative technologies. Partly, it is a loss of third world markets, particularly in the increasingly troubled basic capital goods industries (farm machinery, construction machinery, machine tools) due to debt crises, continued recession and rising bellicosity, which is shifting larger shares of third world GNP into arms purchases. While not all of heavy industries' ills are attributable to military preoccupations - the overvalued dollar and poor management are also at fault - the role of the former looms much larger than is commonly understood.

The differential performance of U.S. arms vs. non-military sectors in international trade is striking (Table 18.2). With the exception of autos, where liberalized trade with Canada kept exports growing, no sector except military-type goods managed to increase its exports appreciably. Furthermore, import penetration was severe in all sectors but arms, which managed to triple its positive trade balance while all other sectors suffered declines. Non-military capital goods industries, in contrast, lost $30 billion to imports with no increase in exports, eliminating two-thirds of their traditionally strong export markets.

18.4 DEFENSE-LED RECOMPOSITION OF WORK

Because the product of military-related industries does embody continual innovation, the workforce in these industries is highly skewed toward those professional and technical occupational categories which are associated with scientific research. In manufacturing as a whole, the jobs going to engineers, engineering technicians, computer scientists, life and physical scientists and mathematicians account for less than 6% of employment. In missiles, they account for 41%, in computers 27%, in communications equipment 22% and in aircraft 19% (Markusen, Hall and Glasmeier, 1986). Studies from the early 1970s (no recent data has been found) show that between 22% of all U.S. electrical engineers, 48-59% of aeronautical engineers, and 23-38% of physicists work on military-related projects.

The current military buildup has created an enormous, unsatiated demand for engineers and related high tech professionals. The current demand for college engineering graduates on the part of California firms cannot be met by Californian universities (Parzen, 1982), resulting in aggressive recruitment campaigns aimed particularly at the better engineering institution in the Midwest. The job opportunities are not lost on young college students worried about their futures - electrical engineering is widely acknowledged to be the most sought after major for entering freshmen at California's top academic institutions.

Blue collar workers in certain skilled categories have traditionally been well-represented in defense manufacturing. In 1970, 54% of tool and die makers, 25% of sheet metal workers, 19% of machinists and airplane mechanics and 14% of draftsmen worked in military production (Dempsey and Schmude, 1971, Rutzick, 1970). But the shares of these occupations within the defense workforce are declining as defense-related production becomes even more high tech. This is occurring for two reasons. First, the quantity of weaponry produced is diminishing as its quality and price rise. Many fewer fighter planes are built now for the same real dollar expenditure than was true in the 1960s. Second, major defence contractors have pioneered the adoption of many types of automation within the production process, supported by Pentagon incentives. There is also some anecdotal evidence that firms are engaging in explicit union-building, again with Pentagon encouragement (Markusen, et al, 1985).

Defense-led innovation has had a strong adverse impact on those members of the workforce engaged in basic commercial manufacturing, a group which has consisted traditionally of white males but which, in the past decade or so, had welcomed increasing numbers of woman and minorities into its ranks. The adversity experienced by this group is in large part a function of the relative neglect of these sectors under the defense-oriented industrial policies at the national level, as well as the relatively poor performance of management in many of these sectors. However, job loss in these sectors has also been hastened by the application of defense-bred technologies to the production process in basic industry. Why defense-oriented research should have produced this particular type of technological change, rather than say the improvement of mass transit systems or better, cheaper-to-build housing, requires some explanation.

Table 18.2 U.S. Trade by major sector, 1980-1984 ($ billions)

Major Sector	1980	1984
Foods, feeds and beverages		
Exports	35.3	30.9
Imports	18.1	21.0
Balance	17.2	10.5
Industrial supplies and materials		
Exports	70.2	61.5
Imports	129.2	122.1
Balance	-59.0	-60.6
Capital goods, not automotive		
Exports	72.6	72.0
Imports	30.5	59.8
Balance	42.1	12.2
Automotive, vehicles, parts, engines		
Exports	16.2	20.9
Imports	28.0	55.1
Balance	-11.8	-34.3
Consumer goods, not food or auto		
Exports	16.2	13.3
Imports	38.4	60.0
Balance	-18.2	-46.6
Military-type goods		
Exports	10.2	24.4
Imports	4.6	7.7
Balance	5.6	16.7
All categories	220.8	223.6
Exports	244.9	325.7
Balance	-24.1	-102.2

Source: Norman Fieleke, "The Foreign Trade Deficit and American Industry", *New England Economic Review*, The Federal Reserve Bank of Boston, July/August, 1985:43-52.

Defense-oriented innovation has ultimately been aimed at the flexibility, maneuverability, invisibility, undetectability and infallibility of lethal weapons and their delivery systems. One of the most outstanding features of the products of this process is their *mobility* whether on land, in the air or on water. This mobile weaponry requires immense investments in computing, electronics, communications and instrumentation equipment to make it perform reliably. Thus, the sizable proportions of output of these industries currently accounted for by DOD, not to mention the leading role of DOD in their inception.

To use some examples from the present Strategic Computing Program, its goals are to "extricate the soldier from hazardous situations" (presumably obviating the hazardous situation itself is impossible), and "to help human decision-makers cope with the pace of conflict". The Pilot's Associate, for instance, the Air force's share of this project, will help the pilot (who is not going to be removed from the hazardous situation as will the tank operator in the autonomous land vehicle) who is "overwhelmed with incoming data and communications". Similarly, the Navy's Battle Management System, its share of the pie, will "generate potential courses of action, evaluate them and give supporting documentation" to naval officers in battle.

The types of automation which are presently driving the restructuring of manufacturing are not products of a natural evolutionary process, but explicit products of defense-driven priorities (Shaiken, 1985; Noble, 1984; Melman; 1983). It is the past research products of just such priorities that have given us the flexible production technologies, whether they be integrated circuits, numerically-controlled machine tools, or CAD|CAM systems, which have proved so potent in reorganizing the process of work (Noble, 1972). Just as these past generations of technological breakthroughs have permitted pilots to rely on auto piloting, they have provided the tools for removing the machinist from his and her lathe. As a result, levels of unemployment appear to have risen in the 1980s and those displaced are having a more difficult time finding new work (Bluestone and Harrison, 1982; Flaim and Seghal, 1985).

But aren't new jobs created in industries which manufacture the robots and automated technologies in turn? Ironically, the commercial applications of these innovations cannot generally be credited to American firms, but to the Japanese. Superior industrial engineering has enabled Japanese firms to develop the automated products which have displaced American workers both in machine tool producing industries and in the industries that use them, a subject explored in the next section.

In sum, then, the increased militarization of manufacturing is shifting job structure dramatically toward professional/technical workers and away from skilled and semi-skilled blue collar workers. Two important corollaries can be drawn. First, since the military-oriented high tech industries are highly labor-intensive, their ascendance over traditional capital-intensive industry may be responsible for a large portion of the productivity slowdown in the last decade. Second, the accompanying recomposition of the manufacturing workforce may be contributing to the observed worsening of the income distribution (Bluestone and Harrison, 1987). This occurs both because enlarged numbers of high-paid engineers appear as manufacturing workers and because the majority of formerly blue-collar workers are forced downward into either unemployment, retirement or lower-paid service sector jobs.

18.5 DEFENSE-LED INNOVATION AS INDUSTRIAL POLICY

The observation that workers in basic manufacturing are being displaced without compensating new job creation raises the question, again, of the efficacy of defense-led innovation as a long-term economic development and employment strategy. Especially if manufacturing workers are displaced at an accelerated rate, an unintended consequence of spin-offs from automating warfare, can defense-underwritten research guarantee that new sectors will be created to ensure the U.S. a niche in international trade and to simultaneously absorb displaced workers? An affirmative answer is assumed by those who espouse this path to an American competitive strategy.

But the evidence is far from reassuring on this front. Indeed, Japanese firms, without the benefit of a huge Defense budget and with a second-rate university research establishment, appear to have been much better positioned to take U.S. defense-generated technologies and apply them to basic industrial production. The ability to translate basic innovations into commercial hardware is a function of several key features of Japanese business organization: an evaluation of innovation in light of overall social goals; an emphasis on certain targetted industries (autos, steel, machine tools, electronics are examples); a cooperative and coordinated planning process between industry and government, between finance and industrial capital and among farms; lifetime job security and superior labor relations, and a system geared to "technology pull" - the concentration of the nation's scientific expertise on production technologies and product quality. In contrast, the domination of many high tech industries by military priorities, the relatively low status of industrial engineers compared with aerospace and electrical engineers and physicists, and the absence of any explicit commitment to targetting certain commercial sectors as the core of the economy, all contribute in the U.S. to a relatively poor record of translation of such innovations into high performance in basic industry.

Ironically, the present explicit shift in emphasis from competing with the Soviets to competing with the Japanese, and the choice of the strategic defense computing initiative as a primary vehicle for doing so, threatens to worsen this process. Each round of defense-bred innovation has, to date, resulted in the successful Japanese entry into a once strong domestic and export sector. This is most recently evident in the deterioration in the domestic machine tool industry, where Japanese sales of small and medium-sized machine tools have deeply penetrated American domestic markets.

A further concentration of U.S. research talent into another round of strategically-oriented research will invite another such translation, perhaps this time in computers. Indeed, the Defense Department is now considering the purchase of some types of applied technology from Japanese firms, because the latter have surpassed American products in performance. In return, the Japanese are asking for access to the newest generation of defense-funded research results, and the Department of Defense appears willing to accede to the bargain. These kinds of arrangements will accelerate the rate at which the major industrial competitor of the U.S. receives and commercializes the latter's high tech inventions, presumably with further depressive effects on basic industrial sectors within the U.S.

Put another way, the enemy of U.S. industrial viability is not the Japanese, but policymakers' proclivity to run faster in the innovation race. Yet, without past generations of defense-generated technologies, the U.S. would not be experiencing the present destabilizing impact on economic restructuring, at least not at the present pace. Indeed, the entire phenomenon of competition in high tech or the innovation race as I like to call it, bears an uncanny yet understandable resemblance to the arms race.

Thus, in addition to being expensive, the use of military Keynesianism to stave off recession in the short run does not serve as a intelligent long-term investment strategy for the U.S. economy. It undermines the more commercially-oriented manufacturing sectors, by draining resources away. It displaces skilled blue collar labor at an accelerated rate, without sufficient alternative opportunities for use of that labor pool, thereby worsening income distribution and increasing the social service burden on the public sector. And finally, and ironically, its lopsided favoring of firms oriented toward the military as clients renders American manufacturing in general less able to capitalize on new defense-generated innovations than are competitor nations with no appreciable military commitment.

18.6 THE REGIONAL CONSEQUENCES OF DEFENSE-LED INNOVATION

A separate question concerns the location of defense-led production and the associated spatial costs and benefits to the U.S. economy. For reasons I have discussed elsewhere, large-scale manufacturing plants associated with military procurement have tended to create new, spatially concentrated industrial complexes (Markusen, 1986b). These complexes are disproportionately located outside the older manufacturing cities, because of the attraction of large parcels of undeveloped land, the proximity of existing military bases, and the presence of a non-unionized workforce. They have created, in the process, new peripheral communities which house disproportionate numbers of skilled workers and are marked by relatively conservative and promilitary attitudes. These communities may be adjacent to older metropolitan centers, but have few interconnections with them, essentially forming their own labor market and cultural ties. Or, they may be in relatively nonurban or out-of-state locations, where one type of activity dominates the local economic structure. In this section, I review the evidence on shifts in U.S. manufacturing location in the postwar period, arguing that defense priorities have been responsible for a striking redistribution of the center of gravity of manufacturing capacity, at a high cost to the nation in resources associated with accompanying population resettlement.

In the postwar United States, defense spending, particularly on weapons development and procurement, has played a major role in the regional restructuring of manufacturing. Indeed, in place of the sunbelt/frostbelt dichotomy, which cannot explain the healthy manufacturing growth of places like Minneapolis and New England, I prefer the metaphor of "defense perimeter" - that area of the country surrounding the industrial heartland, a ring around the nation that extends from New England through Long Island, Florida, Texas, the southwestern states, up through California and Oregon. It is this perimeter which has received the lion's share of the nation's defense procurement and research dollars.

While I can only demonstrate correlation, not causation, the statistics on both manufacturing shifts and regional redistribution of defense dollars are quite compelling. In the postwar period, manufacturing has shifted its locus in the U.S. from the older industrial belt toward the south and west. Manufacturing growth rate differentials among the states were enormous between 1972 and 1984. While total U.S. manufacturing employment grew by a modest 1.1% over that period, states outside the industrial heartland added manufacturing jobs at remarkable rates (Table 18.3). Furthermore, it has been shown that manufacturing growth in the sunbelt has not consisted primarily of relocating manufacturing plants seeking a better business climate, but of a shift in the seedbed function, with the perimeter of the country hosting a disproportionate share of new, innovative industries (Norton and Rees, 1979). Both the sunbelt and New England have been major beneficiaries of high tech growth since the early 1970s. Their attractions appear to include high levels of per capita defense spending as well as selected amenities and business services (Markusen, Hall and Glasmeier, 1986).

These shifts in manufacturing closely parallel postwar shifts in concentration of defense prime contracts (Table 18.4). During the Korean War, as in WWII, the major recipients of defense contracts, measured on a per capita basis, were lower in New England, the Mid-Atlantic states, the industrial midwest and California and Washington. By the mid-1980s, the midwestern states' receipts had plummeted, along with New York and New Jersey, while new states on the periphery, including several in the southeast, the southwest and the plains, jumped from very low levels to higher-than-average ones. Both New England and the Pacific states held and/or increased their per capita receipts. It should be kept in mind that these represent very wide differentials in actual receipts. In real dollars, Illinois' prime contracts fell from $3.4 billion in 1951 to $0.6 billion in 1984, while California increased its receipts from $8.4 billion to $12 billion in the same period.

Table 18.3 Percent change in state manufacturing employment, 1972-84

Region/State	1972-1984	Region/State	1972-1984
NEW ENGLAND		PACIFIC	
New Hampshire	+35.7%	Alaska	38.2
Vermont	26.4	California	33.2
Massachusetts	10.7	Washington	27.0
SOUTH ATLANTIC		MIDDLE ATLANTIC	
Florida	42.9	Pennsylvania	-22.4
Georgia	14.4	New York	-16.9
		New Jersey	-11.9
WEST NORTH CENTRAL		SOUTH ATLANTIC	
South Dakota	55.4	West Virginia	-26.1
North Dakota	43.5	Maryland	-12.3
Kansas	21.2		
Minnesota	20.5		
WEST SOUTH CENTRAL		EAST NORTH CENTRAL	
Texas	34.9	Illinois	-23.1
Oklahoma	23.3	Ohio	-16.4
Arkansas	15.2	Michigan	-13.8
		Indiana	-12.4
MOUNTAIN WEST			
Nevada	114.2		
Arizona	74.5		
Utah	55.7		
Colorado	40.2		
New Mexico	24.3		
Idaho	24.3		

Shapira, 1986, from BLS statistics. Only those states with gains or losses exceeding 10% are shown. In the same period, the U.S. as a whole gained 1.1% manufacturing jobs.

The industrial heartland fares particularly badly in the research category (Table 18.5). The East North Central, which hosts a sizable portion of the nation's corporate R&D labs, received in 1983 only 14% of the national average in research expenditures per capita. The largest concentrations were in the South Atlantic area (both Virginia and Florida in particular), the Pacific states, New England and the Mountain states in that order. Since research is the key to the erection of new "seedbeds of innovation" and helps to tie nascent industries to its existing agglomerations, this highly skewed research spending pattern has presumably played a major role in the redistribution of defense-related production.

Table 18.4 Per capita prime contracts relative to U.S. average*

Region/State	1951	1958	1967	1977	1984
NEW ENGLAND					
Connecticut	440	300	350	310	330
Massachusetts	100	120	130	200	230
New Hampshire	70	50	120	90	130
EASTERN SEABOARD					
New York	190	120	100	120	100
New Jersey	160	120	90	80	80
Maryland	130	130	120	130	180
Virginia	30	50	80	190	160
SOUTHEAST					
Georgia	30	70	140	50	10
Mississippi	10	20	30	10	160
Florida	10	60	70	60	70
PLAINS					
Kansas	80	450	100	80	180
Missouri	60	100	260	230	250
SOUTHWEST					
Texas	40	130	180	100	100
Arizona	30	130	80	110	110
Utah	20	80	90	80	100
Nevada					
PACIFIC					
California	180	350	180	220	210
Washington	140	360	100	220	130
Alaska	-	-	160	150	170
Hawaii	-	-	50	120	100
MIDWEST					
Indiana	240	80	90	70	90
Michigan	200	60	60	70	50
Ohio	140	90	80	50	50
Wisconsin	100	30	50	40	40
Illinois	90	50	50	20	20

*All states not listed have had per capita prime receipts well below the national average.

Expenditure on prime contracts is an imperfect proxy for the impact of government spending on manufacturing. Procurement is heavily manufacturing oriented - some 66% of all private employment attributable to Department spending occurred in manufacturing sectors (Oliver, 1971). Some of this impact is felt by parts, materials and component suppliers, who may be located quite differently from major weapons producers and research labs. Prime contract data do not chart the indirect demands which filter through the economy via subcontracting.[6]

Table 18.5 Per capita prime contracts relative to U.S. average*

	Total Receipts ($ bills)	Procurement Ratio*	Personnel Ratio*	Research Ratio*	Total Ratio
New England	18.21	2.06	0.74	1.57	1.44
Mid- Atlantic	24.18	0.83	0.44	0.61	0.64
East North Central	18.02	0.45	0.44	0.14	0.43
West North Central	15.03	1.07	0.60	0.91	0.85
South Atlantic	53.38	0.97	1.67	2.27	1.35
East South Central	10.05	0.53	0.84	0.41	0.66
West South Central	23.39	0.81	1.08	0.21	0.89
Mountain	18.40	0.96	2.08	1.14	1.47
Pacific	56.69	1.85	1.44	1.75	1.66

*The ratios represent the region's per capita receipts divided by the national per capita expenditure for 1983.

Source: Jay Stein, "U.S. Defense Spending: Implications for Economic Development Planning", Working Paper, Georgia Institute of Technology, City Planning Program, 1985: Table 18.3.

An alternative way of capturing the subcontracting effect is to look at military shipments data from manufacturing industries. When arrayed by location quotient, these do not indicate an appreciably different distribution (Table 18.6). The same perimeter states rank in the top fifteen, while the interior states are heavily represented in the bottom fifteen. (Not all coastal states participate in the defense dividend - Oregon, Louisiana, the Carolinas and Delaware are conspicuously low in defense manufacturing). Some states, like Illinois and Indiana, fall somewhere in the middle, which suggests that a modest feedback effect through metals, machinery and parts is operating. Some states with relatively low per capita receipts, like Colorado, Vermont, Rhode Island and Florida, have a large share of their manufacturing shipments dedicated to the Department of Defence.

The relatively poor performance of the industrial heartland in that current period, even when indirect demand is taken into account, can be attributed to the changing nature of weapons production. Increasingly over the postwar period, weaponry has become more sophisticated with an ever larger component of guidance systems, communications

[6]Several studies have tracked the spatial array of subcontracting by prime contract recipients (Karaska, 1967; Riefler and Downing, 1968; Rees, 1981; 1982). All found that subcontracting tended to reinforce the primacy of the major prime contract regions, specifically the Pacific coast and northeastern regions. Subcontracting, on the basis of these studies, does not appear to have substantially dispersed funds from the military budget. Rees' study shows that the "periphery", defined as the arc of census regions from West North Central through South Atlantic, receives less through subcontracting than do the Pacific and Manufacturing Belt regions.

equipment and scientific instruments. A diminishing share of the price of a fighter plane or a cruise missile consists of materials and machine-made parts. Nor are many machines purchased for use in production. Instead the lion's share of the cost goes to quite labor-intensive design, prototype construction, research, testing and hand-crafted assembly. Furthermore, fewer and fewer units are purchases in any one sale, because the unit cost has gotten so high.

Table 18.6 Highest and lowest ranking states in military shipments, 1983

Rank	Top States	Military Shipments Location Quotient	Rank	Bottom States	Military Shipments Location Quotient
1	California	2.62	50	Arkansas	0.001
2	Connecticut	2.49	49	West Virginia	0.02
3	Missouri	2.49	48	Idaho	0.06
4	Arizona	2.30	47	North Carolina	0.07
5	Utah	2.10	46	South Carolina	0.08
6	Maryland	1.97	45	Kentucky	0.10
7	New Hampshire	1.88	44	Nebraska	0.12
8	Kansas	1.85	43	Alabama	0.26
9	Massachusetts	1.71	42	Montana	0.27
10	Vermont	1.66	41	Wisconsin	0.28
11	Virginia	1.55	40	Tennessee	0.30
12	Rhode Island	1.35	39	Oregon	0.33
13	Florida	1.25	38	Iowa	0.33
14	Colorado	1.23	37	Delaware	0.43
15	Texas	1.17	36	Michigan	0.45

Source: Breandan, O'hUallachain, "Some Implications of Recent Growth in the American Military-Industrial Complex". Working Paper, Northwestern University, Department of Geography, 1986, Table 1, p. 4. Data are from the Department of Commerce's *Current Industry Reports* and were weighted by 1982 employment levels to reflect the lag between production and final delivery.

The regional consequences of this change in production process are significant. The demand for steel and machine tools from the industrial heartland dwindles, while the demand for electronics, telecommunications equipment and instrumentation, heavily concentrated on the defense perimeter, escalates. To the surprise of many industrialists and union officials in the midwest, who have always believed that military build-up was good for their steel and machinery factories, the share of their output accounted for by the Department of Defense has fallen to less than 10%.

In contrast, the share of an industry like telecommunications equipment, the single largest high tech industry in the U.S., will exceed 60% by 1987 (Table 18.1). These industries are heavily concentrated around or near the new weapons-oriented agglomerations, particularly Los Angeles. The new complexes on the defense perimeter have been in the process of construction since the 1930s, when the aircraft industry relocated from the northeast. They received an additional boost after the Korean War with the rise of atmosphere warfare and primacy of the Air Force over the Army and Navy with their land and sea-based techniques.[7]

In sum, then, the defense build-up in the postwar period appears to have been a huge manufacturing redistributor in the United States. It is responsible for initiating entirely new seedbeds of innovation, outside of the traditional American industrial heartland. And since its activities continue to be highly innovation-oriented, these new centers of agglomeration have exhibited tremendous staying power.[8] The single most outstanding example is the Los Angeles area, where the aerospace industry has spawned growth to the north, to the east and to the south as high tech components in communication and electronics sectors cluster around the major weapons makers. A second major one is the greater Boston areas. To the extent that decentralization from these crowded agglomerations occurs, firms' branch plants gravitate toward specialized military facilities, like Colorado Springs space/defense complex or Melbourne, Florida's space corridor, with very little relocation to the industrial heartland. Instead, despite large concentrations of first-rate engineering schools and corporate laboratories, engineers and technicians are recruited in large numbers from heartland institutions, siphoned off toward the defense perimeter. Therefore, a quite select population migration has accompanied skewed defense research and spending patterns.[9]

18.7 CONCLUSIONS

The evidence is rather compelling that the federal government in the United States, through its role in military-related research and development, has become a major director and funder of innovative activity in the postwar period. Its involvement has been primarily driven by the strategic concerns of Cold War. But in a manner, perhaps unintended, it has altered the way the underlying economy operates. Specifically, I have argued that it has staved off severe recessions by underwriting a continual process of innovation, producing a number of economy-transforming inventions. In other words, the U.S. has increasingly come to take the place of Schumpeter's entrepreneur in the process of creative destruction and has artificially induced fundamental inventions which, according to Mensch, had previously bunched up at the troughs of long waves.

If defense-led innovation has mitigated some negative tendencies in capitalist economy, it has nevertheless exacerbated other problems. At least, this appears to be so for the United States. While I have not been able to offer definitive proof of causality in this paper, I have hypothesized, and pointed to supporting correlative data, that six rather new phenomena may be associated with military-driven technology. First, there are the sheer costs of the program itself, and the possibility that it constitutes a rather inefficient way of achieving basic research breakthroughs.[10] The costs have burgeoned into a growing

---

[7]We know very little about the actual regional *development* consequences of defense spending, especially over time. An exception is the ambitious study undertaken by Roger Bolton in the 1960s (Bolton, 1966).

[8]See Markusen, 1986, for an extended discussion of why innovative industries tend to agglomerate.

[9]I have not pursued this latter contention in this paper, but the argument has been made by many. See, for instance, Stowsky, 1986; Fallows, 1982; Kaldor, 1981; and Melman, 1974).

[10]Defense-led innovation does not appear to have fueled interregional shifts in most other advanced capitalist countries with large military commitments, with the notable exception of Germany (Kunzmann,

federal deficit, which the administration has not been able to control despite an adamant commitment to do so.

Second, I have argued that defense-led innovation has had a debilitating effect on commercially-oriented manufacturing, especially in the traditionally strong capital goods industries. By draining research and engineering resources away from these sectors, they have become more vulnerable to international competition. Ironically, the greatest gains appear to have accrued to countries like Japan and Germany, which have no appreciable defense effort themselves, and whose industrial engineering establishments appear to be admirably suited to adapting U.S. military-generated innovations into commercially successful products. Thus, indirectly, military-led innovation may be a significant contributor to the negative U.S. trade balance. Of course, the extraordinarily poor performance of the non-military capital goods industries in the 1980s has deepened the budget deficit problem as well, through depressive effects on tax receipts.

Third, the spin-offs from military-led innovation have been heavily skewed towards automation and communications technologies, which have played a major role in the displacement of workers from conventional manufacturing. As a result, relatively higher levels of sustained unemployment have persisted in the 1980s. This represents an inefficiency in the economy from the point of view of resource utilization. And again, the associated social costs of this group, as well as foregone income taxes, have contributed to the deficit problem.

Fourth, I have suggested that the dramatic compositional shift toward defense-related manufacturing in the postwar period may have contributed, by the late 1970s, to the productivity slowdown in manufacturing. Of course, the automating spin-offs in other spheres of manufacturing should register in gains to productivity. But the high real interest rates of the 1980s, caused by the enlarged deficit, have deterred productivity-enhancing capital expenditures in many industries. For its part, since defense-related manufacturing is so research-intensive, it is extraordinarily labor-intensive. Productivity is unlikely to have improved much in these industries in the last decade.

Fifth, the shift away from traditional blue-collar industries toward defense-related manufacturing has been accompanied by a dramatic shift in the composition of the manufacturing work force toward engineers and technicians. Generally, the latter are paid much higher wages and salaries than are blue-collar workers, even those in oligopolized industries like auto and steel. As a result, I have hypothesized that defense-led activity has contributed to the observed worsening of the income distribution in the U.S. While many blue-collar workers are pushed downward into low-paying service jobs or into unemployment or early retirement, more white-collar engineers, technicians and scientists move up in the income ranks.

Sixth, and finally, I have argued that the extraordinary geographical shifting in the manufacturing center of gravity in the U.S. is a function of defense-led innovation. With the rise of new defense-related high tech agglomerations, like Los Angeles, Silicon Valley and Route 128 in Boston, a societally costly population relocation and community building program has taken place. Not only has much of this relocation been paid for by taxpayers, both in terms of plant construction and labor force migration. But it has also generated underutilized productive capacity and infrastructure in other parts of the country, especially in the traditional industrial heartland.

All of these phenomena may have other causes as well. But the role of defense-led innovation has been largely unexamined, especially since the Vietnam war era. The explicit championing of new programs like the SDI as America's competitive strategy could backfire by exacerbating the competitiveness problem, the productivity slowdown, the deficit, worker displacement, income inequality, and geographical imbalances. It appears reckless to embark on such a strategy without hard research results on the interrelationships between defense-led innovation and performance of the economy.

1985). It has, however, led to new exurban production complexes, as in England's M-4 Corridor west of London (Heim, 1987; Hall et al., 1987).

REFERENCES

Adams. G., 1981, *The Iron Triangle: The Politics of Defense Contracting*, Council of Economic Priorities, New York.

Baldwin, W.L., 1967, *The Structure of the Defense Market, 1955-65*, Duke University Press.

Bluestone, B. and B. Harrison, 1982, *The Deindustrialization of America*, Basic Books, New York.

Bolton, R., 1966, *Defense Purchases and Regional Growth*, Brooking Institution, Washington, DC.

Boretsky, M., 1975, "Trends in U.S. Technology: A Political Economist's View", *American Scientist*, January/February: 70-82.

Dempsey, R. and D. Schmude, 1971, "Occupational Impact of Defense Expenditure", *Monthly Labor Review*, December 1971: 12-15.

Fallows, J., 1981, *National Defence*, Random House, New York.

Fieleke, N., 1985, "The Foreign Trade Deficit and American Industry", *New England Economic Review*, The Federal Reserve Bank of London, July/August: 43-52.

Flaim, P.O. and E. Seghal, 1985, "Displaces Workers of 1979-83: How well have they fared?", *Monthly Labor Review*, June: 3-16.

Gansler, J., 1980, *The Defense Industry*, MIT Press, Cambridge, Ma.

Hall, P., 1982, "Innovation: Key to Regional Growth", *Transaction/Society* 19, 5.

Harrison, B., C. Tilly and B. Bluestone, 1986, "Increasing Inequality in Wage and Salary Income in the U.S. Congressional Joint Economic Committee", *The American Economy in Transition*, Washington, DC, January 16-17.

Heim, C., 1987, *Government Research Establishments, State Capacity and Distribution of Industry Policy in Britain*, Working paper, Department of Economics, Yale University.

Henry, D., 1983, "Defense Spending: A Growth Market for Industry, *U.S. Industrial Outlook*:XXXiX-XLVII.

Kaldor, M., 1981, *The Baroque Arsenal*, Hill and Wang, New York.

Karaska, G., 1967, "The Spatial Impacts of Defense-Space Procurement: An Analysis of Subcontracting Patterns in the United States, Peace Research Society, Papers 8:108-22.

Klare, M., 1984, *American Arms Supermarket*, University of Texas Press, Austin.

Lawrence, R., 1984, *Can America Compete?*, Brookings Institution, Washington DC.

Malecki, E., 1981, "Government-funded R&D: Some Regional Economic Implications", *The Professional Geographer* 33, 1:72-82.

Markusen, A., G. Clark, C. Curtis, S. Deitrick, G. Fields, A. Henny, E. Ingersoll, J. Levin, W. Patton, J. Ross and J. Schneider, 1985, "Military Spending and Urban Development in California", *Berkeley Planning Journal* 2, 1.

Markusen, A., P. Hall and A. Glasmeier: *High Tech America: The What, How, Where and Why of the Sunrise Industries*, Allen and Unwin, London and Boston.

Markusen, A., 1986a, "Defense Spending: A Successful Industrial Policy?", *International Journal of Urban and Regional Research* 9, 1.

Markusen, A., 1986b, "Defense Spending and the Geography of High-Tech Industries" in J. Rees, *Technology, Regions and Policy*, Rowman and Littlefield, Totowa, NJ, 94-119.

Markusen, A., 1987a, "The Economic and Regional Consequences of Military Innovation" in Edwards, P. and R. Gordon, *Strategic Computing: Defense Research and High Technology*, Columbia University Press, New York.

Markusen, A., 1987b, *Regions: The Economics and Politics of Territory*, Rowman and Littlefield, Totowa, NJ.

Melman, S., 1974, *The Permanent War Economy*, Simon and Schuster, New York.

Melman, S., 1983, *Profits without Production*, Alfred A. Knopf, New York.

Mensch, G., 1979, *Stalemate in Technology: Innovations Overcome the Depression*, Ballinger, Cambridge Ma.
Noble, D., 1984, *The Forces of Production*, Alfred A. Knopf, New York.
Norton, R.D. and J. Rees, "The Product Cycle and the Spatial Decentralization of American Manufacturing", *Regional Studies* 13:141-51.
O'hUallachain, B., 1986, *Some Regional Implications of Recent Growth in the American Military Industrial Complex*, Working paper, Department of Geography, Northwestern University.
Oliver, R.P.,1971, "Employment Effects of Reduced Defense Spending, *Monthly Labor Review* 12:3-11.
Parzen, J., 1982, *The Effect of Increased Military Spending in California*, Office of Economic Policy, Planning and Research, Department of Business Development, State of California, May, San Francisco.
Rees, J., 1981, "The Impact of Defense Spending on Regional Industrial Change in the United States" in G.W. Hoffman, (ed.), *Federalism and Regional Development*, University of Texas Press, Austin.
Riefler, R. and P. Downing, 1968, "Regional Effect of Defence Effort on Employment", *Monthly Labor Review* 91, 7:1-8.
Rutzick, M.A., 1970, "Skills and Location of Defense-Related Workers, *Monthly Labor Review* 93, February, 11-16.
Shaiken, H., 1985, *Work Transformed*, Holt, Rinehart and Winston, New York.
Shapira, P., 1986, *Industry and Jobs in Transition: A Study of Industrial Restructuring and Worker Displacement in California*, Unpublished doctoral dissertation, University of California, Berkeley.
Stein, J., 1985, *Defense Spending: Implications for Economic Development Planning*, Working paper, City Planning Program, Georgia Institute of Technology.
Stowsky, J., 1986, *Beating Our Plowshares into Double-Edged Swords: Assessing the Impact of Pentagon Policies on the Commercialization of Advanced Technologies*, Working paper, Berkeley Roundtable on the International Economy, University of California, Berkeley.
Tirman, J, (ed.), 1984, *The Militarization of High Technology*, Ballinger, Cambridge, Ma.
Zysman, J. and S. Cohen, 1987, *Manufacturing Matters*, Basic Books, New York.

CHAPTER 19

Some Reflections on Innovation Stimulating Policy

Hendrik Folmer and Theo Hutten

19.1 INTRODUCTION

Since the early 1970s the contributions of the traditional growth industries to value added and, especially, employment have declined sharply. Various new industries have emerged, although their contribution in terms of production and employment have not (yet) fully compensated the losses incurred from the traditional industries. As they are basic goals, economic policy has primarily focussed on stimulating economic and employment growth during the 1970's and 1980's. At first, the traditional instruments of economic policy were applied. To begin with, direct stimulation of growth has been pursued by means of such instruments as employment programs, direct financial aid to and participation in privately owned companies, Keynesian public expenditure policy and investments in economic, social and physical infrastructure. A second category of instruments has consisted of migration-stimulating measures for labour and capital. A third category has aimed at stimulating economic and employment growth in an indirect way by providing subsidies on capital, labour, land-use, transportation or energy use.

Since the late 1970's economic policy has increasingly begun to stimulate the development of new industries characterized by technical change and innovations. The main reason for the switch of attention from traditional non-innovative industries to innovative industries is that, in the long run, it is believed that the former will decline or stagnate in spite of the substantial public (financial) support whereas (some) industries belonging to the latter category are expected to act as leading industries and, hence, to sustain economic growth. Because of the differences in goals, sustaining and stimulating economic policies are likely to require quite different policy handles. The former obviously implies government intervention, primarily in the form of financial aid. In the case of stimulating the development of innovative industries, the question arises whether or not traditional welfare policy, which relies on the market as a tool of allocation and advocates government intervention only in the presence of externalities, public good or increasing returns to scale, is adequate.

The purpose of the present paper is to devise some policy guidelines to promote technical change and innovation. Attention will be paid to the complete Schumpeterian sequence of invention, innovation and diffusion. The organization of the remainder of the paper is as follows. In Section 2 it will be shown that the stimulation of basic research is an important policy handle. In Sections 3 and 4 attention will be paid to the policy instruments with regard to firm size, market structure, embodiment cost, and education.

In Section 5 regional economic policy with regard to technical change will be considered. The paper ends with a concluding section.

19.2 RESEARCH

With regard to research, a distinction is usually made between basic and applied research. The former consists of original investigations for the advancement of scientific knowledge that do not have specific commercial objectives. Applied research consists of investigations aimed at the discovery of new scientific knowledge which has specific commercial objectives with respect to products or processes. Both basic and applied research are vital for inventions and innovations. Their relationship has been clearly described by Evenson and Kislev (1975). These authors treat applied research as drawing successive samples from a probability distribution of potential yields. The distribution is determined by nature, the state of science, and technology. The research pay-off is the difference between the sample point with the highest yield and the current yield. It is important to note that the applied researcher cannot change the parameters of the distribution of potential yields. The parameters, however, can be affected by basic knowledge and technology. So, basic research is supportive of applied research.

The vast bulk of industrial research is on applied research and development and only a very small percentage is on basic research. Basic research is mainly financed by government and undertaken by public institutions. Hence, because of the support it lands to applied research and its subordinate role in industrial research, an important public policy handle to stimulate inventions and innovations would be the stimulation of basic research.

Not only is research important with regard to the development of inventions and innovations but it also plays a role in the *transfer* of technology and innovations. This follows from Benvignati (1982) who found that innovations, which are of domestic origin, show a tendency to diffuse faster than those of foreign origin. This is because native originators of techniques are likely to be more responsive to domestic market needs than foreign originators for the following reasons. First, the former are in a better position to perceive domestic market needs. Secondly, they receive better feedback from customers, enabling more effective product alterations to be made in the early stages of production.

On the basis of these considerations the provision of publicly funded basic research is a powerful policy handle with regard to the transfer of technology and the diffusion of innovations. Because cultural and geographical proximity is an important factor in the diffusion and adaptation of innovations, an organization per industry similar to that in agricultural research might prove to be quite efficient. Three general patterns of research unit organization in agricultural can be distinguished (c.f. Binswanger et al. 1978, p.180):

(i) A large number of independent, spatially more or less equally spread units, each of which performs screening and minor adaptive research.
(ii) A small number of central institutions for adaptive research, each of which is connected with several branch units of layer (i).
(iii) A group of research institutions for comprehensive scientific as well as adaptive research. Each of these institutions are connected with the units of layers (ii) and (i)

The optimal size and location of research units could depend inter alia on the supply of research skills, the degree to which production factors differ spatially, the proportion of the total production of a region of a given commodity and the possibilities of transfer from neighbouring regions.

The following remarks apply. First, it is important to integrate national research in an international framework so as to further the exchange of basic research results and to facilitate adaptive research and the direct transfer of techniques. Secondly, government plays a major role with regard to basic research via its various institutions for scientific research. By means of the allocation of funds it can influence the direction of basic research and, hence, potentially, applied research and development. Thirdly, given its industry-specific or even firm-specific nature, the possibilities of direct government involvement in applied research and development are quite limited. Its concern with screening and adaptive research activities should usually be induced by the private sector. Another way to influence private research is by means of subsidies for basic and applied research at private institutes and for development costs. This follows for instance, from Binswanger (1974) and Binswanger et al. (1978) who showed that applied research can be viewed as an investment process. The choice of the research investment portfolio depends on present and expected future total factors costs (made up of factor prices and factor shares), research costs and research productivities. Total factor costs and research productivities have a positive impact on research; research costs a negative. So, reducing research costs by means of subsidies would stimulate research. It is also possible, however, that a continued rise in the costs of a given factor leads to a reduction of the research bias of that factor because of the diminishing returns from research and the fact that a research line focussed on the reduction of a given factor usually saves some of all the factors.

It is also important to note that experiences in the Netherlands and Germany show that basic research at applied research institutes stimulates applied research and development (Koumans, 1987). The main reasons are:
(i) Basic research is needed to fill gaps in knowledge with regard to specific projects.
(ii) Basic research may help anticipate future developments.
(iii) Basic research may facilitate the interaction with other research institutes (see also Hotz-Hart, 1989).

This section ends with the remark that for highly complex technology projects government subsidies are a prerequisite because of the extent of the research and development costs. Moreover, in many situations, international competition and government behavior in competing countries make it virtually impossible not to subsidize projects which are considered to be key-projects (Roobeek, 1987). In this regard, it is also important for the government to stimulate cooperation between government research institutes and private R&D institutes and take part in the risks involved in this kind of project.

19.3 MARKET STRUCTURE AND LICENCE SYSTEM

The Evanson-Kislev model referred to in the preceding section has been extended by Binswanger (1974) and Binswanger et al. (1978). From this extended model the conclusion has been drawn that an increase in the scale of output will tend to increase applied research efforts, unless research is constrained by a research budget. This observation is consistent with the neo-Schumpeterian hypothesis that applied research is a function of size and monopoly power. This hypothesis is partly supported by the empirical finding that applied research has been concentrated within large firms (e.g. in terms of number of employees). In the Netherlands the five largest firms are responsible for more than 75% of the total private expenditure on R&D, which is about 66% of total expenditures on R&D (van Dijk and Kleinknecht, 1985). Similar results hold for the U.K. (Clark, 1986). Moreover, as argued by Galbraith (1952), Demsetz (1969) and Kamien and Schwartz (1982) among others, large firm size and monopolistic industrial structure are essential conditions for innovation for the following reasons:
- The cost of innovation are usually considerable and can only be borne by large firms.
- Because of indivisibilities, innovation must often be on a large scale.
- Some control over the market is needed for innovations to be worthwhile.
- Large firms are in a better position to protect their patents.

- Large firms may be in a better position to attract higher qualified personnel and may use their skills better.
- Innovation leads to actual and potential competition. Large firms are able to defend themselves better against new entries and to enter new markets themselves.

There is no evidence, however, that research *intensity* increases significantly with size or with the degree of concentration in the market (Comanor, 1967; Schrieves, 1978). Moreover, compared with the research of smaller firms, the research carried out by large firms tends to be aimed at modest advances. This observation is consistent with Hamburg (1963) who argues that inventions in large organized research laboratories are not primarily focussed on new products or major improvements in existing products and processes. The reasons for this bias are the desire for quick and reliable results, the desire to concentrate research on products which have already proved themselves commercially, the type of individual attracted to the corporate laboratory, the stifling of individual creativity in an organized environment and the fact that team research involves comprise (see also Varayia, 1989).

Small firms (also) play an important role in the process of technical change. The main characteristics which influence this role are independence and the absence of complex bureaucratic structures, which leads to flexibility, and the identity between management and property, which stimulates motivation. Moreover, small firms can usually operate satisfactorily on relative small markets. This kind of firm, however, often suffers from limited knowledge and experience with regard to research, development and management. Moreover, they have limited access to financial resources and have relatively few possibilities to spread out. Finally, we remark that Kamien and Schwartz (1975) concluded that intermediate firms have the highest research intensity.

Related to the neo-Schumpeterian hypothesis is the thesis that applied research is stimulated by product diversification (Nelson, 1959). Diversified firms are assumed to be more research intensive than single product firms because research may yield unexpected inventions and innovations, and diversified firms are able to produce and market a higher proportion of products resulting from these unexpected discoveries. There is, however, little empirical support for these theses (see, for instance, Grabowski, 1968 and Mansfield, 1968). The empirical evidence on the relation between innovation on the one hand and market structure and firm size on the other is still sketchy and conflicting. The following provisional conclusions with regard to these relationships could be formulated:

- Mergers may help reach a threshold level below which research is not viable or help reap economies of scale in the use of inventions and innovations.
- Large firms or monopolistic industries ar not necessarily more progressive nor do they produce more fundamental technical change than small firms.

As Schmookler (1954) argued, the incentive for research and invention will be greater the greater the likelihood that the innovator will be granted a proprietary or monopoly right by means of a patent. This implies, however, that a competitive industry may be converted into a monopoly and that research is likely to be focussed on areas where competition can be most readily extinguished. Both factors may lead to an artificial scarcity of new techniques and affect the future rate of technical progress adversely. This may be mitigated by a system which makes patents available on licence.

An important issue in technological development, especially with regard to complex and expensive projects, is to be the first to introduce a new product or process and to obtain a substantial market share. This could be furthered by protectionism in the domestic market in the early stages of development. Moreover, competition in the domestic market, especially in small, open economies may be counterproductive to obtaining the minimally required market share.

On the basis of these considerations the following policy handles with regard to market structure and firm size can be derived:

(i) The creation of favorable conditions for the establishment and development of small firms. This could be pursued by financial and organizational arrangements, such as the provision of subsidies and management facilities, management education, information services, cooperation and study clubs.

(ii) The stimulation of cooperation among competing firms, both nationally and internationally. This may require juridical adaptations with regard to mergers and international cooperation among governments.

(iii) The establishment of a patent and licence system which furthers inventions and innovations as well as their dissemination. Compulsory licences could be granted if the patentee does not develop his invention himself or grant a licence to others on reasonable terms.

19.4 INVESTMENT SUBSIDIES AND SCHOOLING

Some technical change is embodied in capital items. This implies that the purchase of a new machine is a prerequisite for using a capital embodied innovation. Technical change which is not embodied in machines can increase the efficiency of the production process even if old machines are used. Examples of this kind of technical change would be improvements in organization, labour or intermediate products.

To realize any technical change, a firm must purchase production factors that embody it or must embody it in the production factor it has already purchased. This implies that an important determinant for technical change and its speed is formed by embodiment costs. Some organizational changes may not require much learning time and may be implemented rapidly. Technically improved intermediate inputs can replace existing inputs as soon as the latter are exhausted. Technical change embodied in fixed capital, however, can usually not be replaced before it has been depreciated. Therefore, this kind of technical change is likely to spread less rapidly than technical change embodied in intermediate inputs and organizations.

Improvement in one production factor usually requires modifications and improvements in the other production factors. This applies, in particular, to technical change embodied in capital, which often requires new organizational structures, and, especially, more highly qualified labour. As it is increasingly felt as a limiting factor, we will pay somewhat more attention to the latter.

The two most important factors improving the quality of labour are learning and education. The learning process refers to the effect of cumulative experience on labour productivity. Since product types constantly change, there is probably no limit to the learning process in the aggregate, although there is with regard to individual products.

Education can be viewed as an investment in human capital. It improves the quality of new workers if educational standards improve. Moreover, it exerts its effect through changes in the overall quality of labour. Education may also speed up the process of the diffusion of technical and organizational innovations by facilitating the understanding and acceptance of new ideas (Gelauff, et al., 1984). Moreover, increases in the stock of education contribute to advances in knowledge which may further basic research, applied research and development.

From this exposition the following policy handles can be derived:

(i) Investment subsidies which might speed up the introduction of technical change embodied in capital. It is obvious that a selective use of this instrument should be made. It should only be applied to investments related ´to genuine innovations with high costs which are expected to stimulate growth and which might not be realized without the subsidies.

(ii) Government expenditure policy which may not only improve the chances to reach a break-even point but may also contribute to the learning process. Moreover, without sufficient domestic demand, technical developments may fade away from the originating country. This applies in particular to small, open economies (Rothwell and Zegveld, 1981). It is obvious that this kind of policy intervention may endanger international competition.

(iii) Improvements in the quality of schooling, especially with regard to subjects related to technical and technological changes, which may speed up the development, introduction and diffusion of innovations. These improvements meet the demand for more highly educated personnel, which is an increasing requirement because of technical changes in capital and organization. In particular, nowadays there exist acute shortages of specific professionals such as biotechnologists, engineers, etc. Moreover, rapid technological and technical change may create new shortages in other professions in relatively short intervals in the (near) future. These acute shortages, which may seriously hamper technical change, can be mitigated by adaptive, and, preferably, anticipating changes in schooling programmes and by influencing the inflow of students. It is obvious that all this implies that the organizational structure of the schooling system should allow rapid adaptations to emerging needs.

(iv) Advanced scientific education which may further the possibilities of domestic research and development and facilitate the introduction of foreign scientific discoveries.

(v) Additional schooling and facilities for on-the-job-training programmes because technological and technical changes are likely to obsolete the skills and knowledge of individuals whose period of employment exceeds a given limit.

(vi) Increases in the quantity of schooling which may increase the minimum level of education and, hence, the quality of the labour force. As mentioned above, there are shortages of qualified labour in various industries. By increasing the quantity of education this problem may be mitigated. It should be observed that education is often also of great importance with regard to the demand side. Various consumption expenditures depend inter alia to a large extent on education. An example is the consumption of personal computers.

It is obvious that the conditions formulated above require a highly flexible educational system both with regard to organizational structure, staff, curriculum and equipment. In particular, various feed-back mechanisms between industry and the schooling system are needed.

19.5 TECHNICAL CHANGE AND REGIONAL DEVELOPMENT POLICY

An important problem in regional economic policy is to what extent and in what direction technical change affects regional economic development. More specifically, in what ways can peripheral regions with high unemployment rates benefit from technical change and how can this be enhanced by regional economic policy?

The starting point with regard to this problem is the notion that product inventions and innovations tend to be concentrated to metropolitan areas. The main reasons are that technical expertise, skilled labour, a ready market and information flows are most readily available in these areas (see, among others, Anderstig and Karlsson, 1989, and the references therein). When products reach the later life cycle stages, production is shifted to more peripheral regions characterized by low-cost production factors. A similar hierarchical patterns holds for process innovations in as far as they are of high complexity and focussed on differentiating production. Process innovations of low complexity aimed at standardizing production do not show a hierarchical diffusion pattern. (For further details see Anderstig and Karlsson, 1989.)

From these brief remarks on the spatial diffusion of innovations and the various aspects of technical change outlined in the preceding sections the following policy handles for regional development policy can be derived.

(i) The stimulation of inventions, innovations and the initial adaptation of innovations *in general*. As mentioned above, these activities tend to be concentrated to metropolitan areas. At a later life cycle stage, they tend to trickle down to more peripheral regions. Hence, stimulation in general tends to benefit not only metropolitan areas but also peripheral regions. It should be observed that the latter kind of effect is likely to lag substantially behind the former.
Another argument to promote regional development by stimulating technical change in general follows from Howells (1989). This author states that with the increasing fluidity of research and technological linkages both on an intra- and inter-corporate basis it is becoming increasingly common for innovations generated in one (metropolitan) are to be developed and commercially exploited in another (non-metropolitan) area. In as far as these other regions coincide with peripheral regions, regional development may be enhanced.
Finally, we observe that it is likely to be counter productive to stimulate invention and innovation-generating activities to locate in other regions than their "natural" ones.

(ii) Improvement of invention and innovation generating conditions in peripheral regions. As mentioned above, metropolitan areas are characterized by inter alia technical expertise, skilled labour and substantial information flows. The provision of such conditions, e.g. in the form of research and educational facilities such as (technical) universities, may stimulate the generation of inventions and innovations in peripheral regions.

(iii) The stimulation of process innovations aimed at lowering production costs by such instruments as schooling and education, investment subsidies, etc. It should be observed that this kind of policy at first glance may seem to be detrimental to the main goal of regional policy, (i.e. the stimulation of employment growth) because standardization may go together with substantial reductions in employment. However, given interregional competition, both at the national and the international level, this may be the only way to prohibit a given industry from disappearing from the region.

19.6 CONCLUSIONS

On the basis of the foregoing selection of items from the literature on technological and technical change the following conclusion can be formulated:

- The main policy handles for innovation-stimulating policy are the provision of basic research, schooling and (permanent) education, specific investment subsidies, public expenditure policy, the provision of patents, the maintenance of market structures and the stimulation of national and international cooperation between firms.
- The most important policy handles at the regional level are the stimulation of innovations in general, improvement in peripheral regions of innovation-generating conditions and of process innovations aimed at lowering production costs.

REFERENCES

Anderstig, C., and C. Karlsson, 1989, Spatial Diffusion of Information Technology in Sweden, This volume.

Benvignati, A.M., 1982, "The Relationship Between the Origin and Diffusion of Industrial Innovation", *Economica*, 313-323.

Binswanger, H.P., 1974, "A Micro-Economic Approach to Induced Innovation", *The Economic Journal*, 84:940-958.

Binswanger, H.P., V. Ruttan and others, 1978, *Induced Innovation, Technology Institutions and Development,* The John Hopkins University Press, Baltimore.

Clark, R., 1985, *Industrial Economics*, Basil Blackwell, New York.

Comanor, W.S. 1967, "Market Structure, Product Differentiation and Industrial Research", *Quarterly Journal of Economics*, 81:639-57.

Demsetz, H., 1969, "Information and Efficiency: Another Viewpoint, *Journal of Law and Economics*, 1-22.

Dijk, A. van, and A. Kleinknecht, 1985, *Aanzetten voor een schaalgericht industrieel vernieuwingsbeleid*, University of Limburg, Maastricht, The Netherlands.

Evenson, R. and Y. Kislev, 1975, *Agricultural Research and Productivity*, Yale University Press, New Haven.

Galbraith, J.K., 1952, *American Capitalism*, Houghton Mifflin, Boston.

Grabowski, H.G., 1968, "The Determinants of Industrial Research and Development. A Study of the Chemical, Drug and Petroleum Industries", *Journal of Political Economy*, 76:292-306.

Gelauff, G.M.M., A.H.M de Jong and A.R.M. Wennekes, 1984, *Een putty-clay model met 5 produktiefactoren en deels endogene technische ontwikkelingen*, CPB monografie 32, Den Haag, The Netherlands.

Hamburg, D., 1963, *Invention in the Industrial Research Laboratory*, Cited in C. Kennedy and A.P. Thirlwall

Hotz-Hart, B., 1989, Innovating Behavior of Swiss Industry - Findings and Policy Conclusions, This volume.

Howells, J., 1989, Research and Development, Corporate Organisation and Industrial Location: Prospects for Regional Development, This volume.

Kamien, M.I. and N.L. Schwartz, 1975, "Market Structure and Innovation: A Survey", *Journal of Economic Literature* 13:1-37.

Kamien, M.I. and N.L. Schwartz, 1982, *Market Structure and Innovation*, Cambridge University Press, Cambridge, UK

Kennedy, C. and A.P. Thirlwall, 1972, "Surveys in Applied Economics: Technical Progress", *The Economic Journal*, 82:12-72.

Koumans, W.A., 1987, *De rol van de kennisinfrastructur in het technologietraject*, E.S.B., 72:474-478.

Mansfield, E., 1968, *The Economics of Technological Change*, Norton, New York.

Nelson, R.R., 1959, *The Simple Economics of Basic Scientific Research*, Cited in C. Kennedy and A.P. Thirlwall.

Roobeek, A.J.M., 1987, *De rol van de overheid in de technologische ontwikkeling*, Research Memorandum 8711, University of Amsterdam, Amsterdam, The Netherlands.

Rothwell, R. and W. Zegveld, 1981, *Industrial Innovation and Public Policy: Preparing for the 1980's and 1990's*, Francis Pinter, London.

Schmookler, J., 1954, *The Level of Inventive Activity*, Cited in C. Kennedy and A.P. Thirwall.

Schrieves, R.E., 1978, "Market Structure and Innovation: A New Perspective", *Journal of Industrial Economics*: 26:329-47.

Varayia, P., 1989, Productivity in Manufacturing and the Division of Mental Labor, This volume.

CHAPTER 20

New Issues in Science and Technology Policy: Discontinuities in the Process of Knowledge Generation, Knowledge Transfer and Knowledge Transformation

Walter Zegveld

20.1 INTRODUCTION

For many years the dominant model of the transfer of fundamental knowledge has been that in which it was produced in the universities and subsequently diffused to other sectors of society, including industry. The application of this knowledge led eventually to new technological and social developments. This is the model of the "pool of knowledge": knowledge, like water, flows under "gravitational" forces along contours in the socio/economic landscape to areas where inevitably it leads to growth and progress. During the past decade, however, there have been certain developments which make it necessary to seek now concepts of knowledge generation, transfer and transformation. This paper is designed to stimulate the process of developing these new concepts. Its purpose is to determine what the new issues of science and technology policy are and, in the light of the analysis, question whether the current infrastructure and the traditional manner in which politics, science and industry deal with knowledge (patents, privacy rewards, recognition, etc.), are appropriate.

The first section of the paper deals with recent developments in our understanding of the innovation process, developed and published mainly by economists. Secondly the lines of developments and discontinuities in the industrial structure, in business-corporations and in the universities, as well as their interrelationships are delineated. These developments are in particular determined by scientific and technological progress which, through their growing political and economic significance, are leading to radical changes throughout the economy and society. These developments are not confined to the industrial sector, but can also be observed in agriculture, in commercial services and in health care. The paper concludes by both posing a set of questions and reaching a number of conclusions. They are related to new issues in science and technology policy. These issues are knowledge infrastructure and priority setting, R&D-expenditure philosophy, the role of the government, human resource development and the international aspects of R&D-policy.

20.2 CHANGING VIEW ON INNOVATION PROCESSES

We can obtain an important contribution to the development of the new concepts of knowledge generation, transfer and transformation through considering changing views on the innovation process developed in economics.

Economic theory building has focused mainly on the relationship between invention, innovation and economic change (mainly productivity growth). Neoclassical economic models seek connections between technical progress and economic growth measured, in the first instance, in terms of Gross National Product (Salter, 1960; Denisen, 1968). More detailed studies in this tradition have originated from the more general, linear models of the innovation process which stress the causal relationship between basic science and technological progress in social sectors. Studies in this tradition can be divided into two categories. In the first, emphasis is put on impulses from, and the momentum of, the market (demand-pull). In the second category emphasis is placed on the "enabling" role of new technology (technology push).

According to the technology push model, fundamental scientific research leads to technological developments, which inevitably results in a stream of new products and processes in the market-sector. From the mid-sixties, the technology push view of the innovation process began to change. A growing number of empirical innovation studies led to a model with a strong emphasis on the role of demand. According to this model (the market-pull model), innovations originate and develop as a consequence of a perceived market demand.

In the last ten years the technology-push and market-pull models have increasingly been considered as extreme and unrepresentative examples of a more general process of interaction between science, technology and the market place. The demand-pull and technology-push studies provide no answer to the question of "why?" and "when?" certain technologies originate and others do not. In addition, the relative importance of technology-push and market-pull can vary over the different phases of an industrial cycle. In market-pull studies, science is considered implicitly as a black box which is capable of changing and responding according to the nature of demand. Overemphasis on market needs can lead to a regime of incremental innovation and lack of radical innovation (Hayes and Abernathy, 1980). Studies which emphasize technology-push rarely take into account the influence of economic change on the direction of technological activities (Mowery and Rosenberg, 1979).

The above critique of the one-dimensional, linear technology-push and market-pull models, combined with the results of studies of new forms of innovation processes emerging as a consequence of the development of radical new technologies, combined again with observed changes in the production sector, lead to a more representative model of innovation which is embedded more firmly in a proper industrial context. This is the so-called interactive model of innovation. In this interactive system-model, innovation is considered as a logical, sequential, but not necessarily continuous or linear process that can be depicted as a series of functionally separate, but interacting and mutually reinforcing stages, i.e. research, development, manufacturing, marketing, etc. According to this model, the innovation process can be considered as a complex network of communication links, through which the various functions encompassed inside a business company are interconnected and connected also to external developments in the scientific and technological infrastructure and the market -place.

Freeman, in his study on the 'Long term theories of innovation', developed the concept of 'new technology systems'. This concept refers to clusters of mutually connected innovations, scientific inventions, technologies and markets, based on radical changes in science and technology, and which are associated with rapid economic growth and the emergence of new industries (Freeman, 1982).

Hughes, in his historical study on the development of the electric-supply-network in the United States of America, also uses the 'system' concept. Centrally directed, interacting institutions and technical components (e.g. the electric transmission system) comprise such a system. Hugh stresses the characteristics of interconnectedness, i.e., a change in one component impacts on the other components of the system (Hughes, 1983).

Nelson and Winter have elaborated the analogy between innovation processes and biological processes in their evolutionary theory of economic growth (and in their formal theory building used the Markov-process model, which is extensively used in physics and is based on probabilities). Nelson and Winter developed the concepts of 'natural trajectories' and 'selection environments' in order to explain the observed changes in the pattern of invention and innovation. These concepts refer to an inner logic and self-impulse which guide the development of an innovation or a set of innovations. A natural trajectory is the pathway along which a successful technological stream has developed. Natural trajectories are specific for a particular technology. Nelson and Winter place strong emphasis on the social character of technological developments, and innovations are conceived as a network of social relations. A specific technology develops under the influence of several selection mechanisms, which constitute the selection environment. The selection-environment consists not only of economic, but also of social, institutional, political, technical and cognitive components.

Dosi has elaborated the theoretical concepts of Nelson and Winter. He explicitly stresses the technical components of a trajectory: the technological trajectory. He states that the selection environment has a strong influence on the technological trajectory. In addition to the market, he considers the R&D laboratory as an important component of the selection environment (Dosi, 1982). The work of Nelson and Winter, amplified by the Dosi 'refinement', lays the foundations of a useful theory of innovation.

The concept of technological trajectories aids us in taking not so much a 'macro', but rather a 'meta' view of technological change. It is important also to note that public institutions and public policy are themselves potentially important elements of the selections environment (Rothwell and Zegveld, 1985).

Although, I do not claim to give an exhaustive overview of the field of theory-development I believe that the new concepts and taxonomies in theory building on the innovation process emphasize the interactive character of this process. The conceptualisation of the innovation process will develop further under the influence of the increased attention, mainly influenced by developments in Japan, on quality and in particular on the business logistics of the Japanese innovation process.

I have elaborated these new insights into the innovation process in detail because I believe that the new concept of the knowledge-infrastructure is closely related to them. The interactive character of the new form of innovation process must be taken into account in the new concept of the knowledge-infrastructure.

Before ending the theoretical part of this paper I will remark on the role of government in the theory building on the innovation process. The studies mentioned above have focused mainly on innovations and technological developments in the market sector of the economy and the role of government generally has been regarded only as a side issue. Nelson and Winter provide some attention to the role of government as a means of stimulating and regulating innovation and, as such, as part of the selection-environment of a natural trajectory. This explicit attention by Nelson and Winter is, however, the exception.

Because of the role of the government in the innovation process in areas which are not directly related to the market mechanism, but which are of social or national interest, a plea is to be made for more explicit attention to be given to the role of government in future theory building of the innovation process.

20.3 CHANGES IN THE INDUSTRIAL STRUCTURE

A number of fundamental changes have taken place in the structure of trade and industry, and in their relationship both nationally and internationally which have consequences for the processes of knowledge-generation, knowledge-transfer and knowledge- transformation. These changes can best be illustrated by the changing strategies of large corporations, in the origin of functional relations between big and (often new) small enterprises, in the internationalization of trade and industry and in the role of financial organizations (venture capital).

In the 'fifties and 'sixties the dominant corporate vision was for the creation of large multinational enterprises. This picture has, however, now changed. The focus then apparent on developing a variety of product-market-technology combinations has been replaced by a strategy of remaining as close as possible to the 'core' strengths of the company. In order to react as flexibly as possible to market changes and to new scientific and technological developments, side-activities are being rejected. We can detect new forms of co-operation between big and small enterprises in pursuit of dynamic complementarities. First in Japan, later in the U.S. and now also in Europe, we can see a tendency towards more and more co-operative connections between small, often high-tech enterprises and large enterprises. The smaller enterprises can operate more flexibly and find their way, via the bigger enterprises, to the market place. New High-tech enterprises are an important element in the process of knowledge-transformation and knowledge-transfer (the diffusion of CAD-technology being an example of the latter). As a company among many others, Philips strives to form close relationships with a select group of suppliers. In this new form of co-operation the competitive principle is replaced by a formula fruitful for mutual support: co-makership. This is based on one of the strongest characteristics of the Japanese production system; an elaborated logistic with regard to demand, production capacity and supply. In Japanese hierarchical relations there are, in the vertical company-column, 'long-life contractorships' between the large final producer and many small clients. Along with the strong emphasis on product quality, this is one of the important factors which has contributed to the enormous growth of Japanese exports.

The internationalization of trade and industry is a consequence of the comparative location advantages of the various company-functions such as R&D, production and marketing. The internationalization of the R&D component can best be illustrated by the expenditures of American multinational enterprises: in 1960 2% of their total R&D was carried out in foreign countries; today this is 10%. Philips on the other hand, carries out 50% of its R&D in the Netherlands , while the percentage of sales in the Netherlands is only 6.

Other examples of the internationalization of trade and industry are: the acquisition of foreign (mainly U.S.) small-new-high-tech enterprises by European companies and the joint-ventures with Japanese companies undertaken by a number of European companies. Rapid technological development decreases patent-activities, which means that in the future the amount of international joint-ventures for R&D, production and marketing will increase.

The internationalization of trade and industry accompanies and influences the internationalization of public policy. Various national governments are in the process of tuning their policy to accomodate international developments. They are more or less forced to do this by countries which have passed very quickly through a number of stages in developing the use of new technologies.

20.4 RESTRICTIONS ON TECHNOLOGY-TRANSFER FROM THE UNITED STATES OF AMERICA

According to Lester Thurow, whether the existing level of world economic integration will be maintained depends, among other things, on the strategy the U.S. chooses to solve her national economic problems. The U.S has a productivity problem and if there is no solution in the near future, she will increasingly withdraw from international competition and disconnect her economy from world-trade.

Protectionist steps with specific regard to the export of high-tech products derive from both military-strategic and economic-commercial considerations. The national-security policy of the U.S. overlaps with regard to its technology policy on more than just military technology (Dickson, 1984). This has consequences for international knowledge-transfer and for the position of Europe and Japan in scientific-technological developments.

The U.S. Department of Defense (DoD) plays an important role in technological developments. The expenditures of the DoD on R&D in 1985 (apart from the SDI-funding) were Dfl.100 billion, which is approximately 15 times the total Dutch R&D-expenditures. In 1981 approximately 12% of all engineers and scientists in the U.S. were working on behalf of the DoD. According to the Pentagon this percentage will be 14 in 1987.

In the past, the DoD has talked about civil and military applications of technological developments and about being assured of a production base that satisfies military needs. During the past years more has been said about civil applications of technology and the importance of a competition-lead over other countries in general and Japan in particular.

Dr. Richard Delauer, under-secretary of Defence for Research and Development, in his speech in February 1982 to representatives of the U.S. semiconductor-industry stated: "the n-th generation computer development programme is the U.S. answer to the fifth-generation programme supported by the Japanese government." The U.S. programme Dr. Delauer mentioned is formally called "Strategic Computing and Survivability". In the light of these new objectives, the U.S. DoD can be compared with the Japanese MITI, because DoD now acts more or less as a part of the U.S. industrial and technology policy system.

The fact that scientific and technological developments are considered as national-security matters, can be deduced from DoD measures with regards to the new civil technology and to the adaptation of the industrial base:

- support of R&D that leads to civil and military application,
- support of R&D, that leads only to military applications, but which also gives experience in dealing with advanced apparatus and so has relevance from a civilian point of view,
- allocation of funds to "independent R&D" in areas which are important for the Department of Defense in the form of "overhead"-costs-allowance in the case of defence-contracts,
- support to, and encouragement of, university research,
- implementation of a taskforce to improve the industrial response, with one of the goals being to create a sufficient cadre of trained manpower to fulfill the needs of industry and increase industrial production.

One of the eye catching measures of the Reagan Administration is the restriction on the export of high-technology. There are three important developments in this area: the U.S. government control on the export of so-called dual-use technology (for both commercial and defense applications); admittance to government funded research programs; inclusive access to patents and other restrictions with respect to technology-transfer.

These measures must be seen in the light of the resoluteness of the Reagan Administration to re-establish and expand the military, scientific and industrial-technological leadership of the U.S. As well as the considerations with regard to the economic adaption process, there is also the argument that, in several areas, since civil technology is advanced beyond military technology, it is difficult to control from the view point of international transfer.

Intensified control of technology transfer is not only aimed at obstructing the direct transfer of military-relevant technology to the Soviet Union, but it also effectively leads to the obstruction of West-West transfer. The U.S. measures evoke sharp comments like: "how do we protect an asset that can be destroyed by an act of protection?"

20.5 THE CHANGING ROLE OF TRADE AND INDUSTRY WITH REGARD TO SCIENCE AND TECHNOLOGY

Social activities, including economic activities, are becoming more and more knowledge-intensive. This knowledge-intensive development in commerce and industry is evidenced in new products, in the production process, in expenditures for R&D, in the increase of fundamental research, in the growing demand for technical and other highly educated manpower, in management, in the organization of production, in sales, in marketing and in accountancy. The more the knowledge-intensity of activities in the market sector increases, the greater becomes the need of commerce and industry to carry out in-house R&D, to connect themselves to external knowledge-sources and to employ highly educated manpower with both the technical know-how and an ability to link this to markets needs. The following figures illustrate that the volume of R&D has recently increased in almost all industrialized countries. After 1980 R&D-expenditures, expressed in % GNP, increased in Japan from 1.9% to 2.6%, in FRD from 2.1% to 2.7%, in France from 1.9% to 2.3% (with a planned level in 1987 of 2.6%) and in Sweden from 2.1% to 2.6%.The increases are mainly in industry. The Netherlands, with regard to both the total R&D-expenditures (2.1%) and to industrial R&D falls far behind other large and medium-sized economies.

The increased industrial R&D expenditure has been accompanied by greater efforts in basic scientific research in industrial R&D laboratories. In some scientific areas industry leads the universities in state-of-the-art expertise.

In the framework of the traditional models of knowledge-transfer, knowledge is carried over from universities to industry. In the new situation - for instance in the case of modern production-technologies and new R&D-experimental apparatus - there is now a reversed knowledge-transfer problem with regard to both research and to education. Rapid technological development also has consequences for the system of education and training and in particular it raises the question whether training and education should continue to be performed by traditional, government financed institutions.

In addition to the debate in the Netherlands about the share of vocational and engineering training in so-called second-stage education (i.e. Masters' and Doctoral Degrees) there is now the question of the re-education of the category of employees between the ages of 35-45. Re-education is an important issue in most other industrialized countries. In France and Sweden, for example, the governments have set up regulations which oblige companies to spend a certain percentage of their turnover and wages bill respectively on re-education efforts. In Japan about 5 percent of turnover is spent on re-education (without statutory obligations). Commerce and industry have a growing need for more and differently trained people. In Sweden the Royal Academy of Engineering Science (IVA) is doing research on the question of whether this need is short-term or structural.

Many large, multinational enterprises currently give a great deal of attention to re-education efforts. In the Netherlands, for example, Philips intends to spend 5% of the wage bill on this and intends, in the long term, to increase this further. IBM spends, worlds-wide, 9% on retraining (Eurich, 1985). In the U.S. almost all major enterprises have their own education center, where they also employ visiting university professors and some of these education centers award certified university credits.

In addition to the growing structural importance of new knowledge for the creation of knowledge-intensive products and production processes, management, the organization of production process, sales management, marketing and accountancy are becoming increasingly structured around know-how. An important development in industrial enterprises is the increased connection between R&D-management and other company functions including production, marketing and financing, with strong emphasis being placed on identifying possibilities for technology-market combinations. This so-called 'interface' management has been established to stimulate constant feedback from other company-functions in order to perform R&D as an intensive interaction process.

20.6 THE CHANGING ROLE OF UNIVERSITIES IN SCIENCE AND TECHNOLOGY

Universities in the industrialized countries have been confronted with an enormous increase in the number of student enrollments, which has been accompanied by the introduction of democratic decision-making processes. Relatively little effort has been spent on the quality of training and forgoing linkages with social change processes, in particular the economic change process. The greater part of university studies in the Netherlands offer the students no possibility of gaining practical experience in industry or in other market or social institutions. The recently implemented changes in the structure of higher education, especially the introduction of the so-called 'first-stage' (4 to 5 years), failed to introduce such a possibility.

Research in universities, including research training, is now, more than ever before, focused on the developments and needs of the social sectors, including industry. With regard to higher education, this relates in particular to the postgraduate 'second-stage'. Universities and polytechnics are more or less forced to do this because of financial constraints and new government regulations. During the past few years, governmental expenditures on higher education have decreased to such an extent that universities are more or less compelled to seek alternative finance. The increase in contract-research means that more short-term, applied research is being done. Basic scientific research, as a consequence, enjoys relatively less attention. Industry, in contrast, is increasing its share of basic research (by doing it itself, or, as in the U.S., in research institutions at arms length from the universities).

The fundamental question is, however, whether basic research should be performed in institutions like industry, which are quite vulnerable to changes in external economic, social and political conditions.

In the U.S. the level of R&D-support from industry to university has reached saturation point as reflected in the extensive state subsidies now being given to U.S. universities (e.g. in Texas, Mass. New Mexico).

There are a number of more substantial developments in science and technology which have made an important contribution to the radical changes mentioned in the introduction. The relationship between basic research and technology has intensified strongly, which can best be detected in the increase of R&D-instrumentation in scientific research. In addition to an integration between basic and applied research, this also means that the institutional barriers between university and industrial R&D-laboratories are fading. There is also a strong increase in interest in areas of high multi- and interdisciplinarity. This is linked to an increasing emphasis on solving certain social problems (e.g. ecology-studies) as well as with the fact that, to-day, many new innovative developments and insights are expected to derive from the boundary between biology, chemistry and physics. We can also distinguish a number of new courses in higher education which have a strong interdisciplinary character (e.g. management studies, environmental studies etc.).

The interactive, integrative character of recent innovations is, however, best illustrated in a number of the 'new' technologies; information-technology (IC, telecommunication,

CAD), in biotechnology and, as we might expect, in new materials. These developments illustrate the changing position of the universities in the innovation process. According to Stankiewics, the existing symbiosis between education and science must change to make room for technology-generation and technology-transfer (Stankiewics, 1984).

20.7 CONCLUSIONS

In the previous sections I have described how current developments in university and industry are qualitatively different from those characteristic of the preceding three decades. The relative roles of universities and industry in basic and applied research and in manpower training are changing; there are changes in industrial structure, in the relation between small and large enterprises and in the financing-structure for industry; we are at the height of internationalization in the world economy. We can say, through the interaction between these developments and with the rapid developments occurring in science and technology, that currently we stand at the beginning of a qualitatively new era.

I shall end this analysis by addressing a number of issues which, internationally, are important aspects of the current debate on science and technology policy.

20.7.1 Knowledge Infrastructure and Priority Setting

Which institutions do we need in the future for knowledge/generation, transfer and transformation? Should this be done in institutions which differ with respect to culture, motivation and involvement and which are also at arms length from industry (e.g. University, Max Planck Gesellschaft, Frauenhofer, A.I.F.)? What is the best way to institutionalize the interface between academic and industrial R&D, with regard in particular to priority setting tasks?

Which organizations (research councils, ZWO, TNO) are most appropriate to the setting of priorities in strategic research and to performing this research? (Strategic research is a part of basic research that is not performed and financed out of pure curiosity, but because of some expectation on the use and contribution to the development of new technologies) (Irvin and Martin, 1984)? Which institutions are most appropriate to carry out the strategic research; universities, 'Max Planck Gesellschaft-like' institutes or industrial laboratories stimulated and financed by government?

What is the relationship between the several institutions, each with its unique character, but each of which is part of the knowledge-infrastructure and higher education? What should be the financial and substantive input from industry to the government-financed knowledge-infrastructure?

The debate on this issue is determined by the question of whether the above mentioned developments are structural or whether they represent short-term perturbations on the road back to equilibrium.

20.7.2 R&D-Expenditures Philosophy

Which measuring-stick do we need to use in R&D-expenditures policy? The answer to this question depends on what precisely R&D-expenditures are, and with which expenditures and in which other countries we wish to compare.

The service sector in the Netherlands is rather large, but R&D expenditures in this sector is not included in the national R&D-statistics.

Other countries, in particular the U.S., spend a relatively high percentage of national R&D in the defence sector. This is not the case in the Netherlands. Should we compensate for this, or do we already effectively have these increased R&D expenditures because of our multinational enterprises who carry out R&D on behalf of production- and marketing-activities in other countries?

Sweden had a governmental rule which obliges industry to lodge 10% of their profits in an account at the National Bank. This amount should be immediately put at their disposal if they spend it to R&D inside the company, or training their own employees, on university research, or on any combination of the three. This Swedish measurement implies that R&D a company-based activity and as 'human resource development' are mutually interchangable.

So, we are still left with the question: what precisely are R&D activities? Are there relevant arguments that can be supported by the results of research to justify an increase in government R&D-expenditures? Do we have an overview of R&D-expenditures in companies and social sectors? What criteria and arguments are relevant to justify the financial stimulation of particular sectors? Should we decrease or increase our expenditures in health care? What is the situation in other sectors, which are primarily initiated by government out of social or national interest, such as environment and social security? Do we need to stimulate biotechnology further financially and do we have enough manpower to make this a viable proposition?

20.7.3 The Role of Government

There has recently been a fundamental turn-around in economic policy. The profit-capacity of industry has improved, investment has increased and enterprises now invest more company money in R&D. This has led to the need for a different set of governmental measures with regard to market-focused R&D than those implemented in the seventies.

This new set of measures should contain a number of generic rules that guarantee a certain basic expenditure on R&D; but there should also be rules which have the purpose of encouraging more selective and specific efforts. It is government that should organize this selection process and in turn give incentives and encouragement which lead to R&D-investments in industry.

In the 'sixties and 'seventies' government placed some emphasis in science and technology policy on applied research. The dual consequences of this were that basic research at universities and R&D-institutes decreased and that stronger links between the remaining basic research and developments in the market-sector were forged.

What about to-day. Does the government have to withdraw from applied research and focus more on basic and strategic research? Developments in the U.S., where the N.S.F. and general State-administrations have made big research investments, point to a reinforcement of governmental activities with regard to basic and strategic research. Does the government have to increase its time-horizon on long term research and increase its investments in basic R&D? Should the government, in the process of determing which strategic research should be done, go beyond the point of organizing the priority-setting process? The 'macro-level strategic research forecasting' of Irvin and Martin is perhaps an adequate model for this.

20.7.4 Education, Training and New Technologies

After our experience with industrial innovations and technology policy, do we have to prepare ourselves to move on to develop 'human resource development'-policy? Certainly the rapid development and application of new technologies makes re-education and manpower re-training necessary.

How large is the extent of re-education and re-training in Europe? How is it organized and structured? How is the situation in Europe compared to the U.S. and Japan? Are the regulator education and training institutions able to fulfill the growing need for skilled manpower? Must new education and training institutes be set up in an industrial-context? Should they be organized by the different sectors or be independent and cross-sectoral?

20.7.5 The Internationalization of the Innovation Process

An on-going monitoring of technology protection measures in necessary. If we supply this with an overview of European companies, which acquire small U.S. enterprises, we can obtain a realistic impression of the important role of international technology transfer. With regard to international co-operation at the European level, and since a number of large multinational enterprises have taken the initiative towards a joint European research effort, the question arises whether the internationalization of higher education and strategic and basic research is now inevitable?

Does it go without saying that an international association between national institutes in the area of basic research (e.g. Max Planck Gesellschaft and FOM) and organizations in the area of applied research (e.g. Frauenhofer and TNO) will be formed? It is possible that international, multidisciplinary Centres of Excellence will have an important task in the education of top-scientists for better international co-operation (Prigogin)?

In addition there is the question not only of the need for European co-operation, but also of the conditions and form it should take. Is the Eureka-initiative an example of a suitable form of European co-operation? To what extent can this form of initiative have meaning with respect to the size of the market as well as to the internationalization of the innovation process?

REFERENCES

Van Delden, P., 1985, "co-makership: Philip' Europese Strategie", *Intermediair* 25:31.

Denison, E. and K. Pouillier, 1986, *Why Growth Rates Differ*, Allen and Unwin, London.

Dickson, D., 1984, *The New Politics of Science*, Pantheon Books, New York.

Dosi, G., 1982. "Technical Paradigms and Technical Trajectories", *Research Policy* 11:147.

Eurich, N.P., 1985, *Corporate Classrooms*, Carnegie Foundation for the Advancement of Teaching, Princeton, N.Y.

Freeman, C., J. Clark and L. Soete, *Unemployment and Technical Innovation*, Francis Pinter, London.

Hayes, R.H. and W.J. Abernathy, 1980, "Managing our Way to Economic Decline", *Harvard Business Review* july/aug, 67.

Hughes, T., *Networks of Power*, The Johns Hopkins University Press, Baltimore and London.

Mowery, D. and N. Rosenberg, 1979, "The Influcen of Market Demand upon Innovation: A Critical Review of Some Rent Empirical Studies", *Research Policy* 8:102.

Nelson, R.R. and S.G. Winter, 1977, "In Search of Useful Theory of Innovation", *Research policy* 6:36.

Nelson, R.R. and S.G. Winter, 1982, *En Evolutionary Theory of Economic Change*, Belknop Press (HUP), London.

Pavitt, K., 1984, "Sectoral Patterns of Technical Change: towards a Taxonomy and a Theory", *Research Policy* 13:343.

Rothwell, R. and W. Zegveld, *Reindustrialization and Technology*, Longman, Essex, U.K.

Salter, W.E.G., 1960, *Productivity and Technical Change*, C.U.P.,Cambridge.

Stankiewicz R., 1984, *University-IndustryRelations*, Six Countries Programme.
Thurow, L.C., 1985, "De wereldseconomie bevindt zich op een keerput, *NRC*, 15-6-1985

CHAPTER 21

Micro-Macro Interactions and U.S. Industrial Change

Roland Artle

When less than complete appropriability of the fruits of private sector spending on R&D exists, conventional wisdom holds that government support of the research is justified. The (reverse) case of national defense is much more controversial. How significant are the spillover effects from defense R&D onto the private-civilian sector of the American economy? Will the recent U.S. Technology Transfer Act at all contribute to the diffusion process? Defense spending accounts for nearly seven percent of the U.S. GNP, but as much as half of all U.S. spending on R&D is absorbed by the Department of Defense. Given a similar situation in the Soviet Union, essential questions about the "true burden" of defense spending come to mind. As one ponders America's slipping role in defining and setting the world's technology frontier, the massive absorption of resourses into defense-related R&D cannot be lightly disregarded.

I realize that the issues just posed are not central to this Symposium, but I have felt compelled to raise them. In my concluding remarks, I shall consider a much more narrowly defined issue of government-sponsored R&D, namely as it relates to targeted industrial policy.

Instead I shall provide here a view of the macroeconomic setting within which R&D-linked decisions are made today in the United States. As part of the presentation, I shall discuss the intricate macro-micro interactions through which this settings has evolved recently. Indeed, I believe that the period 1980-1984 marks a watershed in U.S. economic history. The tides of change that were set in motion during those years have already begun to profoundly impact the U.S. economy and the world economy at large.

It is interesting to compare Sweden's "solidaristic" wage policy with the traditional American system of wage formation. The dispersion of wages across (as well as within) industries in Sweden - measured, for example, as the relative variance - is only a very small fraction of the corresponding measure in the United States. The difference is critical with respect to the behavior of declining industries. In America, the workers in a business firm facing stagnant or falling demand have traditionally found their wages stagnant or falling. This has served as a cushioning mechanism for the firms, as well as for the workers, in such an industry. Exits of firms from the industry occur, but at a slow pace. By contrast, the wages of workers in a Swedish declining industry are just about as high, and increase by just about the same amount, as the wages of workers in any other industries. In the face of high costs and slow demand, business firms are speedily forced to exit such an industry. The transfer of resources into expanding firms occurs smoothly, helped by a whole arsenal of governmental - "labor market policy" and "industrial policy" - actions.

Thus, the Swedish system of "pushing" workers out of ailing firms and industries has, generally speaking, reallocated resources at a quicker pace than the traditional American system of "pulling" workers into advancing firms and sectors of the economy.

In this simple comparison, I have used the word "traditional" in describing the American system and its consequences, because the description hardly fits the current scene. Recently, a host of events have occurred whose combined effects show up in a dramatic quickening of the pace at which resources are being transferred and adjusted in the American economy. Japan's "just-in-time" approach to inventory control has not only been imitated and adopted by the typical American corporation. It is being extended and propagated into entirely new forms. Consider, for instance, the notion of the "just-in-time" workforce. As another example of the quickening pace of economic change, American mortgage markets are no longer compartmentalized and cushioned from changes occurring in other financial markets. Due to the "securitization" of mortgages, any interest rate change in the bond markets ripples very quickly through the mortgage markets.

More profoundly, all financial markets in the United States are now very closely linked to financial markets abroad. This is part and parcel of the generally increased "openness" of the U.S. economy. During the 25 years from 1960 through 1984, American imports of goods and services (defined broadly as the "current account" concept) increased their share of the U.S. GNP from 4.7 percent to 11.7 percent. The fact that skyrocketing petroleum imports accounted for only one fifth of the increase in the import share indicates that it was broadly based.

Although the U.S. markets opened up gradually to foreign competition, the first half of the 1980s brought more abrupt change, in a sense to be discussed more thoroughly in what follows. There occurred a causal shift between the elements making up the U.S. balance of payments.

Under the present flexible exchange rate system, the value of the U.S. dollar is determined by supply and demand in the foreign exchange markets. The matching of the supply of and the demand for dollars has an accounting correspondence in the balance of payments: If one adds all the items on the current account to the items on the capital account, the sum is zero (disregarding the troublesome measurement errors, which show up in the accounts as a "statistical discrepancy").

It is noteworthy that although the net flow on the current account precisely matches the net flow on the capital account, the value of the gross flows on capital account is vastly larger. As an example of the empirical use of the concepts, consider the year 1980: U.S. exports of goods and services exceeded imports from foreign countries by $32 billion. Thus, the net outcome of these current account transactions was "excess" demand for U.S. dollars in the foreign exchange markets. Again, disregarding measurement errors, this was precisely compensated by an excess supply of dollars resulting from a net ($32 billion) outflow of financial capital from the United States to foreign countries.

The relationships discussed are illustrated in the *Diagram* on p.293. The measure used on the horizontal axis of the diagram is billions of dollars; the vertical axis measures the number of foreign currency units (for example, the number of Japanese yen or a composite of foreign currencies) per U.S. dollar. Two graphs are drawn on the diagram. One depicts the net flow on current account (the balance between the value of exports and imports, travel and other transportation, as well as dividends and interest and other income earned on U.S. assets held abroad minus the income earned by foreign assets held in the U.S.). The other graph depicts the net flow on capital account (the net balance of U.S. purchases of foreign securities and of foreigners' purchases of U.S. securities, as well as the value of direct investments by Americans abroad minus the value of foreign direct investments in the U.S.). A point on the positive part of the current account curve denotes a dollar value at which there is a corresponding "surplus" on current account (resulting in a net inflow of payments to the U.S.); similarly, a point on the positive part of the capital account curve denotes a dollar value at which there is a net financial capital outflow from the U.S. Each curve shows a zero net balance at the point where it intersects the vertical axis. The reason why the current account curve is negatively sloped is that a higher value of the dollar induces more imports into the U.S. and discourages U.S. exports. The capital account curve is positively sloped because a higher dollar makes it more attractive for Americans to purchase foreign assets. The point of intersection between the two curves on the right-hand side of the vertical axis depicts the U.S. balance of payments situation in the year 1980. For comparison and contrast, consider the situation in 1984: The current account surplus ($32 billion) existing in 1980 had then turned into a large deficit. Also shown is the U.S. balance of payments situation in 1986, when a sharp further deterioration had occurred. (The "tilting" of the current account curve is due to so-called "*J*-curve effects".)

What this *Diagram* tells is an important story. Between 1983 and 1986 there occurred an exceptional increase in foreign purchases of U.S. securities and of foreign direct investment in the U.S. economy. By way of a resulting sharp increase in the value of the dollar, U.S. exports stagnated while imports increased rapidly. Hence, during this period, movements on the U.S. current account, especially in the form of increased imports, responded to movements on capital account. It is noteworthy that we tend to think of the causal chain as normally going in the opposite direction, namely from trade to movements of financial capital.

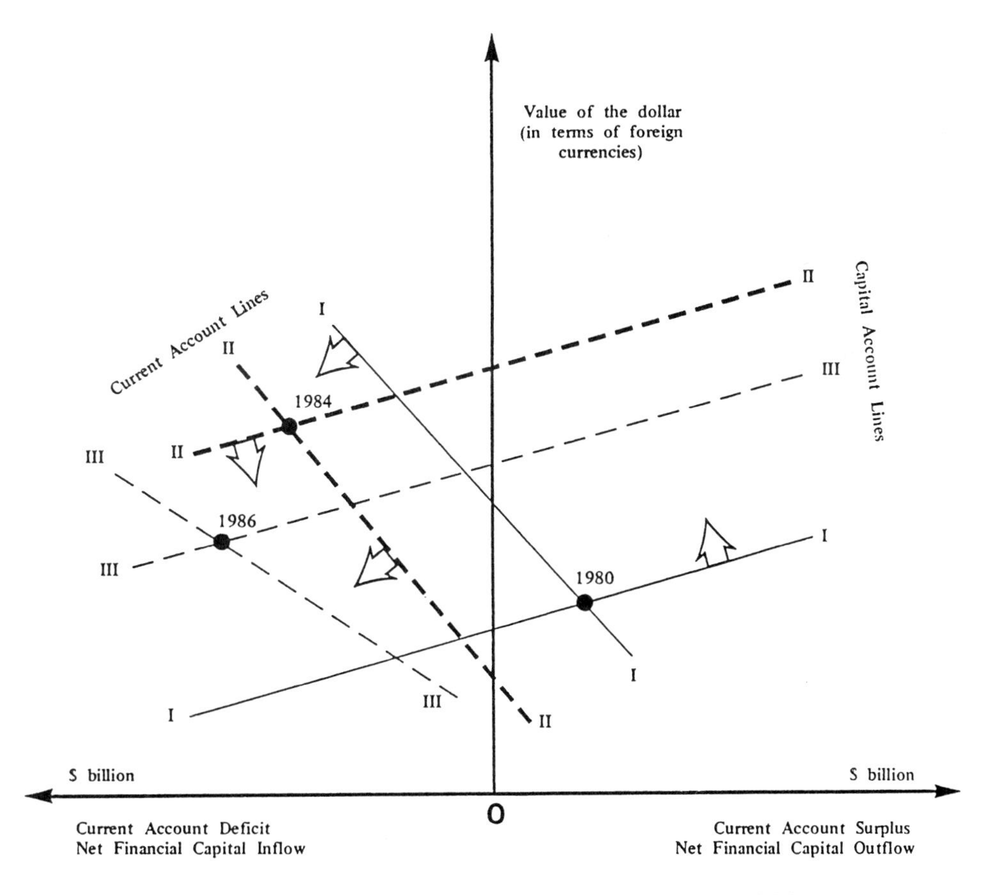

Diagram 20.1 The U.S. balance of payments in 1980, 1984, and 1986

What is the significance of this increased role of foreign financial capital, and how did it come about? To begin with the latter question, the process of deregulating U.S. financial markets initiated in 1980 played a part; so did the lessening of government controls that occurred in the Japanese and British financial markets. A third contributing factor was the gradual drop in energy prices that began in the early 1980s (which benefited the U.S. economy but not foreign petroleum-importing countries because of the simultaneous rise in the value of the dollar). However, the single most important factor was the economic policy mix that prevailed during the first half of the 1980s. Fiscal policy was highly expansionary in the United States, whereas fiscal policy in such countries as Japan and West Germany was quite contractionary in its effects. Apart from brief bursts of accommodation, U.S. monetary policy during the period as a whole had a restraining effect on the economy. This combination of "loose" fiscal policy and "tight" monetary policy contributed strongly to the exceptional elevation of real interest rates in the United States. In turn, the high level of real interest rates was a primary factor in the attraction of foreign financial capital. (In terms of the *Diagram* discussed above, there was a sharp upward shift in the capital account curve.)

It is instructive to examine the microfoundations of these international flows of funds. Take the case of a Japanese government bond and an American government bond of equal maturity. How will an international investor decide between them? Three different considerations will enter the decision process. The reason why a Japanese investor would be attracted to consider an investment in U.S. financial ,markets is the interest rate differential between the usually lower-yielding Japanese bond and the higher-yielding, but otherwise comparable, American bond. The size of that interest rate differential becomes the first consideration in the decision. Second, the prospective investor must form an expectation regarding the risk of a capital loss on the value of the American bond (again, compared with the corresponding risk on the Japanese bond). Third, he must consider the expected rate of depreciation of the U.S. dollar in relation to the Japanese yen during the period considered. Although we have for simplicity used a bond investment, efficiency considerations extend the reasoning to other investment possibilities in the U.S. economy.

Increasingly, during the first half of the 1980s, Japanese as well as other foreign investors decided that on balance the three factors mentioned above weighed heavily in favor of investments in the United States. As the foreign investors exchanged their own currencies for dollars, in order to avail themselves of the American investment opportunities, the value of the dollar was bid up sharply. Several important effects on the U.S. economy followed. First, the real ("purchasing power") income of Americans rose with the value of the dollar. Second, there was an important transfer of income to consumers and away from the American producers of internationally "tradeable" goods and services. Put differently, competition intensified severely both for American exporters and for domestic producers who faced actual and potential new foreign competition. Third, the industrial structure underwent change. Although there were six million more jobholders in 1984 than in 1980, the number of jobs in manufacturing industries dropped by 1.7 million. On the one hand, rising incomes stimulated more demand for services ("high income elasticity") and the rising flow of imports generated new jobs in retail and wholesale trade and other related service industries. On the other hand, output levels stagnated and employment fell sharply in such basic U.S. industries as steel, motor vehicles and textiles. Machine tools, consumer electronics, and semiconductors are examples of other manufacturing industries impacted.

From a different perspective, the sequence of rising U.S. real interest rates, the inflow of foreign financial capital and the rise in the value of the dollar was a manifestation of the growing imbalance between U.S. (real) capital formation, on the one hand, and domestic savings on the other. Very low savings rates among American households, coupled with unprecedentedly high federal government budget deficits, accounted for this imbalance.

Two numbers capture, as if in a nutshell, the linkage. By 1986, the budget deficit had reached five percent of the GNP and the trade deficit 3.5 percent.

In the final phase of the dollar's rise, sheer speculation kept the increase going. The speculative bubble burst in February 1985, and the dollar began to fall. A relaxation in monetary policy contributed to the continuing fall in the value of the dollar. By June 1987, the dollar had dropped by more than 45 percent against the Japanese yen and by generally lesser amounts against other major currencies. The experience gained since the fall of the dollar began in early 1985 suggests that some of the major economic forces set in motion by the earlier sharp rise in the value of the dollar appear to be irreversible.

By any standard of comparison, as we have attempted to show, the U.S. economy has undergone major change since 1980. It is time now to examine how the process of change at the macro-level has been reflected in change at the micro-level.

During the last few years, and in industry after industry, a striking phenomenon has occurred. Larger corporations have shrunk their payrolls, and often drastically so; the elimination of jobs has happened at a pace well beyond what one could normally ascribe to attrition. Nor can the shrinkage in payrolls be attributed to historical trends of increased specialization and increased division of labor in society. Rather, what we have been witnessing is a major restructuring of the American economy, whereby resources, and especially labor, are moving from larger to smaller business firms. It is noteworthy that this restructuring has been accompanied by a falling unemployment rate, albeit at a sluggish pace.

What is emerging has been called a "two-tier" workforce. The regular employees of a company make up its "inside" workforce. In the performance of such tasks as data processing, public relations, advertising, and even legal services, design and engineering, the company is assisted by an "outside" - "shadow" - workforce composed of independent smaller firms as contractors of the work.

The social implications of this emerging two-tier setup are profound - and beyond this paper to consider. Let me just note that the package of benefits typically accruing to employees (health-care and pension plans, paid vacations, etc.) is on the average much slimmer for small-firm employees than it is for employees of larger companies.

Akio Morita, chairman and co-founder of Sony Corporation, coined the alarming words: "American companies have either shifted output to low-wage countries or come to buy parts and assembled products from countries like Japan that can make quality products at low prices. The result is a hollowing of American industry."

I suggest to you that the process of "hollowing" the American corporation has been dual. "Outsourcing" and "globalization", to cite two of the buzzwords, have to a certain extent eroded America's manufacturing base. The emergence of the two-tier work force and the concomitant shift in employment from larger to smaller firms represents a different but equally intriguing form of "hollowing" of the large American corporation.

At the outset, we discussed the long-standing solidaristic Swedish wage policy, which being backed up by governmental labor-market policies, has had the effect of speeding up industrial change in Sweden. A major conclusion of this paper is that macroeconomic change in the United States during the 1980s has wrought an entirely different mechanism through which a quickening of the pace of industrial as well as financial-market change has occurred. But is has come at a price. There is now more insecurity in American labor markets. There is now more uncertainty in American financial markets.

International flows of financial capital have increased rapidly during the 1980s. In particular, we have noted the very steep increase in foreign funds flowing into the U.S. economy. In what ways do such flows connect to the international transfer of technologies and more broadly to the communication of ideas?

Finally, one of the specific industries hurt by economic change during the 1980s has been the semiconductor industry. Leaders of that industry are now seeking federal government support for an ambitious R&D chip manufacturing consortium to be known as Sematech. It raises interesting issues about the effectiveness of targeted industrial policy, about U.S. entrepreneurial individualism and government subsidies, and about the proper role of the U.S. federal government in supporting civilian technology versus defense technology.

CHAPTER 22

Swedish Science Policy: The Government's New Research Bill

Lennart Bodström

Ladies and gentlemen,

I am very pleased and honoured to participate in the opening of this RICE-symposium here in Karlstad. Your choice of theme for the symposium I take as a sign of the importance of looking at research in an economic and industrial perspective.

Distinguished speakers from many countries will address various subjects during the coming days of the conference. I want to extend a particular welcome to the foreign participants. Your participation guarantees that valuable international comparisons can be made and experiences exchanged.

The headings of the lectures indicate the very broad scope of this conference. It will deal with research and development, science and technology, diffusion of technology, structural and industrial change, creativity, innovation, and regional and international perspectives. The fact that the symposium takes place in Karlstad is significant of the regional dimension.

I will of course not try to cover all the possible topics. For example, Mr. Sigvard Tomner will speak about "Innovation and Technology Diffusion in Sweden". I shall concentrate on my area of responsibility in the Government, namely research at the institutions of higher education in Sweden.

The Swedish Government presented a Bill on Research to the Riksdag last February. The Bill was enacted by the Riksdag some weeks ago and the various proposals in it were accepted with very few exceptions. This shows that there is unanimity among the political parties as to the main structure of the research system, the total resources which should be granted to research, the direction of research, and as to various other matters related to research.

I would like to present to you the main issues of research policy as they are formulated in the Bill and the proposals by the Government. Before I do so, however, it is necessary to - very briefly - give you some information about the way Swedish research is organized and the scope and direction of the state grants to research.

- - - - - - - - - -

In an international perspective Sweden is spending a lot of resources on research and development. It amounts to 2,7 per cent of its gross national product which is one of the highest percentages in the OECD area.

Industrial investments in R&D are higher than the state grants but the contributions from industry mainly concern development. For basic or fundamental research state grants are by all means the most important.

As in many other countries basic research at the universities is financed by grants from the Ministry of Education. The Ministry also allocates resources to the three research councils, the Medical Research Council, the Natural Sciences Research Council and the Research Council in the Humanities and Social Sciences.

The Government and the Riksdag grant resources by way of block sums to each faculty at every institution of higher education with permanent research facilities, of which there are 11. They also decide on the total number of professorial chairs and the actual subject matter of each chair. Money may also be granted to the faculties or research councils for research in specific priority areas such as information technology or biotechnology.

For basic research the grants from the Ministry of Education are thus of great importance. For applied research, grants from other ministries may be of equal importance or even more so. This is true, for example, of industrially oriented research where the Ministry of Industry through the National Board for Technical Development is furthering research at the technical universities in areas relevant to industry. A number of other examples could be given of government agencies which commission research of interest to them at the universities.

Here is an important difference between Sweden and most other countries. There are few research institutes or laboratories separate from the universities. Instead, the institutions of higher education are supposed to take on mission-oriented research, and they actually do so to a considerable degree.

- - - - - - - - - -

Let me now turn to the Bill on Research which has recently been debated by the Swedish Riksdag.

International evaluations have become very frequent in recent years, at our universities, research councils and sectorial agencies.

Swedish science and technology policies have also been the object of international review. The OECD undertook an evaluation of these policies a year ago. Some of you may have heard about it, as the OECD reviews often are cited in the international press. The examiners' overall judgement was positive, but they also pointed to some problems in Swedish research and their recommendations were important for the proposals which the Government put forward in the Bill. This is particularly true of those concerning the conditions of postgraduate students and young researchers.

The Bill entails an increase by Skr. 500 million in state grants to research for the next three-year period. This is the greatest increase of resources to research for a very long time. It should be seen in relation to a considerable restraint in the general budget process which the Swedish Government has been forced to exercise in the last couple of years.

The Bill emphasizes university research under the ministry of Education with a particular accent on the general basic resources for research training and research. This emphasis is due to the fact that sectorial and industrial research has increased rapidly in the last decade whereas traditional university research has been fairly stable. In addition, there will be special support allocated to selected areas, such as information technology and biotechnology, but also to the humanities and social sciences.

In addition to the budget proposals other measures aiming at improving the working conditions of research students and researchers were proposed.

While preparing the Bill the Ministry of Education commissioned the National Board of Universities and Colleges, which is the responsible government agency for higher education and research, to work out proposals concerning the academic research environment. The Board was to analyse what factors contribute to a good research environment and favor creativity.

In the course of this symposium some of you will discuss the matter of creativity. Many books have been written about it and there are at least one hundred definitions of the word. There seems to be unanimity, however, that being creative implies bringing about something positive which, to a large extent, is unexpected.

Part of the debate deals with what explains and fosters creativity. Is creativity an individual quality or may the proper environments produce creative individuals?

An interesting point in the paper produced as a result of the assignment to the National Board of Universities and Colleges was that the concept of "critical mass" can be questioned. Other factors such as leadership, seem to play an equally if not more important role. What may be termed "critical mass" seems to vary a lot with the discipline or area of research.

Many of the factors promoting creativity are such that they can only be dealt with in each research environment and by the researchers themselves. What then can a government do to encourage creativity in research?

Well, it can first of all try to improve the working environment of the researchers by granting adequate resources for carrying out research projects. Thus an increase of Skr. 120 million in the general basic resources for the work within the various departments is proposed. There resources are *not* intended to increase the number of posts for researchers but should be used for equipment, travel, conferences, guest lectures from abroad, etc.

Additional resources are proposed for the research councils which play a very important role in the Swedish research system, not the least in the development of high quality research by international standards. Skr. 43 Million are suggested to strengthen the activities of the research councils.

Secondly, the Government can improve the working conditions of research students and researchers by increasing the number of research posts and by improving the financial conditions associated with these posts. Job security is one of the most important factors in fostering creativity, according to studies which have been made.

The Government proposes that the level of students grants which is the most frequent type of study finance be raised by 10 per cent to appr. Skr. 90 000 per year. Extra money is also proposed for the transformation of 700 student grants into post-graduate appointments which are more favorable for the research students as they entail the right to certain social benefits.

Another important change is the proposal to increase the number of research assistantships by 225. Research assistantships are posts for persons who have recently completed their Ph.D. These posts give ample opportunities to carry out research and thereby help the holders to qualify for continued employment in higher education. An increase in the number of these posts allows for better tutoring as well as increasing career possibilities for research students.

The augmentation of research assistantships is also designed to solve a particular problem - or at least reduce its effects - namely the problem of the fairly skewed age structure of the academic research population. Due to the heavy expansion of higher education in the late 1960s and early 1970s the holders of higher education appointments are to a large extent persons who are 40 years or older. Younger generations are represented to a much lesser degree. The proposal for more than 200 new research assistants covers all faculties with some preference for the humanities and the social sciences, the natural sciences and technology.

I have touched upon the issue of inadequate career possibilities in higher education research. Due to the expansion of sectorial research at the universities in the last decade, an expansion which has not been matched by a comparable increase in faculty grants, career prospects for an increasing number of researchers at the universities have been rather gloomy.

In order to tackle this problem - and to strengthen research on a longer term basis - 67 new professorial chairs are proposed. More than half of these posts belong to the humanities and social sciences.

The Government's proposal to the Riksdag also includes special resources to certain priority areas which I mentioned by way of introduction. It would take too long to introduce the various components of the Government's proposal in these. Suffice it to say that they are in the order of Skr. 35 million each for biotechnology and information technology and around Skr. 13 million for research in the so called cultural sciences. A larger proportion of the proposed sums is intended for the research councils which come under the Ministry of Education.

Over and above the usual compensation for rising prices and wage costs, the total sum of new resources amounts to Skr. 360 million for research under the Ministry of Education. To this should be added the Skr. 600 million which the commercial banks have volunteered to reserve during the same three year period, a sum which is primarily intended for the renewal of equipment in higher education.

Taken together, these allocations amount to the greatest increase in research funding for many years. They are intended to remedy the problems of imbalance between the basic resources of the universities on the one hand and the resources for R&D in industry on the other, but also between the basic resources of the universities and sector-based research, which constitutes a very large part of the research carried out at the universities.

Over the 1981-85 period R&D investments in industry doubled at current prices. The Swedish Government naturally welcomes this development. However, it also means that an increasing proportion of Sweden's overall R&D activities is financed and controlled by the business sector. There is thus a risk that long-term and basic research may lose ground. It is therefore most satisfying that the commercial banks have decided to contribute to the financing of basic research at the universities.

When it comes to university research financed by industry it is important that companies do not become so influential in certain departments that research cannot be conducted in an open atmosphere or that it becomes too narrow in its approach. It is the responsibility of the higher education boards to intervene where circumstances so demand. There is a proposal in the Bill that substantial work undertaken for industry by university researchers should be reported to their boards so that an overview regarding cooperation with industry and openness about such contacts may be obtained.

The large increase in sectorial research at the universities may endanger the long-term development of higher education research. This is an issue which has attracted much attention in the Swedish debate. Increasing faculty grants and grants to the research councils allow for more research which is conducted on the researcher's own-programme basis, i.e. research which is not commissioned by an external body. This is another reason for the government's heavy emphasis on basic research. It is also stressed in the Bill that sectorial agencies must finance university research on a longer term basis and pay the full price for commissioned research, something which they are not doing at the moment.

I should now like to turn to a subject which is engaging many people in Sweden today and which may also be of interest to an international audience. This is the localization of research.

Basic higher education exists today at 25 places in Sweden. The Swedish Riksdag has decided not to split the resources for research and research training but concentrate them to eleven institutions of higher education at seven places. There are many people in our country, not the least representatives for higher education institutions without permanent resources for research, who would like to see a permanent research organization at more institutions.

Although we do a lot in the area of research in our country it is difficult to create research environments which are internationally competitive in a country of Sweden's size. Our capacity is simply not big enough. Incidently, one of the recommendations of the OECD examiners was that Sweden should not split its resources for research.

It is of particular importance that post-graduate education is carried out in a stimulating and creative environment. The future teachers and researchers in higher education must get their training in an environment where the international scientific demands can be applied in full and where there are ample opportunities for contacts with active researchers representing different research fields, both nationally and internationally.

Such a research environment cannot be created without rather substantial resources. It probably takes a faculty or an equivalent organization to achieve such an environment. To build up organizations of this kind at several other places is, in the Government's view, not possible within the foreseeable future. It would be detrimental to research and research training as a whole and consequently also to the higher education institutions without permanent resources for research because they too are dependent upon good research for their development and renewal as well as on skillful and well-trained teachers.

Having said this, I want to stress that it is of great importance to make the most of the resources for research and development which actually do exist at the higher education institutions without permanent facilities for research.

When visiting various parts of Sweden I have been struck by the enthusiasm and "pioneer spirit" which exist at these smaller higher education institutions. Considerable and increasing research activities take place on a project basis at almost every such institution. This is very positive both for the teachers themselves and for the vitality of the teaching which they provide but also for the surrounding community and local business.

That is why the Government proposes in the Bill that the smaller institutions of higher education should get more money for research. About Skr. 10 million should be set aside yearly for various activities which make it possible for teachers to do research on a commission basis, for example for local business or municipal authorities or simply to help researchers to conduct research on their own-programme basis.

I mentioned earlier the international evaluations of Swedish research and research policy. It is not by chance that they have taken place. In a small country which cannot conduct research in every possible field it is of particular importance to have contacts and co-operation with researchers in other countries. We know that on the whole Swedish researchers cooperate with foreign colleagues to a large extent.

However, the Government has an important role to play in international research co-operation. This is true not only for the financing of very large international research facilities in the natural sciences, such as CERN or fusion research within the European Community, but also by stimulating international research co-operation on a smaller scale and in many fields.

A couple of years ago Skr. 5 million were set aside for fellowships intended to enable research students and young researchers to do research abroad. Another example is the substantial increases for higher education research in this last Bill which make possible intensified contacts between Swedish researchers and their colleagues abroad.

This conference is good evidence of the interest that Swedish higher education institutions take in international co-operation.

Thank you.

CHAPTER 23

A Ten-Year Review of Science Policy in Sweden

Gudmund Larsson

Mr Chairman of the Municipal Council, Madam Vice-Chairman of the Municipal Executive Board, conference delegates and guests!

I consider it a great honour to have this opportunity to speak to you during your conference here in Karlstad. When I was thinking through what I was going to say to the very distinguished and cosmopolitan gathering at this evening's dinner, it struck me that i would be a natural opportunity to attempt a ten-year review of developments in Sweden, of what has happened in the economic and political spheres and how it has affected the universities and research policy in this country.

Why ten years? Because the most recent major reform of Swedish higher education has been a reality for ten years now. The University of Karlstad is this year celebrating its tenth anniversary. I myself have been actively involved in research and development policy for exactly ten years this month, the last one and a half years of which I have spent in the Cabinet Office. So it seems natural to take a look back.

Before I attempt to give you this ten-year view - and it will be a subjective one - I would like to say a few words about the Government's policy on research. Earlier today the Minister of Education and Cultural Affairs gave you a detailed picture of the proposals in the Government Bill accepted a month ago by the Riksdag with only marginal changes.

So I shall be very brief and simply remind you that it is estimated that central government funding of research will grow by four per cent in real terms over the next three years, which is remarkable in the present budgetary situation. Even more striking, though, is the fact that, within this four per cent rise, university faculty grants will increase by 12 per cent, again in real terms.

This - as the opposition also recognize - is a remarkable increase. The Government Bill gives priority to the same sectors as in other OECD countries, that is to say, information technology, biotechnology, and research aimed at improving the environment. But it should be noted that the Government's proposals also include a strong commitment to the humanities and social sciences. One of the main aims of the Bill, finally, is to strengthen the position of younger researchers, and it therefore provides for a substantial new injection of postgraduate and postdoctoral posts for young research workers.

In the autumn of 1986, as the Government engaged in some tough internal discussion on budget savings and on what sectors were to be provided with greater resources by means of collective savings in other spheres of central government, it was decided that the emphasis should be placed on three areas: improving pensions - a promise made in the last election - regional policy, and research. All the other commitments of central government were restrained in view of the budget deficit. this was a very clear assignment of priorities by the Government.

Research receives special attention from the Prime Minister. The Cabinet Office has a unit specifically to co-ordinate research policy. The Minister for Research is 'second-in-command' to the Prime Minister as regards this area of responsibility, and the main statement introducing the Government Research Bill is the Prime Minister's. This is very unusual. The Prime Minister also has at his disposal a special advisory board whose members include prominent researchers. The board's working committee meets once a month under the chairmanship of the Prime Minister. In other words: research has been brought to the fore as one of the key concerns of Swedish Government policy.

Now I shall turn to my theme, my ten-year review. What has happened in the Swedish economy in the last ten years? What political developments have taken place? Radical changes have occurred. How have they affected the universities and research? Though I myself am a political being and a political appointee, I shall try to be honest and objective, an aim I believe to be particularly important in politics. But I have been an active member of the Social Democratic Party for 23 years and have been employed in the labour movement for almost 15, and I would ask you to bear that in mind.

What economic developments have been seen between 1976 and 1987? Naturally, the developments occurring here in Sweden have not differed fundamentally from those experienced in the industrialized world generally. It is possible to speak of five years of crisis and five years of reconstruction.

During the five crisis years, inflation rose to an annual average of 11-12 per cent. The budget deficit grew to Skr 90 billion a year. The balance on current account took a dramatic turn into the red. Those in employment suffered an erosion of their real earnings, in practice losing a month's pay. Unemployment rose to 3.5 per cent of the labour force - a low level by international standards, but high for Sweden.

It was during these years that Sweden had a coalition Government comprising the conservative Moderate Party, the Liberal Party and the Centre Party, what we in Sweden usually call - in somewhat old-fashioned terms - the 'bourgeois' parties. It was a coalition Government faced with considerable internal tensions and one with very little experience of governing, as the non-socialist parties had been in opposition for 40 years.

But even as steadfast and stubborn a Social Democrat as myself has to concede that three circumstances had a major influence here.

First: these were the years in which the economic recession affecting the West as a whole hit Sweden too, and hit us hard, due to the Swedish economy's dependence on traditional basic industries like steel, forestry and shipbuilding. Any government would have been forced to tackle some very serious problems.

Secondly: one of the reasons the budget deficit grew almost unchecked was that the varying non-socialist coalitions that were in government in the period in question sought to defend a policy of full employment and a system of social security - at least at that time. Naturally this was the result of the long period of Social Democratic government, a continuing hegemony if you like.

The third point that has to be conceded is that our own opposition was not an overwhelming success. It was a new experience for the parties of the centre and right to be in government. It was no less perturbingly new an experience for the Social Democrats to be in opposition.

Let me now pass on to, if not the good years, at least the period characterized by reconstruction of the Swedish economy. The change of Government in 1982 did, after all, bring other changes in its wake.

Sweden devalued its currency sharply - by 16 per cent - in the autumn of 1982. It was a daring policy, but the devaluation came at the right time, when world trade was expanding. A number of other circumstances have meant that this devaluation offensive has had favourable effects to this day. A restrained budgetary policy has cut the budget deficit to only a third of its previous level, and it is still falling. The balance on current account has moved into surplus. We have got rid of the worst structural problems; the shipyards, for instance, have been closed down. This spring, inflation fell to the average level in other OECD countries for the first time for a very long time. Unemployment has now dropped to less than two per cent. This is a uniquely rapid and almost incomprehensible improvement, and for the first time in ten years employed people will be seeing a real rise in their purchasing power.

But remarkable as this recovery is, and however much this simultaneous combination of reduced inflation and full employment may astonish the international community, it has to be admitted that this Government has had as much good luck as the previous one had bad. It should be borne in mind - and this is not so much a matter of luck as a question of political conditions, of confidence in the Social Democratic Government - that the trade unions accepted the strain imposed by the devaluation and did not demand compensation. They have seen their pay remain the same while profits have soared. It is more a question of luck that the exchange rate of the dollar has fallen and that expansion and recovery in our export markets have favoured Sweden's export industries since the devaluation.

To conclude: in the second half of the 1970s many people felt that the growth and stable hopes that had typified the decades leading up to the 70s had been historically unique and that we were now seeing the enduring crisis of the welfare society. Our historical perspective is often as short as this. Today we run another risk: that of viewing the years between 1976 and 1987 as a temporary lapse into insecurity. This is just as dangerous.

What has happened in the political arena in these ten years of economic recession and growing tension? Well, politically, too, illusions have been shattered. 'The Swedish model', which we are quite proud of - with some justification at least - has seemed to be creaking at the joints. Its very foundations have seemed unstable. In the last ten years, Swedish politics and public life have not only been shaken by a brutal and incomprehensible assassination - a shock still powerfully affecting our country - but also by heated and searching discussion about a number of contentious political issues, issues that have divided the nation.

The first of these was the debate on nuclear energy. Never in the whole of my political career have I come across such strong emotions, so much anxiety and fear, such intense involvement on the part of people not normally willing to get involved in specific issues. Our media reverberated with these vehement feelings and conflicting claims. And the more the experts and scientists tried to reassure the anxious, the more worried these people became. Something that was portrayed as being almost absolutely safe came to be viewed as a definite danger to life. The discussion split families and political parties.

The solution was a referendum which gave rise to some unlikely coalitions. Liberals and Social Democrats discovered each other. Communists and Agrarians took to the same barricade. There we perhaps saw the birth of a new party, the Greens. After the referendum in 1980 everyone withdrew exhausted from the battlefield, to return after Chernobyl, in more subdued mood.

The nuclear power issue contributed actively to the downfall of the Social Democratic Government in 1976. Later it led to a split in the non-socialist coalition Government that had taken over.

The other question that caused passions to run high reflected the classical political split between right and left. It was the question of employee investment funds. The issue of collective ownership of industry divided the political map in the way we are accustomed to seeing it divided. On the left, the Social Democrats and trade unions, together with a muttering Communist Party, on the right, the Centre, Liberal and Moderate Parties and the employers. The issue of employee investment fund took on a new lease of life in 1976, and was finally lugged through a reluctant Riksdag in December 1983. But the non-socialist parties are still united in their call for these funds to be scrapped.

If the employee investment fund issue united the non-socialist parties, it caused the labour movement many a headache. To some extent it contributed to the Social

Democratic defeat in 1976, although nuclear energy was probably the decisive question. Today the employee investment funds seem fairly innocuous and are viewed with disappointment within trade union ranks. But these very same funds resulted in Stockholm witnessing one of the most remarkable demonstrations ever seen in Sweden, with a full contingent of business leaders and top executives from the employers' organizations marching to the Riksdag. 75,000 people took part. It was a fantastic sight, and a better-off crowd of demonstrators has never been beheld.

The third major battle in Swedish politics in the past ten years was the last election and the encounter over 'the Swedish model'. It was a discussion mainly between the Moderates, who had espoused the New Liberal thinking of the western world, and the Social Democrats. The discussion appeared to break the Moderate Party and its leaders, but the question is whether this was not just a matter of appearances, for New Liberal thinking is enjoying conspicuous success in Social Democrat-governed Sweden.

How, then - and this is the point I wanted to get to in the end - has all this affected policy on research and the universities in this country? I would say that if it has had an effect, it has been one beneficial to the research community.

The nuclear energy debate undoubtedly left its mark on people's minds, whether those minds were blue, red or green. People drew different conclusions. To many engineers and physicists, indeed scientists in general, the discussion on nuclear power came as something of a shock. They believed themselves to be maligned and misinterpreted, they felt persecuted by the media, and they were worried by all the anti-technology feeling that suddenly flared up. I have met many, physicists especially of course, who were deeply shaken by the way this advance, atomic power, was called into question.

In my ten years of involvement in research policy I have had a strong sense of there being consensus in Swedish society about the importance of research. And not only that - for anything else would surely be astonishing - I have felt there to be common attitudes on how research should be supported. And I would like to give you a brief idea of how I have come to have this impression.

Sweden has one of the strongest trade union movements in the world. When, in 1977, I began working for the Trade Union Confederation - which is the biggest group of unions in Sweden, organizing workers in the private and public sectors - my job was to help evolve a trade union policy programme relating to research. Together with a highly committed fact-finding group made up of representatives of the various member unions, with whom I shared this assignment, I visited most of Sweden's universities and other institutions of higher education. Nowhere did we feel we were unwelcome, wherever we went our interest in research was considered legitimate. We were able to engage in objective discussions.

This has been a characteristic feature of the overall debate on research policy in Sweden during the period I have been talking about. However storm-tossed the seas of public debate may have been over the effects of the economic crisis and how it should be solved, about employee investment funds, nuclear power, or a possible change of system in Swedish politics, discussion on research policy has flowed on as quietly as the Don. Different views as to the right methods of supporting research have been put forward, and time and time again the debate has taken a new turn. But all sides seem to have turned at the same time.

And this is why the research policy pursued by the various non-socialist Governments - though with the same Liberal minister responsible all the time - and the policy pursued by the Social Democratic Government in its 1984 and 1987 Bills do not differ all that much. By and large, the opposition - formerly Social Democratic and now non-socialist - have accepted the Government's policy. It is also striking that all the interested parties outside higher education - the trade unions, industry and so on - have had broadly the same favourable attitude. It goes without saying that this is of value to research in this country. Researchers know they have the nation's support and - hopefully - also know what weighty expectations accompany this support.

Admittedly the management and funding of higher education is now being discussed with some vigour - this is a subject dear to our hearts in this country - but the reformed system of higher education that has been in operation for ten years now is not being fundamentally challenged. The most fascinating question as we look ahead to future Government Research Bills is one currently being discussed as much inside parties as between them, that is to say, the regional distribution of higher education and above all research. In view of the role of research in promoting expansion, I believe everyone can subscribe to the view that it would be desirable to have research going on throughout the country, especially in regions facing major economic problems. But is this possible?

Karlstad is evidence that, to some extent at least, it is possible. I once listened in on a symposium concerned with research on working life here in Karlstad, and was very impressed. This conference will have no less powerful an impact, and the programme for this week is truly impressive. And it is in this town that the 'Karlstad approach' to research has become established, an approach in which a great deal of hope can be placed.

During my years of involvement in the field of research policy, I have taken on several different roles. In the process I have met Lennart Andersson, Vice-Chancellor of the University of Karlstad, a man who has served in the same role throughout this period. I would like to take this opportunity, finally, to thank Lennart for all the valuable co-operation we have shared in, in which his integrity, perseverance and wisdom have been very great assets. I would like to congratulate the Municipality of Karlstad on having so vital a university, and I would like to wish you all a highly successful conference in the week ahead.

LIST OF CONTRIBUTORS

Åke E.Andersson
Institute for Future Studies
Hagagatan 23A, 3 tr.
S-113 47 Stockholm
Sweden

Christer Anderstig
Regional Planning Office
Stockholm County Council
Box 12557
S-102 29 Stockholm
Sweden

Roland Artle
Department of Economics
University of California
250 Barros Hall
Berkeley, CA 94720
USA

Robert Ayres
International Institute for
Applied Systems Analysis
A-2361 Laxenburg
Austria

David F.Batten
Department of Economics
University of Umeå
901 87 Umeå
Sweden

Lennart Bodström
The Swedish Minister for Education

Derek A. Bosworth
Institute for Employment Research
University of Warwick
Coventry
United Kingdom

Evert Jan Davelaar
Department of Economics
Free University
P.O.Box 7161
NL-1007 MC Amsterdam
The Netherlands

Manfred M. Fischer
Inst. for Economic and
Social geography
Vienna University of Economics
Angasse 2-6
A-1090 Vienna
Austria

Hendrik Folmer
Dept. of General Economics
Agricultural University
P.O.Box 8130
NL-6700 EW Wageningen
The Netherlands

Henry Herzog Jr.
Department of Economics
University of Tennessee
Knoxville
Tennessee, 37996-0550
USA

Beat Hotz-Hart
Swiss Federal Institute of
Technology
Hönggerberg
CH-8093 Zürich
Switzerland

Jeremy Howells
Centre for Urban and Regional
Development Studies
The University
Newcastle upon Tyne, NE1 7RU
United Kingdom

Theo Hutten
Agricultural Economic Institute
P.O.Box 29703
NL-2502 LS's-Gravenhage
The Netherlands

Yannis M. Ioannides
Department of Economics
Virginia Polytechnic Institute &
State University
Blacksburg, VA 24061
USA

Charlie Karlsson
Department of Economics
University of Karlstad
Box 9501
S-650 09 Karlstad
Sweden

Kiyoshi Kobayashi
Dept. of Social Systems Engineering
Tottori University
Tottori 680
Japan

T.R. Lakshmanan
Department of Geography
Boston University
675 Commonwealth Avenue
Boston, MA
USA

Gudmund Larsson
Advisor to the Swedish Prime Minister
Cabinet Office

Edwin Mansfield
Department of Economics
University of Pennsylvania
Philadelphia, PA 19104
USA

Ann Markusen
Center for Urban Affirs and
Policy Research
Northwestern University
Evanston, IL 60201
USA

Peter Nijkamp
Department of Economics
Free University
P.O.Box 7161
NL-1007 MC Amsterdam
The Netherlands

Alan M. Schlottmann
Department of Economics
University of Tennessee
Knoxville
Tennessee, 37996-0550
USA

Pravin Varaiya
Dept. of Electrical Engineering
and Computer Sciences
University of California
Berkeley, CA 94720
USA

Kumaraswamy Velupillai
Department of Economics
University of California
Los Angeles, CA
USA

Sam Bass Warner
Department of Geography
Boston University
Boston, MA
USA

R.A. Wilson
Institute of Employment Research
University of Warwick
Coventry
United Kingdom

Michael Wiseman
La Follette Institute of Public Affairs
University of Wisconsin
Madinson, WI
USA

Walter Zegveld
TNO Divison of Policy Research
and Information
P.O.Box 215
NL-2600 AE Delft
The Netherlands